Delmar's
Veterinary Technician Dictionary

D0094337

Delmar's
Veterinary Technician Dictionary

Ray V. Herren
Janet Amundson Romich, DVM, MS

Delmar
Thomson Learning™

Africa • Australia • Canada • Denmark • Japan • Mexico • New Zealand • Philippines
Puerto Rico • Singapore • Spain • United Kingdom • United States

NOTICE TO THE READER

Publisher does not warrant or guarantee any of the products described herein or perform any independent analysis in connection with any of the product information contained herein. Publisher does not assume, and expressly disclaims, any obligation to obtain and include information other than that provided to it by the manufacturer.

The reader is expressly warned to consider and adopt all safety precautions that might be indicated by the activities herein and to avoid all potential hazards. By following the instructions contained herein, the reader willingly assumes all risks in connection with such instructions.

The publisher makes no representation or warranties of any kind, including but not limited to, the warranties of fitness for particular purpose or merchantability, nor are any such representations implied with respect to the material set forth herein, and the publisher takes no responsibility with respect to such material. The publisher shall not be liable for any special, consequential, or exemplary damages resulting, in whole or part, from the readers' use of, or reliance upon, this material.

Delmar Staff:

Business Unit Director: Susan L. Simpfenderfer

Executive Editor: Marlene McHugh Pratt

Acquisitions Editor: Zina M. Lawrence

Developmental Editor: Andrea Edwards Myers

Executive Production Manager: Wendy A. Troeger

Project Editor: Amy E. Tucker

Production Editor: Carolyn Miller

Technology Project Manager: Kim Schryer

Executive Marketing Manager: Donna J. Lewis

Channel Manager: Nigar Hale

Copyright © 2000

Delmar is a division of Thomson Learning. The Thomson Learning logo is a registered trademark used herein under license.

Printed in the United States of America

4 5 6 7 8 9 10 XXX 05 04 03 02

For more information, contact Delmar, 3 Columbia Circle, PO Box 15015, Albany, NY 12212-0515; or find us on the World Wide Web at http://www.delmar.com or http://www.EarlyChildEd.delmar.com

International Division List

Asia:
Thomson Learning
60 Albert Street, #15-01
Albert Complex
Singapore 189969

Japan:
Thomson Learning
Palaceside Building 5F
1-1-1 Hitotsubashi, Chiyoda-ku
Tokyo 100 0003 Japan

Australia/New Zealand:
Nelson/Thomson Learning
102 Dodds Street
South Melbourne, Victoria 3205
Australia

UK/Europe/Middle East:
Thomson Learning
Berkshire House
168-173 High Holborn
London
WC1V 7AA United Kingdom

Thomas Nelson & Sons LTD
Nelson House
Mayfield Road
Walton-on-Thames
KT 12 5PL United Kingdom

Latin America:
Thomson Learning
Seneca, 53
Colonia Polanco
11560 Mexico D.F. Mexico

South Africa:
Thomson Learning
Zonnebloem Building
Constantia Square
526 Sixteenth Road
P.O. Box 2459
Halfway House, 1685
South Africa

Canada:
Nelson/Thomson Learning
1120 Birchmount Road
Scarborough, Ontario
Canada M1K 5G4

Spain:
Thomson Learning
Calle Magallanes, 25
28015-MADRID
ESPAÑA

International Headquarters:
Thomson Learning
International Division
290 Harbor Drive, 2nd Floor
Stamford, CT 06902-7477

Library of Congress Cataloging-in-Publication Data

Herren, Ray V.
   Delmar's veterinary technician dictionary / Ray V. Herren, Janet Amundson Romich.
     p. cm.
   ISBN : 0-7668-1421-1
   1. Veterinary medicine—Dictionaries.   I. Title: Delmar's technician dictionary. II. Romich, Janet Amundson. III. Title.

SF609.H47 2000
636.089'03—dc21

00-021729

Contents

Preface

The role of the veterinary technician has become more technically demanding and varying in scope than in the past. Today's veterinary technician may be employed in private practice, referral practice, laboratory settings, or research facilities. The advancement of science and the increased amount of terminology used in the workplace have made understanding medical language in a precise and concise manner a key to success.

Delmar's Veterinary Technician Dictionary is an ideal tool to assist veterinary technicians in the understanding of veterinary terminology. This resource tool is easy for both the student and professional to carry with them for quick reference of spelling, pronunciation, and defining of terms used in everyday work situations. Terms in this dictionary include medical terms, production terms, and laboratory terms used in various fields related to animal care.

To make *Delmar's Veterinary Technician Dictionary* a quick reference tool, pronunciation keys are easy and straightforward to learn and adapt. Definitions are short and concise. Appendices are included at the back of the dictionary to aid in understanding of medical terms and as quick references for medical information. Appendices include an alphabetical listing of word parts and their definitions (appendix A), guidelines for making plural forms of medical terms (appendix B), abbreviations of medical terms and specialties (appendix C), and a quick reference guide of unit conversions (appendix D). *Delmar's Veterinary Technician Dictionary* is a perfect companion to other references used by students and professionals alike.

Pronunciation Guide for Veterinary Dictionary

Any vowel that has a dash above it represents the long sound, as in

ā	hay
ē	we
ī	ice
ō	toe
ū	unicorn

Any vowel followed by an "h" represents the short sound, as in

ah	apple
eh	egg
ih	igloo
oh	pot
uh	cut

Unique vowel combinations are as follows

oo	boot
ər	higher
oy	boy
aw	caught
ow	ouch

Pronunciation of Vowels

Vowel	Sound	Example
"a" at the end of a word	ah	idea
"ae" followed by r or s	ah	aerobic
"i" at the end of a word	ī	bronchi
"oe"	eh	oestrogen (old English form)
"oi"	oy	sarcoid
"eu"	ū	euthanasia
"ei"	ī	Einstein
"ai"	ay	air
"au"	aw	auditory

Exceptions to Consonant Pronunciations

Consonant	Sound	Example
"c" before e, i, and y	s	cecum
"c" before a, o, and u	k	cancer
"g" before e, i, and y	j	genetic
"g" before a, o, and u	g	gall
"ps" at beginning of word	s	psychology
"pn" at beginning of word	n	pneumonia
"c" at end of word	k	anemic
"cc" followed by i or y	first c=k second c=s	accident
"ch"	k	chemistry
"cn" in middle of word	both c and n c=k n=n	gastrocnemius
"mn" in middle of word	both m and n	amnesia
"pt" at beginning of word	t	pterodactyl
"pt" in middle of word	both p and t	optical
"rh"	r	rhinoceros
"th" at beginning of word		either as in thin or then
"x" at beginning of word	z	xylophone

Acknowledgments

The author and Delmar would like to thank those individuals who reviewed the manuscript and offered suggestions, feedback, and assistance. Their work is greatly appreciated.

Karl Peter, DVM
Foothills College
Los Altos Hills, CA

Carole Maltby, DVM
Maple Woods Community College
Kansas City, MO

Deb Donahue, LATG, Registered Labratory Animal Technologist
Medical College of Wisconsin
Milwaukee, WI

Ron Fabrizius, DVM, Diplomate ACT
Poynette Veterinary Service, Inc.
Poynette, WI

Kelly Gilligan, DVM
Four Paws Veterinary Clinic
Prairie du Sac, WI

John H. Greve, DVM PhD
Iowa State University
Ames, IA

Mark Jackson, DVM, PhD, Diplomate ACVIM, MRCVS
North Carolina State University
Raleigh, NC

Linda Kratochwill, DVM
Duluth Business College
Duluth, MN

Laura L. Lien, CVT
University of Wisconsin
Madison, WI

Sheila McGuirk, OVM, PhD, Diplomate ACVIM
University of Wisconsin
Madison, WI

James Meronek, DVM
Sauk Prairie Veterinary Clinic
Prairie du Sac, WI

Linda Sullivan, DVM
University of Wisconsin
Madison, WI

A

abacterial (ā-bahck-tehr-ē-ahl)–Free from bacteria.

abattoir (ahb-eh-twahr)–A slaughterhouse.

abdomen (ahb-dō-mehn)–The body cavity located between the diaphragm and the pelvis; in lower orders of animals and insects that part of the body that roughly corresponds to the abdomen; belly.

abdominal cavity (ahb-dohm-ih-nahl cah-vih-tē)–A hollow space that contains the major organs of digestion; located between the diaphragm and pelvic cavity; also called the peritoneal (pehr-ih-tō-nē-ahl) cavity.

abdominocentesis (ahb-dohm-ih-nō-sehn-tē-sihs)–A surgical puncture to remove fluid from the abdominal cavity.

abduction (ahb-duhck-shuhn)–To move away from the median plane or midline.

aberration (ahb-eh-ra-shuhn)–In genetics, an irregularity in chromosome distribution during cell division that results in deviation from normal.

abiotic disease (ā-bī-oh-tihck dih-zēz)–A disease caused by an inanimate agent; a nonparasitic disease. Also called physiogenic disease (fihz-ē-ō-jehn-ihck dih-zēz) e.g., a mineral deficiency in animals.

ablactate (ahb-lahck-tāt)–To wean.

ablation (ah-blā-shuhn)–Removal of a part by cutting; used in reference to salvaging the ear canal.

abnormal (ahb-nōr-mahl)–Deviating from that which is typical.

abomasopexy (ahb-ō-mahs-ō-pehcks-ē) –Surgical fixation of the fourth stomach compartment of ruminants to the abdominal wall.

abomasum (ahb-ō-mā-suhm)–The fourth stomach compartment of the ruminant; the true stomach.

aboral (ahb-ōr-ahl)–Away from the mouth.

abort (ah-bōrt)–In animals, to expel the embryo or fetus from the womb prematurely.

abortion (ah-bōr-shuhn)–Termination of pregnancy.

abortive (ah-bōr-tihv)–Defective or barren.

abrasion (ah-brā-shuhn)–An injury in which superficial layers of skin are scraped.

abscess (ahb-sehs)–A localized collection of pus.

absolute age (ahb-sō-loot āj)–Age in years rather than developmental age.

absorb (ahb-sōrb) To assimilate or incorporate one substance into another, as a blotter absorbs ink.

absorption (ahb-sōrp-shuhn)–The process of taking digested nutrients into the circulatory system.

absorption rate (ahb-sōrp-shuhn rāt)–The speed at which a substance enters the body.

abundance (ah-buhn-dehns)–The number of animals per unit area.

acariasis (ahck-eh-rī-eh-sihs)–The condition of animals infested with mites or ticks.

acaricide (ahck-eh-rih-sīd)–A substance, solution, or paste that kills mites or ticks.

Acarid (ahck-eh-rīd)–Any mite or tick of the family Acaridae or order Acarina.

acarine (ahck-eh-rihn)–Any mite or tick.

acclimate (ahck-leh-māt)–To become conditioned to a new climate or different environment.

acclimatization (ahck-lih-mah-tih-zā-shuhn)–Adjustment of an animal to a new environment.

accommodation (ah-kohm-ō-dā-shuhn)–The process of eye adjustments for seeing objects at various distances.

accredited herd (ah-krehd-ih-tehd hərd)–Dairy cattle certified to be free from tuberculosis as a result of two successive tests given under the direction of the United States Department of Agriculture. The term is sometimes erroneously applied to a brucellosis-free herd.

accuracy (of selection) (ahk-yehr-eh-sē ohf seh-lehck-shun)–The correlation between an animal's unknown actual breeding value and a calculated estimated breeding value.

acetabulum (ahs-eh-tahb-yoo-luhm)–The large, cup-shaped, articular socket in the pelvis that holds the ball-shaped head of the femur.

acetic acid (ah-sē-tihck ah-sihd)–An organic acid, CH_3COOH; also important as a product in lactic acid fermentation and, therefore, an important constituent of flavor in many milk products.

acetic bacteria (ah-sē-tihck bahck-tēr-ē-ah)–The bacteria that produce acetic acid from alcohol.

acetic fermentation (ah-sē-tihck fehr-mehn-tā-shuhn)–The process by means of which acetic acid is formed from ethyl alcohol in weak solution through the action of bacterial ferments.

acetone (ahs-ih-tōn)–CH_3COCH_3, a ketone; a clear, rapidly evaporating liquid that is obtained from fermentation of sugar and starch and is present in diabetic urine, breath, blood. Its presence in the breath of a lactating animal (cattle, sheep, milk goats) indicates her failure to oxidize the fatty material of her feed, making her deficient in carbohydrates.

acetonemia (ahs-ih-tōn-ē-mē-ah)–A disease characterized by the presence of excessive amounts of acetone bodies in the blood. Also called ketosis; false milk fever; chronic milk fever; acidosis; acetonuria; pregnancy disease in ewes.

achalasia (ahck-ah-lā-zē-ah)–Inability to relax the smooth muscle of the gastro-intestinal tract.

achondroplasia (ah-kohn-drō-plā-zē-ah)–Skeletal malformation during prenatal development caused by a genetic factor.

acid (ah-sihd)–Property of low pH or increased hydrogen ions. The term applied to any substance with a pH less than 7.0.

acid-fast (ah-sihd fahst)–Property of not being readily decolorized by acids.

acidified silage (ah-sihd-ih-fīd sī-lahj)–Silage preserved by the addition of acid, such as commercial phosphoric acid, sulfuric acid, or hydrochloric acid, or a combination of the latter two.

acidity (ah-sihd-ih-tē)–The measure of how many hydrogen ions a solution contains.

acidophilus (ahs-ih-dohf-eh-lehs)–Refers to organisms that grow well or exclusively in an acidic soil or medium.

acidosis (ah-sih-dō-sihs)–An abnormal condition of low body pH.

acoustic (ah-koo-stihck)–Sound.

acquired character (ah-kwīrd kahr-ahk-tər)–A change in character of an animal, morphological or physiological, due to the environment, which is not passed on to the next generation; not a genetic change.

acquired immunity (ah-kwīrd ihm-yoo-nih-tē)–The ability to resist a disease to which the individual would ordinarily be susceptible. It may be a result of antibodies built up through inoculation with pathogens or having the disease (active immunity), or it may result from the individual who has become immune (passive immunity) by receiving antibodies produced by the dam through colostrum or from another animal's serum or antiserum.

acre per animal unit month (ā-kər pər ahn-ih-mahl yoo-niht muhnth)–The estimated number of acres necessary to provide forage for one animal unit for one month under proper use.

acromegaly (ahck-rō-mehg-ah-lē)–Enlargement of the extremities caused by excessive growth hormone secretion.

acromion (ahck-rō-mē-ohn)–Outer end of the scapula to which the collarbone is attached.

acrosome (ahck-rō-zōm)–A small cap covering the head of the spermatozoa.

active agent (active ingredient) (ahck-tihv ā-jehnt, ahck-tihv ihn-grē-dē-ehnt)–That part of a chemical that has toxic properties to target species.

active immunity (ahk-tihv ihm-yoo-nih-tē)–Immunity following antigen exposure (natural disease or vaccination).

actual analysis (ahck-choo-uhl ah-nahl-ih-sihs)–The composition of a material based on a laboratory chemical analysis rather than a generalized guaranteed analysis.

actual use (ahk-choo-uhl ūs)–The use made of an area by livestock and/or game animals without reference to a recommended utilization; usually expressed in terms of animal units or animal unit months.

acuity (ah-kū-ih-tē)–Sharpness or acuteness; usually used in reference to vision.

acute (ah-kūt)–Short course, sudden onset; implies severe.

acute renal failure (ā-kūt rē-nahl fāl-yər)–Sudden onset of the inability of the kidney(s) to function; abbreviated ARF.

acute toxicity (ah-kūt tohcks-ih-sih-tē)–The potential of a substance to cause injury or illness when given in a single dose or in multiple doses over a period of 24 hours or less.

ad lib (ahd lihb)–As much as desired; abbreviation for ad libitum.

ad lib feeding (ahd lihb fē-dihng)–A system of feeding in which no limit is placed on feed intake.

ad libitum (ahd lihb-ih-tuhm)–The total voluntary intake when feed is available to animals at all times without restriction.

adaptability (ah-dahp-teh-bihl-ih-tē)–The capability of an organism to make changes that make it more fit for its environmental conditions.

adaptation (ahd-ahp-tā-shuhn)–The structures or activities of an organism, or of one or more of its parts, which tend to make it more fit for life in its environment or for particular functions.

additives (ahd-ih-tihvz)–Materials added to food to help manufacture and preserve it and to improve its nutritive value, palatability, and eye appeal. May be classified as emulsifiers, flavors, thickeners, curing agents, humectants, colors, nutrients, or as mold, yeast, or bacterial inhibitors. Amounts used in food are regulated by law.

addled egg (ahd-ld ehg)–An egg in which the yolk has become mixed with the white; sometimes a rotten egg, hence considered inedible.

adduction (ahd-duhck-shuhn)–To move toward the median plane or midline.

adenocarcinoma (ahd-eh-nō-kahr-sih-nō-mah)–A malignant growth of epithelial glandular tissue.

adenohypophysis (ahd-ehn-ō-hī-pohf-ih-sihs)–The anterior pituitary gland; portion of pituitary gland with glandular function.

adherence (ahd-hehr-ehns)–The act of sticking to a surface.

adhesion (ahd-hē-shuhn)–Band of fibers that hold structures together in an abnormal fashion.

adipocyte (ahd-ih-pō-sīt)–A fat cell.

adipose (ahd-ih-pōs)–Fatty or fat.

adjusted weaning weight (Adj. 205-day Wt.) (ahd-juhst-ehd wē-nihng wāt)–The weight of a calf at weaning, adjusted to a standard 205 days of age and adjusted for the age of the dam.

adjusted yearling weight (Adj. 365-day Wt.) (ahd-juhst-ehd yehr-lihng wāt)–The weight of a calf as a yearling, adjusted to a standard 365 days of age and adjusted for the age of the dam.

adjustments (ah-juhst-mehntz)–(1) Range management: changes in animal numbers, seasons of use, kinds or classes of animals, or management practices as warranted by specific conditions. (2) Ecological: the processes by which an organism becomes better fitted to its environment; functional, never structural. See *adaptation*.

adjuvant (ah-jū-vahnt)–(1) Any solid or liquid added to a substance, such as a pesticide or a fertilizer, to increase its effectiveness, e.g., solvents, diluents, carriers, emulsifiers, stickers, spreaders, or sometimes a pesticide to another pesticide or a fertilizer to another fertilizer. (2) A carrier for a biological that releases the biological into the bloodstream over an extended period, thus serving the function of a series of booster shots; consequently, the adjuvant helps lengthen the period of immunity provided by the biological.

adnexa (ahd-nehck-sah)–Accessory structures of an organ.

adrenal cortex (ahd-rē-nahl kōr-tehcks)–The outer portion of the adrenal gland.

adrenal gland (ahd-rē-nahl glahnd)–Gland that secretes hormones to regulate electrolytes, metabolism, sexual function, and the body's response to injury; located near the kidney; also known as suprarenal glands.

adrenal medulla (ahd-rē-nahl meh-doo-lah)–The inner portion of the adrenal gland.

adrenalectomy (ahd-rē-nahl-ehck-tō-mē)–The surgical removal of one or both adrenal glands.

adrenaline (ah-drehn-ah-lihn)–One of the hormones produced by the medulla of the adrenal glands; also called epinephrine; helps in preparing the body for emergency actions.

adrenopathy (ahd-rēn-ohp-ah-thē)–Disease of the adrenal glands.

adventitious sounds (ahd-vehn-tih-shuhs sowndz)–Pathological respiratory sounds; examples are rhonchi and rales.

aerobe (ahr-ōb)–Bacteria or other organisms that live only in free oxygen.

aerogen (ahr-ō-jehn)–Gas-producing bacteria.

aerophagia (ahr-ō-fā-jē-ah)–The swallowing of air.

afebrile (ā-fē-brī-ahl)–Without fever.

afferent (ahf-fər-ahnt)–To carry towards; in the nervous system afferent nerves carry sensory impulses toward the CNS; also called ascending tracts.

aflatoxin (ahf-lah-tohck-sihn)–The substance produced by some strains of the fungus *Aspergillus flavus*, the most potent natural carcinogen yet discovered; a persistent contaminant of corn, other grains, and peanuts.

afterbirth (ahf-tər-bihrth)–The matter that surrounds and attaches the fetus to the uterus. It is expelled after the fetus. Common name for placenta.

agalactia (ā-gahl-ahck-tē-ah)–A failure to secrete milk following the birth of offspring.

agalactic mare (ā-gahl-ahck-tihck mār)–A female horse not producing milk.

age class (ahj klahs)–A descriptive term to indicate the relative age grouping of animals.

aged horse (ahjd hōrs)–Correctly speaking, a horse 8 years of age or over; the term is often used to indicate a horse that is smooth-mouth, that is, 12 years of age or older.

agglutination (ah-gloo-tih-nā-shuhn)–The clumping together of cells or particles.

agglutinins (ah-gloot-n-ihnz)–(1) Antibodies produced in an animal's body in response to an infection or to the injection of microorganisms. They cause agglutination (clumping) of the organisms responsible for their formation. (2) Any antibody capable of clumping the organism that stimulated its production in the animal's body.

agonal breathing (ahg-uh-nuhl brē-thihng)–Respirations near death or during extreme suffering.

agonist (ah-gohn-ihst)–A substance that produces an effect by binding to an appropriate receptor.

agonistic (ahg-oh-nihs-tihck)–A type of animal behavior that involves offensive or defensive activities.

agoraphobia (ahg-ō-rah-fō-bē-ah)–A fear of open places; especially that occurring in animals kept in stalls for too long a period. Among horses, also called picket-line bound and barn rat.

agostadero (ahg-ohs-tah-dehr-ō)–A pasture used only in the summer (southwestern United States).

agouti (ah-goo-tē)–A naturally occurring coat color pattern that consists of dark-colored hair bands at the base of the hair and lighter increments of hair color toward the tip.

agranulocyte (ā-grahn-yoo-lō-sīt)–A cell that does not contain prominent grainlike structures in its cytoplasm.

agrarian (ah-grahr-ē-ahn)–Pertaining to agriculture.

agriculture (ahg-rih-kuhl-chuhr)–The broad industry engaged in the production of plants and animals for food and fiber, the provision of agricultural supplies and services, and the processing, marketing, and distribution of agricultural products.

aids (ādz)–The means by which a rider communicates with a horse (voice, hands, legs, seat, etc.)

air sacs (ahr sahcks)–Spaces in the respiratory tract of birds that store air and provide buoyancy for flight.

alate (ā-lāt)–Winged; having wings.

alba (ahl-bah)–A term meaning white.

albidus (ahl-bihd-uhs)–A term meaning white.

albino (ahl-bī-nō)–An animal with a white coat and pink eyes; devoid of melanin.

albumin (ahl-byoo-mihn)–A water-soluble protein that is coagulated by heat; found in milk, blood, egg white, and muscle.

albuminuria (ahl-bū-mihn-yoo-rē-ah)–The presence of albumin in the urine.

albus (ahl-buhs)–A term meaning white.

alcohol (ahl-koh-hohl)–The family name of a group of organic chemical compounds composed of carbon, hydrogen, and oxygen; a series of molecules that vary in chain length and are composed of a hydrocarbon plus a hydroxyl group, $CH_2=(CH_2)N=OH$; includes methanol, ethanol, isopropyl alcohol, and others.

alcohol test (ahl-koh-hohl tehst)–A test in which equal parts of milk and ethyl alcohol are mixed to discover abnormal milk or milk with unusual salt balances. Normal milk breaks clear of the test tube, while the curd or abnormal milk clings to the glass. The test also detects milk that is apt to coagulate during sterilization of condensed milk.

alcohol-alizarin test (ahl-koh-hohl ah-lihz-ah-rihn tehst)–A test in which alcohol and alizarin are added to milk. A color reaction of lilac-red indicates normal milk, yellow-brown indicates sour milk, and violet indicates mastitis milk.

aleukia (ah-loo-kē-ah)–The absence of leukocytes (white blood cells) in the blood.

alfalfa (ahl-fahl-fah)–A pasture or hay crop perennial having compound leaves with three leaflets. Dry alfalfa has 17.5% crude protein and 23.0% crude fiber.

alimentary tract (ahl-ih-mehn-tahr-ē trahck)–The body system responsible for the intake, digestion, absorption, and elimination of food or nutrients; located from mouth to anus; also called gastrointestinal tract, GI tract, or digestive tract.

alkaline (ahl-kah-līn or ahl-kah-lihn)–The property of high pH or decreased hydrogen ions; pH reading is above 7.

allantois (ahl-ahn-tō-ihs)–The innermost layer of the placenta.

allele (ah-lēl)–(1) The alternative forms of genes having the same place in homologous chromosomes that influence the development of alternative traits or characters. (2) A pair of Mendelian genes at the same locus as a pair of homologous chromosomes. Also called allelomorph.

allergen (ahl-ər-jehn)–A substance that produces an allergic response.

allergy (ahl-ər-jē)–An overreaction by the body to a particular antigen; also called hypersensitivity (hī-pər-sehn-sih-tihv-ih-tē).

allogamy (ahl-lohg-ah-mē)–Cross-fertilization.

allopolyploid (ahl-lō-pohl-eh-ployd)–A polyploid containing genetically different chromosome sets, e.g., from two or more species.

allosomes (ahl-lō-zōmz)–Chromosomes distinguished by peculiarities of behavior or sometimes by a difference in size or shape.

allotment (ah-loht-mehnt)–An area designated for the use of a prescribed number of cattle or sheep, or for common use of both.

alopecia (ahl-ō-pē-shah)–Hair loss resulting in hairless patches or complete lack of hair.

alpaca (ahl-pahck-ah)–(1) *Lama pacos;* a llama domesticated in Peru and adjoining countries. (2) The long wool of the alpaca which is woven into a cloth.

alter (ahl-tehr)–To neuter by castrating or spaying. Also called cut, geld, emasculate.

alternate grazing (ahl-tehr-naht grā-zihng)–Hanging pastures or ranges so that the forage grows back before it is grazed again. Also called rotational grazing.

alternate host (ahl-tehr-naht hōst)–A plant or animal upon which a disease organism exists for only a part of its life cycle, e.g., cedar and apple for rust.

alternation of generations (ahl-tehr-nā-shuhn ohf jehn-ər-ā-shuhnz)–Reproduction in which common characteristics are found only in every second generation, e.g., one generation reproduced sexually and the next asexually.

alveoli (ahl-vē-ō-lī)–A saclike dilation or socket. Usually refers to the small grapelike clusters of air sacs found at the end of each bronchiole; oxygen exchange units; singular is alveolus (ahl-vē-ō-luhs). Also dental sockets and structures of the udder.

amasesis (ahm-ah-sē-sihs)–Incapable of masticating (chewing).

ambidexterous (ahm-bih-dehck-struhs)–Able to use both hands well.

ambient (ahm-bē-ahnt)–Surrounding.

amble (ahm-buhl)–A lateral gait that is different from the pace by being slower and more broken in cadence.

ambler (ahm-blehr)–A horse whose best gait is pacing; a pacer.

amblotic (ahm-bloh-tihck)–Likely to cause abortion.

amblyopia (ahm-blē-ō-pē-ah)–Dimness or loss of sight without detectable eye disease.

ameba (ah-mē-bah)–A protozoan organism with no specific shape that moves in a flowing manner by the extension of a false foot or pseudopod.

amino acids (ah-me-nō or ahm-ih-nō ah-sihdz)–Organic substances from which organisms build proteins, or the end product of protein decomposition.

aminosis (ahm-ih-nō-sihs)–An abnormal condition of the body caused by excessive ingestion of amino acids.

amitosis (ahm-ih-tō-sihs)–Cell division without formation and splitting of chromosomes. See mitosis.

ammonia (ah-mō-nē-ah or ah-mōn-yuh)–NH_3; a chemical compound composed of 82.25% nitrogen and 17.75% hydrogen. At ordinary temperatures, it is a colorless, pungent gas about one-half as heavy as air. Sometimes added to livestock feed to increase nutritional value.

amnesia (ahm-nē-zē-ah)–Memory loss.

amniocentesis (ahm-nē-ō-sehn-tē-sihs)–Surgical punture with a needle through the abdominal and uterine walls to obtain amniotic fluid to evaluate the fetus.

amnion (ahm-nē-ohn)–The innermost membrane enveloping the embryo in the uterus; also called amniotic sac or bag of waters.

amniotic cavity (ahm-nē-oh-tihck kahv-ih-tē)–The space formed by the membrane surrounding the embryo in the uterus.

amniotic fluid (ahm-nē-oh-tihck flūihd)–The liquid found in the amniotic cavity that nourishes the fetus.

Amoeba (ah-mē-bah)–A genus of unicellular protozoan organisms of microscopic size, existing in nature in large numbers; many live as parasites; some species are pathogenic to humans.

amotus (ah-mō-tuhs)–Denoting the hind toe of certain birds, which does not touch the ground.

amphibian (ahm-fihb-ē-ahn)–A cold-blooded animal with legs, feet, and lungs that breathes by means of gills in the early stages and by means of lungs in the later stages of life.

amphigean (ahm-fih-jē-ahn)–Native of both the Old and New Worlds.

amphoteric (ahm-fō-tehr-ihck)–Capable of reacting as either an acid or a base, as casein.

amplitude (ahm-plih-tood)–The intensity of an ultrasound wave.

ampule (ahm-pūl)–A unit of packaging. Frozen semen is now shipped in straws instead of glass ampules.

ampulla (ahm-pūl-lah)–A funnel-shaped structure located at or near the end of a duct; for example, the part of the vas deferens where it joins the urethra in some species.

amputation (ahmp-yoo-tā-shuhn)–The removal of all or part of a body part (usually a limb).

amylase (ahm-ih-lās)–An enzyme secreted by the pancreas and delivered to the small intestine that aids in the digestion of starch.

amylopsin (ahm-ih-lohp-sihn)–A pancreatic enzyme which breaks down starch to sugars; the chief carbohydrate of pancreatic juice.

anabiosis (ahn-ah-bī-ō-sihs)–The resuscitation or revival of an organism after apparent death.

anabolism (ahn-ah-bōl-ihz-uhm or ahn-nahb-ō-lihzm)–A condition of building up; building up of body cells. The opposite is catabolism.

anadipsia (ahn-ah-dihp-sē-ah)–Excessive thirst.

anaerobe (ahn-ahr-ōb)–Organisms (usually bacteria) that live and multiply without free oxygen.

anaerobic (ahn-ah-rōb-bihck)–(1) Living or active in the absence of free oxygen. (2) Pertaining to or induced by organisms that can live in the absence of free oxygen.

anaerobic bacteria (ahn-ar-ō-bihck bahk-tehr-ē-ah)–Bacteria not requiring the presence of free or dissolved oxygen for metabolism. Facultative anaerobes are capable of adapting to an aerobic or anaerobic environment. Strict or obligate anaerobes are hindered or completely blocked by the presence of dissolved oxygen and sometimes by the presence of highly oxidized substances, such as sodium nitrates, nitrites, and, perhaps, sulfates. End-product gases include methane and hydrogen sulfide.

anal (ā-nahl)–Pertaining to the last abdominal segment, which bears the anus.

anal glands (ā-nahl glahndz)–Secretory tissue that are composed of apocrine and sebaceous glands located within the anal sac. The secretion of the anal glands is stored in the anal sacs and may play a role in marking territory, as a defense mechanism, or as a pheromone for sexual behavior.

anal sacculitis (ā-nahl sahck-yoo-lī-tihs)–Inflammation of the pouch(es) located around the anus.

anal sacs (ā-nahl sahcks)–A pair of pouches that stores an oily, foul-smelling fluid secreted by the anal glands located in the skin between the internal and external anal sphincters (located at the four o'clock and eight o'clock position). Each sac has a duct that opens to the skin at the anal orifice and fluid is expressed during defecation, excitement, or social interaction.

analgesia (ahn-ahl-jēz-ē-ah)–Without pain.

analgesic (ahn-ahl jē-zihck)–A substance that relieves pain without affecting consciousness.

analogous (ahn-ahl-oh-guhs)–Structures that differ anatomically but have similar functions.

analysis (ah-nahl-ih-sihs)–The percentage composition of fertilizers, feeds, etc., as determined by chemical analysis, expressed in terms specified by law.

anaphylaxis (ahn-ah-fih-lahck-sihs)–A severe response to a foreign substance; signs develop acutely and may include swelling, blockage of airways, tachycardia, and ptylism. Also called anaphylactic shock (ahn-ah-fih-lahk-tihk shohck).

anaplasia (ahn-ah-plā-zē-ah)–A change in the structure of cells and in their orientation to each other.

anastomosis (ah-nahs-tō-mō-sihs)–A surgical connection between two tubular or hollow structures; plural is anastomoses (ah-nas-tō-mō-sēs).

anatomic position (ahn-ah-tohm-ihck pō-sih-shuhn)–An animal in its normal standing position.

anatomy (ah-naht-ō-mē)–The study of body structures.

ancestor merit (ahn-sehs-tehr mehr-iht)–An unbiased estimate of the production of future daughters of a bull based on an accurate evaluation of the bull's sire and maternal grandsire.

ancestor (ahn-sehs-tehr)–An individual from whom an animal or person is descended.

androgenesis (ahn-drō-jehn-eh-sihs)–Development of offspring with the paternal chromosomes only.

androgynized cow (ahn-drohj-eh-nīzehd kow)–A cow that has received synthetic testosterone (male sex hormone). A bell-shaped marker is attached to the cow's neck to mark cows that are in estrus (heat) as she mounts them.

anechoic (ahn-eh-kō-ihck)–An ultrasonic term for when waves are transmitted to deeper tissue and none are reflected back.

anemia (ah-nē-mē-ah)–A blood condition of less that normal levels of red blood cells or hemoglobin.

anesthesia (ahn-ehs-thē-zē-ah)–The absence of sensation.

anesthetic (ahn-ehs-theht-ihck)–A substance that produces a lack of sensation.

anestrus (ahn-ehs-truhs)–The nonbreeding season; the period of time when a female is not cycling.

aneuploid (ahn-ū-ployd)–An organism or cell that has a chromosome number other than an exact multiple of the monoploid or basic number, i.e., hyperploids=higher; hypoploid=lower.

aneurysm (ahn-yoo-rihzm)–A localized balloon-like enlargement of an artery.

angiocardiogram (ahn-jē-ō-kahr-dē-ō-grahm)–A radiographic study of the blood vessels and heart using a contrast material.

angiogram (ahn-jē-ō-grahm)–A radio-graphic study of the blood vessels using contrast material.

angiography (ahn-jē-ohg-rah-fē)–A radiographic study of blood vessels after injection of radiopaque contrast material.

angiopathy (ahn-jē-ohp-ah-thē)–Disease of the vessels.

angioplasty (anh-jē-ō-plahs-tē)–The surgical repair of blood or lymph vessels.

angiorrhaphy (ahn-jē-ōr-ah-fē)–Suture of a vessel.

angora (ahn-gōr-ah)–A type of long fur on cats, rabbits, or goats.

anhidrosis (ahn-hī-drō-sihs)–A deficiency of sweat in any animal; also spelled anidrosis.

anhydrous milkfat (AMF) (ahn-hī-druhs mihlk-faht)–A product similar in composition to butteroil. However, it is made directly from cream and is used most commonly in milk-deficit countries to combine with nonfat dry milk to make fluid milk, ice cream, and other dairy products.

animal kingdom (ahn-ih-mahl kihng-dohm)–All animals collectively. In the kingdom Animalia there are twelve branches or phyla ranging from I, Protozoa, to XII, Chordata, which includes the class Mammalia in the subphylum Vertebrata.

animal rights (ahn-ih-mahl rītz)–The philosophy that animals have the same rights as humans and that they should not be used for human consumption. See animal welfare.

animal therapy (ahn-ih-mahl thehr-ah-pē)–The use of domestic animals to provide therapy for humans; to relieve lonesomeness, make them feel needed, and soothe the emotional self.

animal type (ahn-ih-mahl tīp)–The combination of characteristics of an animal appropriate to a special kind of use or desirability, such as beef type.

animal unit (ahn-ih-mahl yoo-niht)–A measurement based on the amount of feed eaten and manure produced by an average mature horse, cow, or the equivalent; as two heifers over one year, four calves under one year, seven ewes or bucks, two and one-half brood sows or boars, five hogs raised to 200 pounds each, or 100 hens.

animal unit months (AUMs) (ahn-ih-mahl yoo-niht munthz)–The amount of grazing required by a 1,000-pound (454 kg) cow or equivalent weight of other domestic animal for one month.

animal welfare (ahn-ih-mahl wehl-fahr)–The philosophy that animals should be treated in a kind and caring manner.

anisocoria (ahn-ih-sō-kō-rē-ah)–The condition of unequal pupil size.

anisocytosis (ahn-eh-sō-sī-tō-sihs)–The condition of unequal cells.

ankylosis (ahng-kih-lō-sihs)–The loss of mobility in a joint (may be due to injury, disease, or surgery).

anogenital distance (ā-nō-jehn-ih-tahl dih-stahns)–The area between the anus and genitalia; females have a shorter anogenital distance than males; is used to determine the sex of animals.

anomaly (ah-nohm-ah-lē)–Deviation from normal or deviation from what is regarded as normal.

anophthalmos (ahn-ohp-thahl-mōs)–Without development of one or both eyes.

anoplasty (ah-nō-plahs-tē)–The surgical repair of the anus.

anorectal (a-nō-rehck-tahl)–The distal portion and caudal opening of the large intestine.

anorexia (ahn-ō-rehck-sē-ah)–The lack of or loss of appetite.

anoxia (ā-nohck-sē-ah)–The absence of oxygen.

antagonism (ahn-ta-gohn-ihzm)–The loss of activity of a chemical when exposed to another chemical.

antagonist (ahn-tā-gohn-ihst)–A substance that inhibits a specific action by binding with a particular receptor instead of allowing the agonist to bind to the receptor.

antebrachium (ahn-tē-brā-kē-uhm)–The portion of the forelimb between the elbow and carpal joints.

antefebrile (ahn-teh-fē-brī-ahl)–Before the onset of fever.

antepartum (ahn-tē-pahr-tuhm)–Before the onset of fever.

anterior (ahn-tēr-ē-ər)–The front of the body (in quadrupeds and arthropods the head is on the anterior end of the body; in bipeds the ventral surface is the front of the body).

anterior chamber (ahn-tēr-ē-ōr chăm-bər)–The aqueous-containing space located in the eye behind the cornea and in front of the iris.

anterior pituitary (ahn-tēr-ē-ōr pih-too-ih-tār-ē)–The front part, or lobe, of the pituitary gland, located at the base of the brain. It produces and secretes specific hormones, regulates growth of the body, and regulates and controls the thyroid, adrenal cortex, ovaries, and testicles.

anterior presentation (ahn-tēr-ē-ōr prē-sehn-tā-shuhn)–The normal birth position of mammals; front feet and head presented first.

anterior segment (ahn-tēr-ē-ōr sehg-mehnt)–The cranial one-third of the eyeball that is divided into anterior and posterior chamber.

anthelmintic (ahn-thehl-mihn-tihck)–A substance that works against intestinal worms.

antiarrhythmic (ahn-tih-ah-rihth-mihck)–A substance that controls heartbeat irregularities.

antibacterial (ahn-tih-bahk-teh-rē-ahl)–Any substance that has the ability, even in dilute solutions, to destroy or inhibit the growth or reproduction of bacteria and other microorganisms.

antibiosis (ahn-tih-bī-ō-sihs)–An association between two organisms in which one harms the other.

antibiotic (ahn-tih-bī-oh-tihck)–A substance that inhibits the growth of or kills bacteria.

antibody (ahn-tih-boh-dē)–A disease-fighting protein produced by the body in response to the presence of a specific antigen.

anticarcinogen (ahn-tih-kahr-sihn-ō-jehn)–A substance that inhibits or eliminates the activity of a carcinogen (cancer-producing substance).

anticoagulant (ahn-tih-kō-ahg-yoo-lahnt–A substance that slows and prevents blood clotting.

anticonvulsant (ahn-tih-kohn-vuhl-sahnt)–A substance that prevents seizures.

antidiarrheal (ahn-tih-dī-ah-rē-ahl)–A substance that prevents diarrhea or frequent bowel movements.

antidiuresis (ahn-tih-dī-yoo-rē-sihs)–A decrease in the production and excretion of urine. See diuresis.

antidote (ahn-tih-dōt)–A remedy for counteracting a poison.

antiemetic (ahn-tih-ē-meh-tihck)–A substance that prevents vomiting.

antigen (ahn-tih-jehn)–A substance that the body regards as foreign (such as a virus, bacterium, or toxin).

antihypertensive (ahn-tih-hī-pər-tehn-sihv)–A substance that lowers blood pressure.

antimutagen (ahn-tih-moo-tih-jehn)–A substance that inhibits or eliminates the activity of a mutagen. See mutagen.

antineoplastic agent (ahn-tih-nē-ō-plah-stihck ā-gehnt)–A substance that treats neoplasms; usually used against malignancies.

antioxidant (ahn-tih-ohck-sih-dahnt)–A compound that prevents oxidation. Used in mixed feeds to prevent rancidity or loss of vitamin potency.

antipruritic agent (ahn-tih-pər-ih-tihck ā-gehnt)–A substance that controls itching.

antipyretic (ahn-tih-pī-reh-tihck)–A substance that reduces fever.

antisepsis (ahn-tih-sehp-sihs)–The prevention of infection by the exclusion, inhibition, or destruction of the causative organisms.

antiseptic (ahn-tih-sehp-tihck)–A chemical agent that kills or prevents the growth of microorganisms on living tissue.

antiserum (ahn-tih-sēr-uhm)–A serum that contains specific antibodies extracted from a hyperimmunized animal or an animal that has been infected with the microorganisms that contain the antigen.

antispasmodic (ahn-tih-spahz-mohd-ihck)–Any drug that prevents or counteracts spasms (muscular, intestinal, bronchial, etc.)

antitoxin (ahn-tih-tohcks-sihn)–A specific antiserum aimed against a poison that contains a concentration of antibodies extracted from the serum or plasma of a healthy animal.

antitussive (ahn-tih-tuhs-ihv)–A substance that reduces coughing.

antivivisectionist (ahnt-ih-vihv-ih-sehck-shuhn-ihst)–A person who opposes surgery on live animals for research or educational purposes.

antler (ahnt-lər)–A deciduous, ossified, protrusion originating from the skull.

antrum (ahn-truhm)–The cavity in a hollow organ or structure, such as the cavity inside a developing follicle.

anuria (ā-nū-rē-ah)–The complete suppression of urine production.

anus (ā-nuhs)–The caudal or aboral opening of the gastrointestinal tract.

aorta (ā-ōr-tah)–The main artery that originates from the left ventricle of the heart.

aortic semilunar valve (ā-ōr-tihck sehm-ē-loo-nahr vahlv)–The membranous fold located between the left ventricle and the aorta; semilunar means half moon.

apathy (ahp-ah-thē)–Indifference.

aperture (ahp-ər-chər)–An opening.

apex (ā-pehcks)–The tip.

aphakia (ah-fahk-ē-ah)–The absence of a lens.

aphtha (ahf-thah)–A vesicle or sac containing a thin fluid on the udder, inside the mouth, or sometimes between the toes of a cloven-footed animal. The condition is characteristic of foot-and-mouth disease.

aplasia (ā-plā-zē-ah)–The lack of development of an organ or tissue.

apnea (ahp-nē-ah)–Without or the absence of breathing.

aponeurosis (ahp-ō-nū-rō-sihs)–A fibrous sheet that gives attachment to muscular fibers and serves as a means of origin or insertion of a flat muscle.

appendage (ah-pehn-dihj)–An attachment to something larger; a leg or limb.

appendicular skeleton (ahp-ehn-dihck-yoo-lahr)–The bones of the extremities, including the shoulder and pelvic girdle.

appositional (aph-ō-sih-shuhn-ahl)–Placing side to side.

approach (ah-prōch)–In surgery, the specific procedure by which an organ or part is exposed.

apterium (ahp-tehr-ē-uhm)–Areas or tracts of skin without feathers or down; plural is apteria (ahp-terh-ē-ah).

apterous (ahp-tehr-ohs)–Wingless.

aqueous humor (ah-kwē-uhs hū-mər)–The fluid that nourishes the intraocular structures in the anterior segment of the eye.

arachidonic acid (ah-rahck-ih-dohn-ihck ah-sihd)–A substance essential to body tissues built by the animal body from the simpler fatty acids that are derived from the food fats.

arachnid (ah-rahck-nihd)–A group of arthropods having four pairs of legs and one or two body segments; mites, ticks, and spiders are examples.

arachnoid (ah-rahck-noyd)–Cobwebby; of slender entangled hairs.

arachnoid membrane (ah-rahck-noyd mehm-brān)–The second layer of the meninges.

arch (ahrch)–A curve.

archetype (ahr-kih-tīp)–In biology, the antecedent of a group of plants or animals from which certain typical characteristics have been inherited; a progenitor.

arena (ah-rē-nah)–The sphere of action, such as a livestock arena.

areole (ah-rē-ōl)–(1) A small area, especially the open space between anastomosing veins.

ark (ahrk)–A small house with ridged roof, often used for a small number of poultry rabbits.

arrector pilus (ah-rehck-tər pī-luhs)–A tiny muscle attached to the hair follicle that causes hair to stand up; plural is arrector pili (pī-lī).

arteriectomy (ahr-teh-rē-ehck-tō-mē)–The surgical removal of part of an artery.

arterioles (ahr-tē-rē-ōlz)–The smaller branches of arteries.

arterioplasty (ahr-tē-rē-ō-plahs-tē)–The surgical repair of an artery.

arteriosclerosis (ahr-tē-rē-ō-sklehr-ō-sihs)–The hardening and thickening of the arteries.

arteriotomy (ahr-tē-rē-oh-tō-mē)–An incision of an artery.

artery (ahr-tər-ē)–A blood vessel that carries blood away from the heart.

arthralgia (ahr-thrahl-jē-ah) –Joint pain.

arthritis (ahr-thrī-tihs)–An inflammatory condition of the joints.

arthrocentesis (ahr-thrō-sehn-tē-sihs)–A surgical puncture of the joint space to remove synovial fluid.

arthrodesis (ahr-thrō-dē-sihs)–The fusion of a joint or the spinal vertebrae by surgical means.

arthrodynia (ahr-thrō-dihn-ē-ah)–Joint pain.

arthrogram (ahr-thrō-grahm)–A film record of a radiographic examination of a joint after injecting a contrast material.

arthrography (ahr-throhg-rah-fē)–A radiographic examination of a joint after injecting contrast material.

arthrology (ahr-throh-lō-jē)–The study of joints.

arthropathy (ahr-throhp-ah-thē)–Any disease affecting a joint.

arthropod (ahr-thrō-pohd)–An individual of the plylum Arthropoda which includes insects, spiders, and crustacea; characterized by a coating which serves as an external skeleton and by legs with distinct movable segments or joints.

arthrosclerosis (ahr-thrō-skleh-rō-sihs)–A hardening or stiffening of the joints.

arthroscope (ahr-thrō-skōp)–An endoscopic instrument used to visually examine the internal structure of a joint.

arthroscopy (ahr-throhs-kō-pē)–Endoscopic examination of the internal structure of a joint.

articulate (ahr-tihck-yoo-lāt)–To join or to come together in a manner that allows motion between the parts; also means to speak clearly.

articulation (ahr-tihck-yoo-lā-shuhn)–The connection between two bones.

artificial hormone (ahr-tih-fih-shahl hōr-mōn)–A manufactured substance that is used in place of a naturally produced hormone.

artificial insemination (ahr-tih-fih-shahl ihn-sehm-ihn-ā-shuhn)–A breeding method in which semen is collected, stored, and deposited in the uterus or vagina without copulation; abbreviated AI.

artificially acquired immunity (ahr-tih-fih-shahl-ē ah-kwīrd ihm-yoo-nih-tē) –Immunity that comes about as a result of a vaccination.

artiodactyla (ahr-teh-ō-dahck-tih-lah)–The zoological order of mammals that includes hoofed animals with an even number of toes; includes cattle, sheep, and swine.

ascending (ā-sehnd-ihng)–Progressing upward or cranially.

ascites (ah-sī-tēz)–An abnormal accumulation of fluid in the peritoneal cavity.

ascorbic acid (vitamin C) (ahs-kōr-bihck ah-sihd)–A chemical compound, $C_6H_8O_6$, that occurs in fruits and vegetables and prevents scurvy in mammals.

asepsis (ā-sehp-sihs)–A state without infection.

aseptic technique (ā-sehp-tihck tehck-nēck)–The precautions taken to prevent contamination of a surgical wound.

as-fed basis (ahs fehd bā-sihs)–The amount of nutrients in a diet expressed in the form in which it is fed; sometimes referred to as air-dry.

asper (ahs-pehr)–Rough.

asphyxia (ahs-fihck-sē-ah)–An abnormal condition resulting from lack of oxygen intake.

asphyxiation (ahs-fihck-sē-ā-shuhn)–An interruption of breathing resulting in lack of oxygen; also called suffocation.

aspirate (ahs-pər-āt)–The act of inhaling; removing fluid or gas via suction.

assay (ahs-ā)–An assessment or test to determine the number of organisms, cells, or amount of a chemical substance found in a sample.

assess (ah-sehs)–To evaluate.

assessment (ah-sehs-mehnt)–An evaluation.

assimilation (ah-sihm-ih-lā-shuhn)–Absorption.

ass (ahs)–(1) *Equus asinus;* the donkey; a domesticated beast of burden used in the breeding of mules by crossing an ass with a female horse. (2) Any of several species of *equus;* e.g., wild asses of Asia and North Africa.

asthenia (ahs-theh-nē-ah)–Debility, or the lack of body strength.

asthma (ahz-mah)–A chronic allergic disorder.

astomous (ah-stō-muhs)–Mouthless.

astringent (ah-strihn-jehnt)–A drug, such as tannic acid, alum, and zinc oxide or zinc sulphate, that causes contraction of tissues.

astrocyte (ahs-trō-sīt)–A star-shaped cell; a type of neuroglial cell.

astrocytoma (ahs-trō-sī-tō-mah)–A malignant intracranial tumor composed of astrocytes (a type of neuroglial cell).

asymmetrical (ā-sihm-meht-rih-kahl)–Without proper proportion of parts; unsymmetrical.

asymptomatic (ā-sihmp-tō-mah-tihck)–Without signs of disease.

asystole (ā-sihs-tō-lē)–Without contraction; lack of heart activity.

atavism (aht-ah-vihzm)–Reversion; reappearance of a characteristic or disease after a lapse of one or more generations.

ataxia (ā-tahck-sē-ah)–The inability to voluntarily control muscle movement; "stumbling."

atelectasis (aht-eh-lehk-tah-sihs)–The incomplete expansion of the alveoli or collapse of alveoli resulting from resorption of air.

atherosclerosis (ahth-ər-ō-skleh-rō-sihs)–Hardening and narrowing of the arteries resulting from fatty deposits.

atlas (aht-lahs)–Cervical vertebra 1.

atom (aht-ohm)–The smallest unit of an element to retain the chemical characteristics of that element. It consists of negatively charged particles called electrons orbiting around the nucleus and, within the nucleus, protons (which are positively charged and balance the extranuclear electrons) and neutrons (which have no charge).

atomization (aht-ohm-ih-zā-shuhn)–The process of breaking a liquid into fine spray.

atonic (ā-tohn-ihck)–Lacking muscle control.

atopy (ah-tō-pē)–Hypersensitivity reaction in animals involving pruritus with secondary dermatitis; commonly called allergies.

atraumatic (ā-traw-mah-tihck)–Pertaining to, resulting from, or caused by a noninjurious route.

atresia (ah-trez-ah)–An occlusion or the absence of a normal body opening or tubular organ.

atrioventricular (ā-trē-ō-vehn-trihck-yoo-lahr)–The atrium and ventricle.

atrioventricular node (ā-trē-ō-vehn-trihck-yoo-lahr nōd)–A small mass of tissue located in the right atrium that transmits electric impulses from the sinoatrial node to the bundle of His.

atrioventricular valve (ā-trē-ō-vehn-trihck-yoo-lahr vahlv)–The membranous fold located between the superior and inferior chambers of the heart; described as left or right.

atrium (ā-trē-uhm)–The superior chamber of the heart; plural is atria (ā-trē-ah).

atrophy (aht-rō-fē)–The wasting of tissue; "without growth."

attaint (ah-tānt)–A wound on a horse's leg caused by its own hooves.

attenuate (ah-tehn-ū-āt)–To dilute, thin down, enfeeble or reduce (the virulence of an organism).

attenuation (ah-tehn-yoo-ā-shuhn)–The loss of intensity of the ultrasound beam as it travels through tissue.

atypical (ā-tihp-ih-kahl)–Disagreeing with the form, state, or situation usually found under similar circumstances; not typical.

aubin (aw-bihn)–An irregular gait of a horse; slower than a gallop but resembling it.

auditory (aw-dih-tōr-ē)–Hearing.

auditory ossicles (aw-dih-tōr-ē ohs-ih-kulz)–The three little bones of the middle ear that transmit sound vibrations.

aural (awr-ahl)–The ear.

aural hematoma (awr-ahl hēm-ah-tō-mah)–A collection or mass of blood on the outer ear.

auscultation (ahws-kuhl-tā-shuhn)–Listening to body sounds (usually involves the use of a stethoscope).

autoclave (aw-tō-klāv)–An apparatus for sterilizing by steam under pressure.

autogenous vaccine (aw-tohj-eh-nuhs vahck-sēn)–A vaccine made from the patient's own bacteria instead of from stock cultures.

autoimmune disease (aw-tō-ihm-yoon dih-zēz)–A disorder in which the body makes antibodies directed against itself.

autoimmunity (aw-tō-ihm-yoo-nih-tē)–Production by an animal of an allergic reaction to its own tissues, which may produce clinical disease.

autolysis (aw-tohl-ih-sihs)–Self-digestion. The natural softening process of fruits or vegetables after picking, or meat after slaughtering.

autonomic nervous system (aw-tō-nah-mihck nər-vuhs sihs-tehm)–The portion of the peripheral nervous system that consists of nerves that control involuntary action; abbreviated ANS.

autosome (aw-tō-zōm)–A non-sex-determining chromosome.

autotrophic (aw-tō-trohf-ihck)–Refers to organisms that supply their own food source.

autumn lag (aw-tuhm lahg)–The period during the autumn and early winter when rabbits and pigeons slow down or stop reproducing.

available energy–See metabolizable energy.

average daily gain (ADG) (ahv-ər-ihj dā-lē gān)–The calculation of an animal's postweaning gain figured by dividing the weight gain by the days on feed.

avian (ā-vē-ahn)–Refers to birds; the class Aves.

aviary (ā-vē-ehr-ē)–A place for keeping birds; pigeon loft.

avicide (ā-vih-sīd)–A substance used to kill birds.

aviculture (ā-vih-kuhl-chuhr)–The caring for and raising of birds.

avirulent (ā-vihr-ū-lehnt)–Without the ability to produce disease.

avitaminosis (ā-vī-tah-mih-nō-sihs)–A condition or disease of an animal resulting from a vitamin deficiency.

avoidance (ah-voyd-ahns)–A response whereby organisms prolong their dormancy, thereby achieving less vulnerability to environmental stresses.

avulsion (ā-vuhl-shuhn)–Tearing away of a part.

axenic (ā-zehn-ihck)–Germ free.

axial feather (ahck-sē-ahl fehth-ər)–The short feather in the middle of the wing of a bird, which separates the primary from the secondary feathers.

axial skeleton (ahcks-ē-ahl skehl-eh-tohn)–The skull, hyoid bones, vertebral column, ribs, and sternum.

axillary (ahcks-ihl-ār-ē)–The armpit.

axis (ahcks-ihs)–Cervical vertebra 2.

axon (ahcks-ohn)–The structure of the neuron that extends away from the cell body and conducts impulses away from the nerve cell.

axonal (ahck-son-ahl)–Axon.

azoospermia (ā-zō-ō-spər-mē-ah)–The absence of sperm in the semen; also called aspermia (ā-spər-mē-ah).

azotemia (ā-zō-tē-mē-ah)–The presence of urea or other nitrogenous elements in the blood.

azoturia (ahz-ō-tū-rē-ah)–Excessive urea or other nitrogen compounds in the urine.

B

Babcock test (bahb-kohk tehst)–A test of the fat in milk that involves certrifugation with sulfuric acid. It was developed by Dr. S. M. Babcock at the University of Wisconsin in 1890. Also known as Babcock's test.

bacillus (bah-sihl-luhs)–A morphologically descriptive term for bacteria meaning rod-shaped. Plural is bacilli (bah-sihl-ī).

back (bahck)–Trotting in reverse.

backcross method of mating (bahck-krohs meth-ohd ohf mā-tihng)–The breeding of F_1 females to a male of the same breed as the F_1's parents.

bacteremia (bahck-tēr-ē-mē-ah)–A disease caused by the presence of bacteria in the blood.

bactericide (bahck-tēr-ih-sīd)–A substance that destroys bacteria; a germicide.

bacterin (bahck-tər-ihn)–A killed bacterial vaccine.

bacteriolysis (bahk-tēr-rē-ohl-ih-sihs)–The destruction or dissolution of bacteria inside or outside the animal body.

bacteriophage (bahck-tēr-rē-ō-fāj or bahk-tē-rē-ō-fāhj)–A viruslike bacteria-destroying agent that can propagate itself only in the presence of young, active, susceptible bacteria.

bacteriostat (bahck-tē-rē-ō-staht)–A compound that inhibits growth and reproduction of or kills certain bacteria.

bacterium (bahck-tē-rē-uhm)–A microscopic, unicellular organism; plural is bacteria (bahck-tē-rē-ah).

bacteriuria (bahck-tē-rē-yoo-rē-ah)–The presence of bacteria in the urine.

bad mouth (bahd mowth)–A malocclusion where the top and bottom teeth do not meet.

bag (bahg)–The udder; the scrotum; a sack; the total kill in a hunter's possession.

bag up (bahg uhp)–The development of mammary glands or udder near parturition; also called bagging up.

balance (bahl-ehns)–(1) The skeletal and muscular makeup of an animal, which gives the animal visual appeal. A well-balanced animal's body parts appear to fit together and blend harmoniously and symmetrically. (2) A device for determining weights. (3) To put in proper proportions, such as balancing a feed ration. (4) A term used to express the ratio between the resources of land, labor, capital, and management that attains the optimum use of each resource in the production of crops and livestock to maximize financial returns and to maintain or improve soil productivity.

balanced ration (bahl-ehnsd rah-shuhn)–The daily allowance of livestock or fowl feed; mixed to contain suitable nutrients required to promote normal development, maintenance, lactation, gestation, etc.

bald face (bahld fās)–A wide white marking that extends beyond both eyes and nostrils.

baldy (bahl-dē)–An animal that is naturally polled or hornless.

balk (bahlk)–The act of an animal stopping and refusing to move.

ball-and-socket joint (bahl ahnd sohck-eht joynt)–A type of synovial joint that allows free movement in all planes; also called enarthrosis (ehn-ahr-thrō-sihs).

balling gun (bahl-ihng guhn)–A tool used to administer pills or magnets to livestock.

ballottement (bahl-oht-mehnt)–A diagnostic technique of hitting or tapping the wall of a fluid-filled structure to bounce a solid structure against its wall; used for pregnancy diagnosis and to determine abdominal contents.

band (bahnd)–A large group of range sheep.

band cell (bahnd sehl)–An immature polymorphonuclear leukocyte.

bandage (bahn-dahj)–(1) To cover by wrapping or the material used to wrap. (2) A cloth used for dressing wounds.

banding (bahn-dihng)–(1) A style of mane that is sectioned and fastened with rubber bands; seen in Western show horses. (2) The marking of animals or birds by fastening bands to their legs or ears or by painting their feathers for identification. (3) A method of castration in which a tight rubber band is placed around the scrotum; circulation is stopped and the testicles atrophy.

bar (bahr)–(1) The mouthpiece of a bridle bit. (2) The part of a horse's hoof that is bent inward, extending toward the center of the hoof. (3) Ridges in the roof of a horse's mouth. (4) The space in front of a horse's molars where the bit is placed.

bar pad (bahr pahd)–A leather covering for a horse's hoof to prevent slipping. It extends to each heel over the bars of the hoof and is fitted as part of a short shoe having a soft pad to cover the frog. See bar shoe.

bar shoe (bahr shoo)–A horseshoe that has a bar plate consisting of a flat piece of metal across the heel opening to protect the frog of the hoof. See bar pad.

barb (bahrb)–(1) One of the parallel filaments projecting from the main feather shaft (rachis); forms the feather vane. (2) A breed of horses related to the Arab and native to Barbary; probably introduced into Spain by the Moors and probably smaller

and coarser than the Arab; its strain is evident in all known present breeds. (3) Mucous membrane projections for the openings of the submaxillary glands under the tongue of horses and cattle.

barbering (bahr-bər-ihng)–A behavior disorder in animals where dominant animals bite or chew the fur from subordinate animals.

barbule (bahr-byoo-uhl)–One of the small projections fringing the edges of the barbs of feathers; attach to adjacent barbules to give the vane rigidity.

bareback (bār-bahck)–(1) A poultry term denoting a chicken that is not fully feathered out on the back. (2) To ride bareback is to ride a horse without a saddle.

barium sulfate (bār-ē-uhm suhl-fāt)–A contrast material; also called barium.

barn (bahrn)–A farm building used for storage of hay, grain, farm implements, etc., or for housing domestic animals.

barn book (bahrn book)–The record book in which breeding and management data on livestock are kept. Commonly included are breeding dates, birth dates, health data, and weight data.

barnyard (bahrn-yahrd)–Land immediately adjacent to a barn, fenced in to enclose livestock or fowls; also called barn lot.

barny (bahr-nē)–A term used in the judging of dairy products to designate an off flavor characterized by cow-stable odor and its persistence on the palate.

barred (bahrd)–A trait in chickens involving the alternating white and dark stripes on the feathers.

barrel (bahr-ahl)–The circumference of the chest or trunk of equine; used as a guideline for feed capacity.

barren (bār-ehn)–Not able to reproduce.

barren mare (bār-ehn mār)–An intact female horse that was not bred or did not conceive in the previous season; also called open mare.

barrier sustained (bār-ē-ər suh-stānd)–Gnotobiotic animals that are maintained under sterile conditions in a barrier unit.

barrow (bār-ō)–(1) Male pig that was castrated when young. (2) In New Zealand, to shear or partially shear a sheep.

bars (bahrz)–A support structure that angles forward from the hoof wall to keep it from overexpanding; also the gap between a horse's incisors and molars; also the side points on the tree of a saddle.

barton (bahr-tuhn)–(1) A poultry yard or farmyard. (2) Farm buildings attached to a farmhouse; common in New England.

basal feed (bā-sahl fēd)–A feed used primarily for its energy content.

basal metabolism (bā-sahl meh-tahb-ō-lihzm)–The energy necessary to sustain the vital functions of cellular change, respiration, and circulation when an organism is at rest.

base exchange capacity (bās ehcks-chānj kah-pah-siht-ē)–The maximum capacity of the blood to exchange base (alkali) for acid in maintaining a physiologic balance.

base narrow (bās nahr-ō)–The term used to describe a horse that stands with front or rear feet close together, yet standing with legs vertical.

base wide (bās wīd)–The term used to describe a horse that stands with front or rear feet wide apart, yet with legs vertical.

basic (bā-sihck)–On the alkaline side of neutral pH (above 7.0).

basophil (bā-sō-fihl)–A class of granulocytic leukocyte that promotes the inflammatory response.

basophilic (bā-sō-fihl-ihck)–Things that stain readily with basic, or blue dyes in many commonly used stains such as H & E, Giemsa's, and Wright's.

battery (baht-eh-rē)–In poultry husbandry, a series of cages used for raising chickens without a hen, for fattening chickens and broilers for market, and for increasing egg production in laying hens.

bay (bā)–A horse with a body color ranging from tan to red to reddish-brown. The mane and tail are black, and the lower legs are usually black.

bay roan (bā rōn)–A coat color of a horse that is brown with gray.

beak (bēk)–A modified epidermal structure that covers the rostral part of the maxilla and mandible.

beard (behrd)–(1) Hairy appendages on the face of an animal, such as the whiskers of a goat. (2) A group of feathers hanging from the throat of certain breeds of chickens. (3) A tuft of bristly hair projecting from the upper part of a turkey's breast. (4) The part of a horse's lower jaw that carries the curb of the bridle.

bearing rein (behr-ihng rān)–A rein pushed against the neck of a horse toward the direction of turn; also called neck rein.

beast (bēst)–A four-footed animal.

beat (bēt)–The time when the foot (or feet if simultaneous) touches the ground.

bed (behd)–(1) A place for animals to sleep. (2) The hauling platform of a wagon or truck.

bedding (behd-ihng)–The material used to cushion the animal's shelter.

beef (bēf)–(1) Any cow, bull, or steer that is fattened for slaughter for food. (2) The meat derived from cattle nearly one year old or older.

beget (bih-geht)–To procreate, as a sire.

bellwether (behl wheth-ər)–A wether (a male sheep castrated before sexual maturity) with a bell tied around his neck who serves as a flock leader.

belly (beh-lē)–Common term for an animal's abdomen.

belly band (behl-ē bahnd)–Common term for an abdominal wrap; circumferentially wrapping the abdomen with bandages to apply pressure to this area.

belt (behlt)–A band of hair or skin of another color, often white, around an animal's body, as in Dutch belted cattle.

belted (behl-tehd)–Any animal having a band of different color around its body.

benign (beh-nīn)–Causing little or no harm; not malignant.

bezoar (bē-zōr)–Stones, hair balls, wool balls, etc. found in the alimentary canals of ruminants and sometimes other animals. At one time these were thought to have antipoison properties. See hair ball, trichobezoar.

biceps (bī-sehpz)–A muscle with two heads.

bicolor (bī-kuhl-hər)–Having two colors.

bicornuate (bī-kōrn-yoo-āt)–Having two horns (as in uterus); also called bicornate (bī-kohr-nāt).

bidentate (bī-dehn-tāt)–Having two teeth.

bifurcation (bī-fər-kā-shuhn)–Dividing in two.

bilateral (bī-lah-tər-ahl)–Two sides.

bilateral symmetry (bī-lah-tər-ahl sihm-eh-trē)–Similarity of form, one side with the other.

bile (bī-uhl)–A digestive fluid produced by the liver.

biliary (bihl-ē-ār-ē)–Bile.

bilirubin (bihl-ē-roo-bihn)–A pigment produced from the destruction of hemoglobin.

billygoat (bihl-ē gōt)–A term used incorrectly to describe a male goat; should be buck.

binocular (bī-nohck-yoo-lahr)–Both eyes.

binomial (bī-nō-mē-ahl)–By international agreement, all plants and animals have two Latin names: genus and species.

bioaccumulation (bī-ō-ah-kyoom-yeh-lā-shuhn)–The process by which animals accumulate substances, especially pollutants, that may not be injurious to that organism but may injure other organisms that eat them.

biogenesis (bī-ō-jehn-eh-sihs)–(1) Formation by the action of organisms. (2) The doctrine that all life has been derived from previously living organisms.

biohazard (bī-ō-hahz-ahrd)–A substance that is dangerous to life.

biological (bī-ō-lohj-ih-kehl)–Products derived from a living process or living matter, such as sera, vaccines, bacterins, and antitoxins, etc. Also called biologics.

biological control (bī-ō-lohj-ih-kehl kohn-trōl)–A method of pest control by the use of predatory insects, fungi, or viruses; as contrasted to control by chemical pesticides. See biophage.

biologist (bī-ohl-oh-jihst)–A person who studies living organisms.

biology (bī-ohl-oh-jē)–The field of study dealing with living organisms. It may be divided into the study of plants (botany) and of animals (zoology).

biophages (bī-ō-fāj or bī-ō-fahj)–Organisms that obtain nourishment from other organisms, e.g., predators, parasites, and pathogens. See biological control.

biopsy (bī-ohp-sē)–The removal of living tissue for examination.

biotechnology (bī-ō-tehck-nohl-eh-jē)–Technology concerning the application of biological and engineering techniques to microorganisms, plants, and animals, sometimes used in the narrower sense of genetic engineering.

biotic (bī-oht-ihck)–Pertaining to life; biological.

biotic potential (bī-oh-tihck pō-tehn-shahl)–(1) The maximum reproduction power or ability. The inherent ability of an organism to reproduce and survive in greater numbers. (2) The ability of an organism to reproduce in an optimum, unrestricted, and noncompetitive environment.

biotype (bī-ō-tīp)–Groups of animals primarily distinguishable on the basis of interaction with relatively genetically stable varieties or clones of host plants; animal species.

bipolar (bī-pō-lahr)–Two opposite ends.

birth (bərth)–See parturition.

birth date (bərth dāt)–For racing or showing, a foal's birthday is considered as January 1 (regardless of the actual month it was born).

birth weight (bərth wāt)–The weight of an animal taken within 24 hours after birth. Heavy birth weights tend to be correlated with calving problems, but the conformation of the calf and the cow are contributing factors.

bishoping (bihsh-ohp-ihng)–An artificial altering of teeth of an older horse to sell it as a younger horse. The term is derived from a horse dealer named Bishop.

bit (biht)–(1) The part of a horse bridle that is placed in the mouth; usually made of steel. (2) An earmark for cattle.

bitch (bihtch)–An intact female canine (dog, fox, wolf, etc.).

bitting (biht-ihng)–Teaching a horse to give (yield) to the bit before it is used for riding.

black roan (blahck rōn)–The term used to describe a horse whose coat is black and white, rather uniformly mixed. Also called blue.

blacksmith (blahck-smihth)–A mechanic who welds and fashions iron and other metals, sharpens plowpoints, horseshoes, etc.

blade (blād)–(1) A retail cut of meat from the forequarter or shoulder. (2) The hind part of a fowl's single comb.

blastoma (blahs-tō-mah)–A neoplasm composed of immature undifferentiated cells.

blastula (blahs-tū-lah)–A mass of cells with a cavity that results from the division of a fertilized egg. From this stage the cells begin to differentiate.

blaze (blāz)–A broad white stripe on the face of a horse.

bleed (blēd)–To remove the blood from an animal during the slaughter process.

blemish (blehm-ihsh)–An unattractive defect that does not interfere with performance.

blend price (blehnd prīs)–The price paid a milk producer based on the proportion of milk utilized in each price class, such as fluid milk and manufacturing milk.

blepharectomy (blehf-ār-ehck-tō-mē)–The surgical removal of all or part of the eyelid.

blepharitis (blehf-ah-rī-tihs)–Inflammation of the eyelid.

blepharoplasty (blehf-ār-rō-plahs-tē)–The surgical repair of the eyelid.

blepharoptosis (blehf-ah-rō-tō-sihs)–Drooping of the upper eyelid.

blepharorrhaphy (blehf-ār-ō-rah-fē)–Suturing together of the eyelids; also called tarsorrhaphy (tahr-soh-rah-fē).

blepharospasm (blehf-rō-spahzm)–Rapid, involuntary contractions of the eyelid.

blepharotomy (blehf-ahr-ah-tō-mē)–Incision of the eyelid; also called tarsotomy (tahr-soh-tō-mē).

blind gut (blīnd guht)–The cecum; also called bung.

blind quarter (blīnd kwahr-tər)–The quarter of an udder that will not secrete milk during lactation or one that has an obstruction in the teat.

blind teat (blīnd tēt)–A teat on a sow that is not connected with the milk gland and will not produce milk; also called blind nipple.

blinders (blīnd-ərz)–Small pieces of material, usually leather, fastened to a bridle to keep a horse from seeing anything not straight ahead; also called blinkers.

blindness (blīnd-nehs)–The inability to see.

blissom (blihs-ohm)–A ewe in heat.

blistering (blihs-tər-ihng)–The application of an irritating substance to treat a blemish.

bloat (blōt)–An excessive gas accumulation in the rumen, abomasum, stomach, or cecum; in ruminants usually implies accumulation of excessive ruminal gas resulting in distension.

block (blohck)–To trim fleece to enhance the appearance of a sheep.

blocky foot (blohck-ē foot)–A horse with a noticeably more upright slope to the hoof than is normal.

blood (bluhd)–(1) The fluid that circulates through the heart and vessels that carries nutrients, various chemicals, and oxygen to the body cells; contains 55% liquid plasma and 45% formed elements. (2) A market term referring to the fineness of wool, as in three-eighths blood, quarter blood, low quarter blood, half-blood, blood classification.

blood horse (bluhd hohrs)–A purebred horse.

blood pressure (bluhd prehs-shər)–The tension of blood on the walls of the arteries.

bloodline (bluhn-līn)–The family or family lines of breeding.

bloody milk (bluhd-ē mihlk)–Milk from an animal suffering from ruptured blood vessels and injured inner gland tissues.

bloom (bloom)–A shiny coat for show horses.

blow out (blō owt)–To walk or exercise a horse either to loosen its muscles for further exercise, or to prevent chilling and stiffening after a hard workout.

blue roan (bloo rōn)–A horse with black hair interspersed with white, giving the impression of a bluish or grayish color.

bluebag (bloo-bahg)–A gangrenous form of mastitis that causes a bluish discoloration of the udder; most frequently seen in cattle, sheep, and goats.

blue-gray (bloo grā)–A horse having steely-blue hairs scattered through a gray coat.

blunt (bluhnt)–Dull, not sharp.

boar (bōr)–An intact male porcine or guinea pig.

board (bōrd)–To house.

bob (bohb)–(1) Veal from calves slaughtered when less than four to six weeks old; also called slunk. (2) The docked tail of a horse.

bobtail (bohb-tāl)–An animal with a short tail.

body (bohd-ē)–The physique of an animal.

body capacity (boh-dē kah-pah-siht-ē)–The heart girth and barrel.

bog spavin (bohg spahv-ihn)–In horses, a distention of the joint caused by excess fluid on the inner surface of the hock.

bold (bōld)–(1) Designating the assured gait of a horse. (2) Designating assurance of character in an animal.

bolster (bōl-stehr)–The padded undersection of a saddle.

bolt (bōlt)–To eat rapidly or startle.

bolus (bō-luhs)–A rounded mass of food, large pharmaceutical preparation, or to give something rapidly.

bone (bōn)–Connective tissue that forms the framework that supports and protects the body.

bone meal (bōn mēl)–The product of drying and grinding animal bones, not previously steamed under pressure, and used as a stock feed.

bone plate (bōn plāt)–A flat metal bar with screw holes that is used in bone fracture repair.

bone screw (bōn skroo)–A screw that compresses bone fragments together to repair bone fractures.

bone spavin (bōn spahv-ihn)–A bony growth in the hock that lames a horse. Also called jack spavin, spavin, spavined hock. See spavin.

boning (bō-nihng)–The process of making the hair stand out on a beef animal's leg. The purpose is to make the leg and/or hind quarter appear thicker in the show ring.

boot (boot)–Profuse feathering on the shank and toes of fowls.

booted (boot-ehd)–Fowls that are feathered on shanks and toes and having vulture hocks are said to be booted, as in booted and sultans.

borborygmus (bohr-bō-rihg-muhs)–Gastrointestinal gas movement that causes a rumbling noise.

Bos indicus (bohs ihn-dih-kuhs)–The scientific name for domestic humped cattle common to the tropical countries; Zebu or Brahman.

Bos taurus (bohs tohr-ehs)–The scientific name for domestic cattle common to the temperature zones; Hereford, shorthorn, Angus, etc.

bosal hackamore (bohs-ahl hahck-ah-mohr)–A bitless bridle with a braided noseband (bosal) that controls a horse with leverage and pressure on the nose and jaw.

boss (bohs)–(1) A lumpy formation on an animal or plant. (2) Metal stud on either side of the bit of a horse's bridle. (3) The dominant individual among a group of animals. See pecking order.

bossy (bohs-ē)–An affectionate name for a cow.

bots (bohtz)–The larvae of the bot fly, *Gastrophilus;* occur in the stomach.

bottom (boht-ehm)–Stamina in a horse.

bougie (boo-zhē)–An instrument used for insertion into natural passages of a body for dilation or medication, e.g., a teat bougie; usually constructed of steel, rubber, or plastic.

bovine (bō-vīn)–The species name for cattle.

bovine somatotropin (BST) (bō-vīn sō-mah-tō-trō-pihn)–A naturally occurring hormone that aids in stimulating the production of milk in cows.

bow-line knot (bō-līn noht)–A type of nonslippable knot.

bowl (bōl)–A vat or tank in which wool is scoured.

Bowman's capsule (bō-mahnz kahp-suhl)–The cup-shaped structure of the kidney that contains a glomerulus.

box stall (bohcks stahl)–An enclosure where a horse can move freely.

boxlock (bohcks-lohck)–A movable joint of any ringed instrument.

brace (brās)–Any device designed to strengthen or support, such as a corner-post brace, wall brace, etc. (2) A pair, as a brace of pheasants.

brace bandage (brās bahn-dahj)–A resilient bandage on the legs of horses worn in some cases in an effort to support lame legs, and in other cases to protect a horse from cutting and skinning its legs while racing.

brachial (brāk-ē-ahl)–The arm.

brachial plexus (brā-kē-ahl plehck-suhs)–A network of intersecting nerves that arise from the last three or four cervical and first one or two thoracic spinal nerves (species dependent) that innervate the front limb.

brachium (brā-kē-uhm)–The proximal portion of the forelimb between the shoulder and elbow joints.

brachycephalic (brā-kē-seh-fahl-ihck)–A short, wide head; examples include Pug and Pekingese.

brachydont (brā-kē-dohnt)–A permanently rooted tooth.

Bradford system (brahd-fōrd sihs-tehm)–The system of combining and spinning wool commonly used in the United States. The longer wools are used for this system of manufacture. Bradford system yarns are used to make worsted fabrics.

bradycardia (brād-dē-kahr-dē-ah)–An abnormally slow heartbeat.

bradypnea (brāhd-ihp-nē-ah)–An abnormally slow respiratory rate.

braid (brād)–The coarsest grade of wool.

brain (brān)–(1) The nerve tissues enclosed in the skull or cranium of vertebrates. (2) Calf brains, sold for human consumption.

brain stem (brān stehm)–The portion of the brain that consists of medulla oblongata, pons, and midbrain.

braining (brān-ihng)–(1) Fowl slaughter, in which a knife is inserted through the roof of the mouth to the rear of the skull, piercing the medulla oblongata. (2) Removing the brain from a carcass.

branch (brahnch)–(1) A lateral stem arising from the main stem. (2) Part of a horseshoe from the first nailhole to the end of the heel.

brand (brahnd)–(1) A method of permanently identifying an animal by scarring the skin with heat, extreme cold, or chemicals. (2) The steel or iron die used to mark the animal. (3) The design of the mark of identification commonly used to identify foods and feedstuffs.

brand book (brahnd book)–An official record of ranch brands used to mark cattle.

brawn (brawn)–(1) Muscles, especially of the arm and leg. (2) Boar's flesh, especially the boiled and pickled meat. (3) A pig fattened for its meat. (4) Hardened skin.

bray (brā)–The cry of a donkey or mule.

breachy (brēch-ē)–Describing an animal who passes over, or through, fences.

break in (brāk ihn)–To discipline; to train, as in a horse.

break joint (brāk joynt)–In grading sheep carcasses, the metacarpal (forearm bone) is struck with a heavy object. If the resulting break occurs at the center of the joint, it is referred to as a break joint, and the carcass is classified as a lamb carcass. If the break occurs above the joint and the joint does not separate cleanly, it is termed a spool joint. The carcass is then classified as a mutton carcass.

breaker (brāk-ər)–Utility-grade slaughtered cattle in which some degree of marbling is in the meat. The loins and rounds are broken out and sold as steaks.

breast (brehst)–(1) The front of the body between the neck and the abdomen. (2) The breastbone and flesh surrounding it, especially in fowl, lamb, or veal.

breast band (brehst bahnd)–A strap, used in place of a collar, that extends over the forechest of a horse; also called breast collar. See breast collar.

breast collar (brehst kohl-ahr)–An alternative to a neck collar for a horse. The breast collar is a wide horizontal strap around the lower part of the neck to which traces are attached for traction. See breast band.

breastbone (brehst-bōn)–A common term for the sternum.

bred (brehd)–An animal that is mated and is pregnant.

bred heifer (brehd hehf-ər)–A pregnant, young cow that has not previously borne a calf. See heifer, open heifer.

breech (brēch)–(1) Abnormal presentation of a fetus. (2) That part of an animal from the base of the tail to the hocks. (3) Coarse wool from the breech and hind legs of a sheep.

breech presentation (brēch preh-zuhn-tā-shuhn)–A birth in which the hind feet and rump present first.

breeching (brē-chihng)–The part of a harness that passes around the rump of a harnessed horse. It allows a horse to stop or back a wagon.

breed (brēd)–(1) A group of animals that are genetically similar in color and conformation and that, when mated to each other, produce young identical to themselves. (2) To improve, through control, characteristics in plants and animals.

breed association (brēd ah-sō-sē-ā-shuhn)–A group organized to safeguard the purity of a livestock or poultry breed and to promote the breed. It sets the standards and maintains registers for the breed.

breed character (brēd kahr-ahck-tər)–The details of conformation or color, e.g., shape of horns in cattle, size and shape of ear in swine, etc., that distinguish a breed of animals.

breed out (brēd owt)–To eliminate undesirable characteristics by selective breeding.

breeder (brē-dər)–(1) A specialist in breeding. (2) The owner of the dam when she was bred. The definition holds true in registering all classes of livestock. (3) An animal used for breeding.

breeder tom (brē-dər tohm)–A male turkey used for breeding, which is marketed between April and August.

breeding chute (rack) (crate) (brē-dihng shoot, rahck, krāt)–A structure built to confine a cow, mare, sow, and other female domestic animal at the time of mating to aid in the mating act.

breeding class (brē-dihng klahs)–Designated animals used for breeding purposes.

breeding herd (brē-dihng hərd)–The livestock retained to provide for the perpetuation of the herd or band.

breeding hobbles (brē-dihng hohb-uhlz)–A fetter fastened to the feet of a mare to prevent her from kicking the stallion during mating.

breeding house (brē-dihng hows)–(1) A building, divided into pens about 6–10 feet in size, used for breeding poultry. One male and approximately a dozen females are placed in each pen. (2) Any structure used for breeding.

breeding program goals (brē-dihng prō-grahm gōlz)–The objective or direction of breeder selection programs.

breeding season (brē-dihng sē-zohn)–The period of the year when animals normally reproduce.

breeding unit index (brē-dihng yoo-niht ihn-dehcks)–The measure of a breeding herd, including the total number of female animals capable of giving birth, weighted by the production per head in a base period.

breeding unsoundness (brē-dihng uhn-sownd-nehs)–Malformation of the genitals, cryptorchidism, scrotal rupture, and other abnormalities that disqualify animals for procreation though not for other purposes.

breeding value (brē-dihng vahl-ū)–The ability of an animal to transmit the genetic capability to produce meat, milk, eggs, or other economically important products; the value of an animal as a parent.

breedy (brē-dē)–Cows with a high degree of femininity; bulls with strong masculine features.

bridle (brīd-l)–(1) The part of a harness that includes the bit, reins, and headstall. (2) A short rope with hooks on either end.

bridle wise (brīd-l wīz)–Describing a horse accustomed to a bridle.

bright bay (brīt bā)–A horse with a light, glossy, tan coat.

brindle (brihn-dl)–An animal that is tawny, gray, or brown, having irregular, dark streaks or spots.

bring forth (brihng fōrth)–To produce; to give birth.

brisket (brihs-kiht)–A mass of connective tissue, muscle, and fat covering the cranioventral part of the ruminant chest between the forelegs.

bristle (brihs-ehl)–Stiff, sharp hair or hairlike parts that grow on animals and plants.

broad-spectrum antibiotic (brohd-spehck-truhm ahn-tih-bī-oh-tihck)–An antibiotic that attacks both gram-positive and gram-negative bacteria, and which may also show activity against other diseases.

broad-spectrum pesticide (brohd spehck-truhm pehs-tih-sīd)–A pesticide that kills a wide variety of insects. It may kill many beneficial insects, fish, birds, and mammals as well as target pests.

broiler (broy-lər)–A young chicken approximately 8 weeks old weighing 1.5 kg or greater; also called fryer or young chicken. Usually 8–12 weeks old.

broke (brōk)–Describing a gentled horse that can be harnessed or ridden.

broken coat (brō-kehn kōt)–In rabbits, a condition in which the guard hair is broken or missing in spots, exposing the undercoat.

broken crest (brō-kehn krehst)–A heavy neck in a horse, which breaks over and falls to one side.

broken ear (brō-kehn ēr)–A distinct break in the cartilage of a rabbit's ear, which prevents erect ear carriage; also called lop ear.

broken wind (brō-kehn wihnd)–A hereditary, incurable respiratory disease of old horses, in which the air cells of the lungs are ruptured or dilated. Also called heaves.

bronc (brohngk)–A wild or unbroken horse. Also called bronco.

bronchiectasis (brohng-kē-ehck-tah-sihs)–Dilation of the branches of the trachea.

bronchiole (brohng-kē-ōl)–The smallest branch of a bronchus; also called bronchiolus (brohng-kē-ō-luhs); plural is bronchioli (brohng-kē-ō-lī).

bronchitis (brohng-kı-tihs)–Inflammation of the bronchi. In horses, it sometimes results in broken wind. Also called infectious bronchitis, chick bronchitis, and gaping disease.

bronchoconstrictor (brohng-kō-kohn-strihck-tər)–A substance that narrows the bronchi.

bronchodilator (brohng-kō-dī-lā-tər)–A substance that expands the bronchi.

bronchopneumonia (brohng-kō-nū-mō-nē-ah)–An abnormal condition of the bronchi and lung (usually involves inflammation and congestion of the bronchi and lung).

bronchoscope (brohng-kō-skōp)–An instrument used to visually examine the branch of the trachea going to the lung.

bronchoscopy (brohng-kohs-kō-pē)–Endoscopic examination of the branch of the trachea going to the lung.

bronchospasm (brohng-kō-spahzm)–Sudden involuntary contractions of the bronchial muscles.

bronchus (brohng-kuhs)–Branch of trachea going to the lung; plural is bronchi (brohng-kī).

brood (brood)–(1) The young of any animal. (2) To incubate eggs or protect young from the weather. (3) Of, or pertaining to, an individual kept for breeding, as a brood mare, brood sow, brood cow. (4) A hen inclined to set on eggs, as a brood hen.

brood hen (brood hen)–A hen used to hatch eggs and rear young chicks.

brood mare (brood mār)–An intact female horse used for breeding. Also broodmare.

brood sow (brood sow)–A sow reserved for breeding.

brooder (broo-dər)–A housing unit for rearing birds after hatching; any enclosure for young animals.

brooder house (broo-dər hows)–A structure in which chicks or baby animals are raised without a mother; also called brooder.

brooder ring (broo-dər rihng)–A small wall (usually of cardboard) placed immediately around the brooder house during the first couple of weeks of the bird's life to keep them from wandering off and becoming chilled.

broodiness (broo-dē-nehs)–Describing a hen that attempts to sit on eggs constantly. Also used to describe female animals (especially swine) that have characteristics that indicate reproduction efficiency.

brooding (broo-dihng)–Raising newly hatched chicks in a protected environment.

broody (broo-dē)–(1) A livestock-judging term used to describe an animal, especially a gilt or sow, that possesses characteristics that indicate she will be reproductively efficient. (2) A bird manifesting a desire to sit on a clutch of eggs.

broody coop (broo-dē koop)–An enclosure used to confine a broody hen and hasten her return to normal egg production.

brow band (brow bahnd)–The part of the bridle that goes around the brow of the horse's head.

brown-ticked gray (brown tihckd grā)–The color of a horse with very small spots of brown scattered over a background of gray.

browse (browz)–Leaves, small twigs, and shoots of shrubs, seedling, and sapling trees, and vines available for forage for livestock and wildlife.

brush (bruhsh)–(1) Shrub vegetation and trees that do not produce timber. (2) The lower part of the tail of the cow or horse. (3) A device with set-in bristles for grooming the coat of an animal; used with a curry comp.

brushing (bruhsh-ihng)–The habit of a horse of striking the fetlock with the other hoof, which may result in either roughing the fetlock hair or an actual injury.

bruxism (bruhck-sihzm)–Involuntary grinding of the teeth.

buccal surface (būk-ahl sər-fihs)–The tooth surface facing the cheek; also known as the vestibular surface (vehs-tih-buh-lahr).

buck (buhck)–(1) An intact male rabbit or goat or deer. (2) Wool from rams. (3) Of a horse, to quickly leap with back arched and head held low.

buck herd (buhck hərd)–A group of rams.

bucked knees (buhckd nēz)–A term used to describe a horse or cow with knees bent out.

buckeyed (buhck-īd)–Describing a horse's eye with a convex cornea, which appears to protrude beyond the eyelids. The horse is shortsighted and shies easily.

bucking chute (buhck-ihng shoot)–An enclosure where broncos are saddled.

bucking range (buhck-ihng rānj)–In certain localities, range selected for placing rams with ewes.

buckling (buhck-lihng)–Male goat between one and two years of age.

buckskin (buhck-skihn)–(1) A term for a grayish-brown horse. (2) The skin of a buck deer. (3) A horse that is light brown and has a black mane and tail.

buckstrap (buhck-strahp)–A saddle strap used for a hand-hold on a rearing horse.

buffalo (buhf-eh-lō)–The American buffalo, which is better known as the American bison. The bison has been crossed with cattle to produce cattalo or beefalo.

buffer (buhf-ər)–(1) Animals which serve as food for predators, thus reducing danger to game. (2) A tool used to cut clinches from horseshoe nails before removing the shoe; also called clinch cutter. (3) A system of chemicals that resist changes in pH.

buffy coat (buhf-ē kōt)–The layer of leukocytes and thrombocytes located between erythrocytes and plasma when a blood sample is separated into components by weight.

buggy (buhg-ē)–In the United States, a light, four-wheeled, horse-drawn carriage. In the United Kingdom, the buggy usually has two wheels.

bulb (buhlb)–The upper part of a horse's heel.

bulbourethral glands (buhl-bō-yoo-rē-thrahl glahndz)–The glands that secrete a thick fluid that aids in motility of sperm located on either side of the urethra; also called Cowper's glands; not present in canine.

bulk bin (buhlk bihn)–A large storage bin used for storing feed. The feed is put in through the top using an auger from a feed truck or a feed mill. The feed is usually discharged by an auger from the bottom and is moved to the animal's trough or feeder by an auger or conveyer.

bull (buhl)–An intact male bovine or llama.

bull pen (buhl pehn)–(1) A small wood or metal enclosure for confining bulls. (2) The sale ring at a livestock auction.

bulla (buhl-ah)–(1) A circumscribed skin elevation that is greater than 0.5 cm in diameter and filled with fluid. (2) Any space, as in tympanic bulla; plural is bullae (buhl-ā).

bulldog (buhl-dohg)–(1) A steel clip fitted into the nostrils of a bull, used to control him; also called bull holder. (2) To throw a steer or cow to the ground by grasping its horns and twisting its neck (United States).

buller (buhl-ehr)–A cow in heat for a long time; a cow difficult to get with calf.

bulling (buhl-ihng)–Presenting signs of being ready for mating (heifers and cows).

bullock (buhl-ohck)–A castrated bull. See ox.

bunch (buhnch)–In pigs, an improperly healed wound following castration.

bunker silo (buhnck-ər sī-lō)–A silo for storing silage, consisting of a wide trench constructed in the side of a hill from which surface water has been diverted, such as by a diversion terrace.

bunodont (boon-ō-dohnt)–Animals with teeth that have worn rounded surfaces, e.g., swine.

buphthalmos (boof-thahl-mōs)–An abnormal enlargement of the eye.

burglar (behr-glehr)–A horse, good in appearance, but with a slight defect that can be remedied temporarily when offered for sale. Also called robber.

burn (bərn)–Tissue injury caused by heat, flame, electricity, chemicals, or radiation.

burro (buhr-ō)–A donkey; ass.

burry (buhr-ē)–Describing wool with excessive quantities of burrs, birdseeds, chaff, etc. in the fibers.

bursa (bər-sah)–Fluid filled sac that acts as a cushion to ease movement in areas subject to friction.

bursitis (bər-sī-tihs)–Inflammation of a bursa.

bush (buhsh)–(1) The brush of an animal; the tail. (2) To force a seller to accept a lower price for a horse than was bid in the auction ring.

buster (buhs-tehr)–One who trains animals for saddle or work.

butcher run (booch-ər ruhn)–Unsorted and ungraded rabbits purchased for slaughter.

butterfat (buh-tər-faht)–The fat in milk; also called milkfat.

buttermilk horse (buh-tər-mihlk hōrs)–A roan with a coat color between red and blue.

button (buht-n)–(1) A nipple, especially of a hog. (2) A stunted or immature horn growth, as on a calf. (3) Cartilage on the chine bone of cattle. (4) Any shell-like bone construction of the body; also called concha. (5) A leather ring for adjusting a horse's bridle.

butt (buht)–To strike with the head or horns, as do cattle, goats, or rams.

by (bī)–Begotten by a male; sired.

by-pass protein (bī pahs prō-tēn)–A protein that is heat or chemically treated so that it does not get altered in the runimant stomach.

by-product (bī proh-duhckt)–A product of significantly less value than the major product. In beef cattle, the major product is meat; by-products include the hide and other items.

C

caballada (kahb-ahl-lah-dah)–A herd of horses or mules (southwestern United States).

cachexia (kah-kehcks-ē-ah)–General ill health and malnutrition; used in describing the condition of cancer patients.

cadence (kā-dehnz)–The rhythm of a horse's gait.

cage battery (kāj baht-eh-rē)–A system of keeping poultry and rabbits in cages in a controlled environment throughout their life.

cage operation (kāj ohp-ər-ā-shuhn)–A method of raising chickens in which the hens are kept in confinement as they produce eggs.

cake (kāk)–A residue from the pressed kernels of cottonseed, linseed, or soybean, etc., from which the oil has been removed. It is pressed into hard cakelike masses and is used as a feed.

caked udder (kākd uh-dər)–(1) Physiological edema of the udder most often associated with recent freshening. (2) A doughy, swollen udder.

calamus (kahl-ah-muhs)–A hollow shaft at the proximal end of the feather shaft.

calcification (kahl-sih-fih-kā-shuhn)–A deposit of calcium within tissue; making the tissue hard.

calculus (kahl-kyoo-luhs)–An abnormal mineral deposit; plural is calculi (kahl-kyoo-lī).

calendar year (kahl-ehn-dər yehr)–The period from January 1 through December 31 of the same year.

calf (kahf)–(1) A young bovine, usually under one year of age. (2) The hide of such a calf.

calf crop (kahf krohp)–The number of calves produced by a given number of cows, usually expressed as the percent of calves weaned of cows bred.

calf pen (kahf pehn)–A small cubicle in which a calf is raised. The pens are raised off the floor or are movable. The purpose is to lessen the chance that the calf will contract a disease such as scours.

calf puller (kahf puhl-ər)–A device used to assist a cow in delivering a calf. A chain or cord is attached to the calf's front foot, and the calf is drawn from its mother.

calico (kahl-ih-kō)–(1) A cat having three colors of fur (black, orange, and white). (2) A spotted horse.

calk (kawk)–A downward-pointed part of a horse's shoe, which prevents slipping. Also called caltrop.

calking (kawk-ihng)–Injury to the coronary band by the shoe of the horse. Usually incurred by horses whose shoes have calks or by horses that are "roughshod" for walking on ice.

calks (kawckz)–Grips on the heels and the outside of the front shoes of horses.

callus (kahl-uhs)–A bulging deposit around the area of a bone fracture that may eventually become bone; also means thickening of the skin and hypertrophy of the horny skin layer.

calvarium (kahl-vār-ē-uhm)–The top of the skull.

calves (kahvz)–Cattle of either sex under one year of age.

calving (kahv-ihng)–Giving birth to bovine; may also be referred to as freshening.

calving ease index (kahv-ihng ēz ihn-dehcks)–A rating for calving ease that combines cleaving ease scores for heifers, second calves, and older cows, as well as birth weight as reported to breed associations and published in their annual sire summaries. It is also called the Ease of Calving Index. The index is used to select replacement heifers.

calving interval (kahv-ihng ihn-tər-vahl)–The amount of time between the birth of a calf and the birth of the next calf from the same cow.

calving season (kahv-ihng sē-szohn)–The time of year when the calves of a herd are born; occurs at different times in different parts of the country.

calyx or calix (kā-lihcks)–A cuplike organ or cavity.

canal (kahn-ahl)–A tunnel.

canaliculus (kahn-ah-lihck-yoo-luhs)–A duct; plural is canaliculi (kahn-ah-lihck-yoo-lī).

cancellous bone (kahn-sehl-uhs bōn)–The inner, spongy portion of long bone that houses the red bone marrow; also called spongy bone.

cancer (kahn-sehr)–A malignant tumor.

cancer eye (kahn-sehr ī)–The common name for squamous cell carcinoma (SCC).

candle (kahn-dl)–To examine an egg in front of a light to observe internal characteristics associated with edible quality or hatchability, such as air cell size, yolk shadow position, presence of blood or meat spots, and presence or lack of germ development.

candling (kahn-dlihng)–The process of shining a light through an egg to check embryo development.

canine (kā-nīn)–Species name for dog (other canines are foxes, wolfs, jackals, etc.)

canine teeth (kā-nīn tēth)–Long, pointed teeth located between the incisors and premolars; also called cuspids; abbreviated C in dental formula.

canker (kānk-ər)–(1) An ulcer in the mouth. (2) A serious disease of the horse's hoof. The frog frequently discharges a stinking fluid, and ultimately the sole and the frog rot.

canner (kahn-nehr)–(1) A marketing classification for low-grade beef. (2) A wild horse, from the western range of the United States, slaughtered for its meat.

cannibalism (kahn-ih-bahl-ihz-uhm)–Devouring one's own species.

cannon bone (kahn-nohn bōn)–The common name for metacarpal III and metatarsal III in livestock.

cannula (kahn-yoo-lah)–A hollow tube. Cannulas are usually inserted into a body cavity to allow escape of fluids or gas and through which liquids may be introduced into the body.

canter (kahn-tər)–A slow, restrained, three-beat gait in which the two diagonal legs are paired. Also called slow gallop.

canthectomy (kahn-thehck-tō-mē)–Surgical removal of the corner of the eyelid.

canthoplasty (kahn-thō-plahs-tē)–Surgical repair of the palpebral fissure.

canthotomy (kahn-thoh-tō-mē)–An incision into the corner of the eyelid.

canthus (kahn-thuhs)–The angle where the upper and lower eyelids meet; corner of the eye; plural is canthi (kahn-thī).

cap (kahp)–A milk tooth of an animal.

cape (kāp)–The short feathers underneath the hackle of chickens.

capillaries (cahp-ih-lār-ēz)–Blood vessels only one epithelial cell thick. Singular is capillary.

capon (kā-pohn)–A castrated young chicken or domestic fowl.

caponette (kā-pohn-eht)–A male chicken that is neutered before sexual maturity by the implantation of female hormones.

capped hock (kahpd hohck)–In cattle or horses, an enlargement of the point of the hock; usually caused by bruising.

caprine (kahp-rīn)–The species name for goats.

capsule breeding (kahp-suhl brē-dihng)–An older term for artificial insemination. See artificial insemination.

capsule (kahp-suhl)–A gelatin covering for medicines.

caracul (kahr-eh-kehl)–A type of pelt produced by young lambs of Karakul breeding.

carapace (kahr-ah-pāc)–The dorsal region of a turtle shell.

carbohydrate (kahr-bō-hī-drāt)–Any of certain organic chemical compounds of carbon, hydrogen, and oxygen, which include sugars and starches.

carbon (kahr-bohn)–An essential chemical element component in plants and animals.

carbon dioxide (kahr-bohn dī-ohcks-īd)–CO_2; a colorless, odorless gas.

carbonizing (kahr-boh-nīz-ihng)–A process for extracting vegetable matter from wool by treating it with acid or some other chemical.

carbuncle (kahr-buhng-kuhl)–A cluster of furuncles.

carcass (kahr-kuhs)–The body of an animal after it has been slaughtered; usually the head, hide, blood, and offal are removed.

carcass evaluation (kahr-kuhs ē-vahl-ū-ā-shuhn)–Techniques of measuring components of meat quality and quantity in carcasses.

carcass merit (kahr-kuhs mehr-iht)–The desirability of a carcass relative to the quantity of components (muscle, fat, and bone), USDA quality grade, plus potential eating qualities.

carcass quality grade (kahr-kuhs qwahl-ih-tē grād)–An estimate of the palatability of meat within a beef carcass. Quality grade is based primarily on marbling and maturity. See marbling, maturity.

carcass quantity (kahr-kuhs qwahn-tih-tē)–The amount of salable meat (muscle) a carcass will yield. Cutability is an estimate of carcass quantity.

carcass weight (kahr-kuhs wāt)–The weight of an animal carcass after the hide, head, feet, and entrails have been removed.

carcinogen (kahr-sihn-ō-jehn)–A chemical, physical, or biological agent that increases the incidence of cancer.

carcinoma (kahr-sihn-nō-mah)–A malignant growth of epithelial cells.

card (kahrd)–A machine used to separate the wool fibers by opening the locks or tufts of wool. The machine contains multiple rolls with teeth. Hand cards are used chiefly in the fitting of show sheep.

cardia (kahr-dē-ah)–Juncture of the esophagus and stomach.

cardiac (kahr-dē-ahck)–The heart.

cardiac catheterization (kahr-dē-ahck kahth-eh-tər-ih-zā-shuhn)–A radiography study in which a catheter is passed into a blood vessel and is guided into the heart to detect pressures, blood flow, and heart anomalies.

cardiac tamponade (kahr-dē-ahck tehm-pō-nād)–The compression of the heart due to fluid or blood collection in the pericardial sac.

carding (kahr-dihng)–The process of separating wool fibers.

cardiologist (kahr-de-ohl-ō-jihst)–A specialist who studies the heart.

cardiology (kahr-dē-ohl-ō-jē)–The study of the heart.

cardiomegaly (kahr-dē-ō-mehg-ah-lē)–Enlargement of the heart.

cardiomyopathy (kahr-dē-ō-mī-ohp-ah-thē)–Disease of the heart muscle.

cardiovascular (kahr-dē-ō-vahs-kyoo-lər)–The heart and blood vessels.

carditis (kahr-dī-tihs)–Inflammation of the heart.

carnassial tooth (kahr-nā-zē-ahl tooth)–The large, shearing cheek tooth; the upper fourth premolar and lower first molar in dogs and the upper third premolar and lower first molar in cats.

carnivore (kahr-nih-vōr)–An animal that is able to sustain life by only eating animal tissue.

carpus (kahr-puhs)–The joint of the distal forelimb located between the radius/ulna and the metacarpals.

carriage (kahr-ihj)–The manner or bearing of an animal; its physical control and behavior.

carrier (kahr-ē-ər)–(1) An animal that harbors disease without clinical signs and serves as a distributor of infection. (2) An animal or person in apparently good health who harbors pathogenic microorganisms.

carrion (kahr-ē-ohn)–A putrefying carcass.

carry (kahr-ē)–(1) To keep an animal on a maintenance ration without obtaining any weight gain or any products from it. (2) To bear, as a pregnant cow carries a calf.

carrying capacity (kahr-ē-ihng kah-pah-siht-ē)–The maximum number of individual animals that can survive the greatest period of stress each year on a given land area.

cartilage (kahr-tih-lihdj)–A form of connective tissue that forms the temporary skeleton of the embryo, provides for bone growth, and covers the joint surfaces of bone.

caruncle (kahr-uhng-kuhl)–(1) A maternal, placental fleshy mass that attaches to the fetal cotyledon; part of the placentome. (2) The fleshy, unfeathered portion of skin on the turkey's neck.

casein (kā-sēn)–The chief protein of milk; a phosphoprotein that occurs in the milk of different animals. Also called casseinogen.

casing (kā-sihng)–The cleaned sections of hog, cattle, or sheep intestine used as sausage skins.

cast (kahst)–(1) A fibrous or protein material found in the urine with renal disease. (2) An external stabilization device for bone fracture repair. (3) To be caught in a recumbent position and unable to rise. (4) To throw off a horseshoe, as by a horse when the nails become worn, old, etc. (5) The number of young produced.

casting (kahs-ting)–A restraint method using ropes to place an animal in lateral recumbency.

castor (kahs-tehr)–A callus on a horse's leg.

castrate (kahs-trāt)–To remove the testicles or to destroy their use; to geld.

castration (kahs-trā-shuhn)–Surgical removal of the testes.

catabolism (kah-tahb-ō-lihzm)–The breaking down of body cells and substance. The opposite is anabolism.

catalase (kaht-ah-lās)–An enzyme in plants and animals; splits hydrogen peroxide into water and gaseous oxygen.

catalepsy (kaht-ah-lehp-sē)–Rigidity of muscles accompanied with the animal having a trancelike state.

catalyst (kaht-l-ihst)–An agent that promotes interaction between or among other chemical substances without itself being changed.

cataract (kaht-ah-rahkct)–Cloudiness or opacity of the lens.

catarrh (kah-tahr)–Excessive mucus secretion by mucous membranes.

catch (kahtch)–(1) A sheep whose fleece is cut after the day's work has stopped (New Zealand). (2) To conceive after breeding, as a mare.

catgut (kaht-guht)–The tough string manufactured from the intestines of sheep and other animals that is used in surgical sutures, musical instruments, etc.

cat-hammed (kaht hahmd)–A term used to describe a horse having long, relatively thin thighs and legs.

cathartics (kah-thahr-tihcks)–Drugs that cause the evacuation of the bowels.

catheter (kahth-eh-tər)–A tube that is inserted into a body cavity to inject or remove fluid.

cattle (kah-tuhl)–Greater than one member of the genus *Bos*.

cattle feeding (kah-tuhl fē-dihng)–The finishing of cattle for market by feeding them mainly corn and other feed concentrates for fattening. Cattle feeding is most extensive in the corn belt states of the United States.

cattle leader (kah-tuhl lēd-ər)–A rod, or rod and rope, attached to a ring fastened in the nostril of a cow, bull, or steer for restraining or leading the animal.

cattle marker (kah-tuhl mahrk-ər)–A metal or composition tag with identifying numbers or characters, fastened to the ear, neck, or horns. Also called cattle tag.

cattle prod (kah-tuhl prohd)–A device that delivers a harmless electrical current to the end of a rod; used to prod cattle or other animals into moving. Also called stock prod.

cattle run (kah-tuhl ruhn)–(1) The number of cattle sent to market within a stated period, e.g., a day, a week, etc. (2) An alley or passageway for cattle.

cauda equina (kaw-dah ē-kwī-nah)–A collection of spinal roots at the lower part of the spinal cord; has a "horse's tail" appearance.

caudal (kaw-dahl)–Toward the tail end.

caudal fold (kaw-dahl fohld)–A fold of skin at the junction of tail and body.

caudocranial projection (kaw-dō-krā-nē-ahl prō-jehck-shuhn)–An x-ray beam that passes from caudal to cranial; used to describe extremity radiographs; also called posterioanterior projection (poh-stēr-ē-ō-ahn-tēr-ih-ər prō-jehck-shuhn).

caul (kawl)–(1) The fat around the stomach of cattle, sheep, and swine. (2) In embryology, a part of the thin membrane or web surrounding the fetus.

causal organism (kaw-sahl ōr-gahn-ihzm)–The organism (pathogen) that produces a specific disease; also causative agent.

cauterization (kaw-tər-ī-zā-shuhn)–The destruction of tissue using electric current, heat, or chemicals.

cautery (caw-tər-ē)–The application of a burning substance, hot instrument, electric current, or other agent to destroy tissue.

cavesson (kah-vehs-sohn)–A head stall with a noseband (often quite large) used for exercising and training horses.

cavity (kahv-ih-tē)–A hollow space within the body or organ.

cavy (kā-vē)–Any South American rodent of the family Caviidae of which the guinea pig is the main specie.

cecum (sē-kuhm)–The proximal part of the large intestine that forms a pouch.

celiac (sē-lē-ahck)–The abdomen.

cell (sehl)–The ultimate functional unit of an organic structure, plant, or animal.

cellulitis (sehl-yoo-lī-tihs)–Inflammation of connective tissue.

cementum (sē-mehn-tuhm)–A mineralized substance that covers the root of teeth.

centrals (sehn-trahlz)–The two central incisors of a horse. Also called pincers.

central nervous system (sehn-trahl nər-vuhs sihs-tehm)–A portion of the nervous system that consists of the brain and spinal cord; abbreviated CNS.

central test (sehn-trahl tehst)–A location where animals are assembled from several herds to evaluate differences in certain performances traits under uniform management conditions.

central venous pressure (sehn-trahl vē-nuhs prehs-shər)–The pressure of blood in the right atrium; abbreviated CVP.

centrifugal separation (sehn-trihf-yeh-gahl sehp-ahr-ā-shun)–The separation of substances of different densities by means of centrifugal force, as the separation of fat from milk.

centrifuge (sehn-trih-fūj)–A machine that spins samples very rapidly to separate elements based on weight.

centriole (sehn-trih-ōl)–A small organelle located near the nuclear membrane of cells that divides during mitosis and forms the centers toward which the chromosomes move upon division of the cell.

centromere (sehn-trō-mēr)–A small structure located on a chromosome that appears to form an attachment to the spindle fibers during cell division.

centrosome (sehn-trō-sōm)–A minute protoplasmic body sometimes held to be the dynamic center of mitotic activity.

cephalic (seh-fahl-ihck)–The head or toward the head.

cerclage wire (sihr-klahg wīr)–A band of metal that completely (cerclage) or partially (hemicerclage) goes around the circumference of bone that is used in conjunction with other stabilization techniques to repair bone fractures.

cere (sēr)–Thickened skin at base of external nares of birds; may be different colors in some birds to denote sex.

cerebellum (sehr-eh-behl-uhm)–The second largest part of brain. It receives and relays messages from the periphery to the brain; located at the caudal end of the head.

cerebral cortex (sər-ē-brahl kōr-tehcks)–The outer portion of the largest portion of the brain.

cerebrospinal fluid (sər-ē-brō-spīn-ahl flū-ihd)–A clear, colorless liquid that nourishes, cools, and cushions the brain and spinal cord; abbreviated CSF.

cerebrospinal fluid tap (sər-ē-brō-spī-nahl flū-ihd tahp)–Removal of the fluid surrounding the central nervous system; also called CSF tap.

cerebrum (sər-ē-bruhm)–The largest portion of the brain involved in thought and memory.

cerumen (seh-roo-mehn)–Ear wax.

cervical vertebrae (sehr-vih-kahl vər-teh-brā)–The first set of vertebrae that are found in the neck region; abbreviated C in vertebral formula.

cervicitis (sihr-vih-sī-tihs)–inflammation of the cervix.

cervix (sihr-vihckz)–The neck; in relation to the reproductive tract, the cervix is the neck of the uterus.

cesarean derived (sē-sā-rē-ahn)–An animal is delivered via cesarean section into a sterile environment to avoid possible contamination.

cesarean section (sē-sā-rē-ahn sehck-shuhn)–The delivery of offspring through an incision in the maternal abdominal and uterine wall; also called C-section and cesarean birth.

chalaza (kah-lā-zah)–A ropelike structure that holds the yolk to the center of the egg.

chalazion (kah-lā-zē-ohn)–Localized swelling of the eyelid resulting from the obstruction of a sebaceous gland of the eyelid.

champ (chohmp or chahmp)–To chew noisily, as a horse champs at the bit.

champignon (shahm-pihn-yehn)–Inflammation of a horse's spermatic cord.

character (kahr-ahck-tər)–(1) One of the details of structure, form, substance, or function that make up a species and distinguish it from others. (2) A judging term for desirable, necessary, or attractive qualities of an animal or product. (3) The evenness and distinctness of crimp in wool fibers.

check (chehck)–(1) An egg that has a cracked shell, with the inner membrane intact. When the crack is naturally mended, it is called a blind crack. Also called crack, dent. (2) To retard growth.

check ligament (chehck lihg-ah-mehnt)–One of two ligaments to the digital flexors of equine; functions to maintain the limbs in an extended position during standing.

check rein (chehck rān)–A leather strap fastened to the backpad of a harness to prevent a horse from lowering its head. Also called bearing rein. See check bit.

check strap (chehck strahp)–A part of a harness that passes between a horse's forelegs. It is fastened to the collar on one side and to the bellyband on the other. It prevents the collar from rising when the horse is backing up or stopping.

cheek (chēk)–The fleshy portion of either side of the face, forming the sides of the mouth and continuing rostrally to the lips.

cheek pouch (chēk powch)–A space in the oral cavity of hamsters that carries food and bedding.

cheek teeth (chēk tēth)–The premolar and molar teeth; the large teeth at the side of the mouth.

cheeking (chēk-ihng)–Grasping the cheekpiece of a bridle to pull the head of the horse toward the rider.

chelated (kē-lā-tehd)–Bound to and precipitated out of solution.

chemical (kehm-ih-kehl)–(1) A substance obtained by or used in a chemical process. (2) In dairying, an uncommon flavor defect that occurs when cheese has been made from contaminated milk.

chemical castration (kehm-ih-kahl kahs-trā-shuhn)–The process of injecting a chemical into the testicles of an animal. The chemical causes the testicles to atrophy.

chemical dehorning (kehm-ih-kahl dē-hōrn-ihng)–The application of chemicals to the horn buttons of young calves to stop horn growth.

chemonucleolysis (kē-mō-nū-klē-ō-lī-sihs)–The process of dissolving part of the center of an intervertebral disk by injecting a foreign substance.

chemosterilant (kē-mō-stehr-ih-lahnt)–A chemical that can prevent reproduction. See chemical castration.

chemotherapy (kē-mō-thehr-ah-pē)–The use of chemicals to treat infectious diseases.

chest (chehst)–The part of the body between the neck and abdomen; the thorax.

chest tube (chehst toob)–A hollow device inserted into the thoracic cavity to remove fluid or gas.

chestnut (chehst-nuht)–Horny growths on the medial surface of the equine leg, either above the knee in the front limb or toward the caudal area of the hock in the rear limb.

chestnut roan (chehst-nuht rōn)–A horse on which white hairs are mixed with the basic chestnut color. Sometimes called strawberry roan.

chevon (shehv-ohn)–Goat meat.

chiazm (kī-ahzm)–Crossing.

chick (chihck)–A young parrot or chicken.

chick feathers (chihck feh-thərz)–The feathers from chickens, a by-product of the poultry industry, gathered for use in the millinery trade.

chick starter (chihck stahr-tər)–Balanced feed for the quick growth of baby chicks, consisting of ground grains, meat scraps, leaf meal, soybean meal, dried milk, limestone, iodized salt, vitamins, antibiotics, and other items.

chicken (chihck-ehn)–*Gallus domesticus,* family Phasianidae; the domestic fowl, whose ancestry is attributed to *Gallus gallus* and other species of the jungle fowl of India and the Malay peninsula.

chill (chihl)–To reduce the temperature of a product, such as a meat carcass or milk, to cure it, improve the flavor, prevent spoilage, and increase market life.

chin ball marker (chihn bahl mahrk-ər)–A device worn beneath the chin of a surgically altered bull or androgenized cow used to detect and mark cows in heat. The cows are marked by the device when they are mounted.

chin glands (chihn glahndz)–The secretory organs located on the throats of turtles; also called mental glands.

chine (chīn)–The backbone of an animal carcass.

choana (kō-ā-nah)–The posterior naris; the cleft in the hard palate of birds.

choke (chōk)–An acute condition in livestock brought about by a food mass lodged in the esophagus.

choke down (chōk down)–To break in or control an animal by putting a rope around its neck so that it chokes the animal if it struggles.

choker (chōkr)–A rope with a noose for pulling logs.

cholecystectomy (kō-lē-sihs-tehck-tō-mē)–Surgical removal of the gallbladder.

cholecystic (kō-lē-sihst-ihck)–The gallbladder.

cholecystitis (kō-lē-sihs-tī-tihs)–Inflammation of the gallbladder.

choledochus (kō-lehd-uh-kō-uhs)–The common bile duct.

cholesterol (kō-lehs-tər-ohl)–One type of lipid that circulates in the blood; it plays an important role in the synthesis of bile, sex hormones, and vitamin D.

cholinesterase (kō-lihn-ehs-tehr-ās)–A chemical catalyst (enzyme) found in animals that helps regulate the activity of nerve impulses.

chondromalacia (kohn-drō-mah-lā-shē-ah)–Abnormal softening of cartilage.

chop (chohp)–Animal feed of coarsely crushed or finely ground cereal grains.

choppers (chohp-ehrz)–Aged ewes in medium flesh not good enough to grade as fat.

choriomeningitis (kōr-ē-ō-meh-nihn-jī-tihs)–Inflammation of the meninges and choroid plexus.

chorion (kōr-ē-ohn)–The outermost layer of the placenta.

choroid (kōr-oyd)–The opaque middle layer of the eyeball that contains blood vessels; contains the iris, pupil, lens, and ciliary muscles.

choroid plexus (kōr-oyd plehck-suhs)–Vascular folds of the pia mater in the third, fourth, and lateral ventricles of the brain that secrete cerebrospinal fluid.

chroma (krō-mah)–Color.

chromatid (krō-mah-tihd)–One strand of a double chromosome seen in the prophase and metaphase of mitosis.

chromoseres (krō-mō-sehrz)–The smallest particles identifiable by characteristic size and position in the chromosome thread. They are minute subdivisions of chromatin arranged in a linear, beadlike manner on the chromosome.

chromosome (krō-mō-sōm)–A structure in the nucleus composed mainly of DNA that transmits genetic information.

chondropathy (kohn-drohp-ah-thē)–Disease of cartilage.

chronic (krohn-ihck)–Long course, progressive onset; persisting for a long time.

chronic renal failure (krohn-ihck rē-nahl fāl-yər)–The progressive onset of the inability of the kidney(s) to function; abbreviated CRF.

chronological age (krohn-eh-lohj-ih-kehl āj)–The actual age of an animal in days, weeks, months, or years.

chuck (chuhck)–A clamping device for holding a drill bit.

chute (shoot)–(1) A mechanical device that is used to restrain cattle. (2) A stampede of animals (western United States).

chyle (kī-uhl)–A milky fluid consisting of lymph and triglyceride from the digestion of food.

chylothorax (kī-lō-thōr-ahcks)–The presence of chyle in the chest cavity.

chyme (kīm)–A semifluid, creamy substance resulting from the digestion of food by gastric juice.

cilia (sihl-ē-ah)–Thin hairs or eyelashes; singular is cilium (sihl-ē-uhm).

ciliary body (sihl-ē-ār-ē bohd-ē)–The thickened extension of the choroid that assists in accommodation or adjustment of the lens.

ciliary muscles (sihl-ē-ār-ē muhs-uhlz)–Fibers that adjust lens shape and thickness.

ciliate (sihl-ih-āt)–Bearing cilia; fringed with hairs; bearing hairs on the margin.

cinch (sihnch)–The part of a saddle used to hold it onto the horse; placed around the girth area.

circulation (sihr-kū-lā-shuhn)–The pulsatory movement of blood in the body.

circumscribed (sehr-kuhm-skrībd)–Contained within a limited area.

cirrhosis (sihr-rō-sihs)–A degenerative disease that disturbs the structure and function of the liver.

cisterna chyli (sihs-tər-nah-kī-lē)–The origin of the thoracic duct and saclike structure for the lymph collection.

cisterna magna (sihs-tər-nah mahg-nah)–The subarachnoid space located between the caudal surface of the cerebellum and the dorsal surface of the medulla oblongata that is commonly used to sample cerebrospinal fluid or to measure intracranial pressure.

claiming pen (klām-ihng pehn)–A pen or small enclosure within a pen in which an ewe and her newborn lamb are placed until the ewe accepts the lamb.

clamp (klahmp)–(1) An instrument used to secure or occlude things. (2) To castrate an animal by severing the sperm cords by pinching them with a clamp.

clarified milk (klahr-ih-fīd mihlk)–Milk that has been cleared of solid impurities by passing through a centrifugal separator.

clarify (klahr-ih-fī)–To remove undesirable, solid substances from a liquid, such as milk or fruit juice, by ordinary or by centrifugal filtration.

class (klahs)–A division of the plant or animal kingdom lower than a phylum and higher than an order, e.g., the class Insecta.

class of animal (klahs ohf ahn-ih-mahl)–The age and/or sex groups of a kind of animal, e.g., calves, cows, does, ewes, fawns, yearlings, etc.

classification (klahs-eh-fih-kā-shuhn)–The forming, sorting, apportioning, grouping, or dividing of objects into classes to form an ordered arrangement of items having a defined range of characteristics. Classification systems may be taxonomic, mathematical, or other types, depending on the purpose to be served.

classify (klahs-eh-fī)–(1) To systematically categorize animals according to a set scheme. (2) To sort individuals into groups having common characteristics or attributes.

clavicle (klahv-ih-kuhl)–The slender bone that connects the sternum to the scapula (not present in all species); also called the collarbone.

claw (klaw)–(1) The sharp nail on the toe of an animal or bird. (2) A device on a milking machine to which the stanchion tubes are connected.

claybark dun (klā-bahrck duhn)–A coat color of animals, especially horses. Dun imposed on a sorrel chestnut background. Also called copper dun.

clean wool (klēn wool)–A term that usually refers to scoured wool, but occasionally is used to describe grease wool that has a minimum amount of vegetable matter.

cleaning a cow (klēn-ihng ah kow)–A common term for the removal of a retained placenta; also called cleansing a cow.

clear egg (klehr ehg)–An infertile egg.

cleavage (klēv-ihj)–In animal reproduction, the splitting of one cell into two identical parts. Each resulting daughter cell matures and may divide again.

cleft palate (klehft pahl-aht)–A congenital fissure of the roof of the mouth that may involve the upper lip, hard palate, and soft palate; also called palatoschisis (pahl-ah-tohs-kih-sihs).

cleft-footed (klehft foot-ehd)–An animal having a divided hoof or foot, as a cow or hog. Also called cloven-footed.

clefts (klehftz)–In blacksmithing, cracks in the heels of horses.

clinch (klihnch)–The part of a nail bent over to keep it from being pulled out, as in horseshoeing, box manufacture, etc.

clinch cutter (klihnch kuht-tər)–A tool used to remove horseshoes.

clinical (klihn-ih-kahl)–Visible, readily observed, pertaining to treatment.

clinical evidence (klihn-ih-kahl ehv-ih-dehns)–Any symptom of disease that can be determined by direct observation, such as fever, lack of appetite, swellings, paralysis, etc.

clip (klihp)–(1) A semicircular metal piece extending from the outer surface of the horseshoe at the toe or side to prevent the shoe from shifting on the hoof. (2) Shears. (3) (a) An inclusive term for shorn wool. (b) The process of shearing wool. (c) The year's production of wool. (4) (a) To shear the hair of an animal close to the skin, as to clip a dog or cow's flank and udder. (b) To remove fleece from a sheep or goat. Also called shear. (c) To cut the feathers from a fowl's wing to prevent it from flying.

clipping (klihp-ihng)–Trimming the wings of birds to alter their flight; also referred to as to pinion (pihn-yehn).

clitoris (kliht-ō-rihs)–An organ of sensitive, erectile tissue located near the vaginal orifice in female.

cloaca (klō-ā-kah)–A common passage for fecal, urinary, and reproductive systems in birds and lower vertebrates.

clog (klohg)–(1) A heavy, wooden block attached to a horse's hind pasterns by a strap to prevent it from kicking when put out to pasture. (2) An impediment, encumbrance, or restraint.

close breeding (klōs brēd-ihng)–A form of livestock inbreeding that involves very close relatives.

close cropping (klōs krohp-ihng)–The grazing by animals of grasses and other forage dangerously close to the crown of the plant, which can ruin grazing lands. Also called close grazing.

closed breed (klōzd brēd)–A breed of livestock where registration is restricted to progeny of animals themselves registered in that breed.

closed formula (klōzd form-ū-lah)–The statement of ingredients on a tag attached to a sack of feed that gives the guaranteed percentages of protein, fat, and fiber. It does not give the amount of each ingredient.

closed herd (klōzd hərd)–A group of animals that restricts the entrance of new animals.

closed pedigree (klōzd pehd-ih-grē)–Hybrid combinations of unrevealed inbreds that have been put together by privately owned organizations in the seed business. Also called unpublished pedigree.

closed side (klōzd sīd)–The right side of a beef carcass.

closed-faced (klōzd fāsd)–A sheep that has considerable wool covering about the face and eyes. This often leads to a condition known as wool blindness.

close-feathered (klōs fehth-ahrd)–Said of a fowl in which the feathers are held closely to the body, i.e., at no perceptible angle to the body.

clot (kloht)–(1) A coagulum; a semisolid mass, as a clot of cream or a blood clot. (2) To coagulate, as blood.

clouding (klow-dihng)–An undesirable translucent condition of clear liquids.

cloven-footed (klō-vehn foot-ehd)–See cleft-footed.

club foot (kluhb foot)–A deformity in an animal's foot in which the foot turns inward to resemble a club.

cluck (kluhck)–The call of a brooding hen to her chicks.

clutch (kluhch)–(1) A nest of eggs in a hatchery. (2) A brood of chickens. Also called cletch.

coagulant (kō-ahg-yoo-lahnt)–(1) A substance that acts on a liquid to coagulate it, as renin. (2) An agent that aids the clotting of blood.

coagulation (kō-ahg-yoo-lā-shuhn)–The process of clotting.

coaptation (kō-ahp-tā-shuhn)–An act of approximating.

coat (kōt)–The outer covering of an animal, i.e., the fur of a beaver, the feathers of a fowl.

cob (kohb)–(1) A stocky, short-legged horse used for pulling light carts. (2) Male swan.

coccidiostat (kohck-sihd-ē-ō-staht)–Any of a group of chemical agents mixed in feed or drinking water to control coccidia (a protozoa).

coccus (kohck-uhs)–Round.

coccygeal (kohck-sihd-jē-ahl)–Of, or pertaining to, vertebrae of the tail.

coccygeal vertebrae (kohck-sihd-jē-ahl vər-teh-brā)–The fifth set of vertebrae that are found in the tail; abbreviated Co or Cy in vertebral formula; also called caudal vertebrae and abbreviated Cd.

cochlea (kōck-lē-ah)–A spiral-shaped passage in the inner ear.

cochlear duct (kōck-lē-ahr duhckt)–A tube filled with fluid located in the cochlea that vibrates when sound waves strike it.

cock (kohck)–An intact male parrot or chicken.

cocked ankle (kohckd ahnk-uhl)–Dislocation of the fetlock in a horse caused by contraction of the tendon. Also called knuckling over.

cockerel (kohck-ər-ahl)–An immature male chicken.

cod (kohd)–The remnants of the steer scrotum.

codominance (kō-dohm-ih-nahns)–A kind of gene action where one allele does not exhibit complete dominance over the other.

codominant (kō-dohm-ih-nahnt)–Animal species that shares dominance with other species in an area.

codominant genes (kō-dohm-ih-nahnt jēnz)–Genes that are neither dominant nor recessive.

coefficient of digestibility (kō-eh-fihsh-ehnt ohf dī-jehs-tih-bihl-ih-tē)–The amount of a particular nutrient digested and absorbed by an animal; expressed as a percentage of the amount that was in the animal's feed.

coffin bone (kawf-ihn bōn)–The common name for the third phalanx of livestock.

coffin joint (kawf-ihn joynt)–The distal interphalangeal joint between phalanx II and phalanx III in livestock.

Coggins test (kohg-ihnz tehst)–A test for diagnosing equine infectious anemia. It was developed by Dr. Leroy Coggins, Cornell University.

coincidence (kō-ihn-sih-dehns)–In genetics, a term used in estimating the distance between two genes. It is the ratio between the actual percentage of double crossing over and the percentage expected on the assumption that each crossing over is an independent event.

coitus (kō-ih-tuhs)–Copulation; sexual intercourse; service; mating.

cold carcass weight (kohld kahr-kuhs wāt)–The weight of a carcass after the carcass has cooled and shrunk.

cold collar (kohld kohl-ər)–(1) A freshly harnessed horse (southwestern United States). (2) A balky horse.

cold manure (kohld mahn-oo-ər)–Farmyard manure that does not ferment and heat unduly while in storage, such as cattle manure.

cold sterilization (kohld stehr-ih-lī-zā-shuhn)–(1) The use of a cathode ray or an electron beam gun in food processing to kill bacterial life. (2) Chemical sterilization of instruments.

cold backed (kohld bahckd)–A horse that humps his back and does not settle down until the saddle has been on a few minutes. Some cold-backed horses will merely tuck their tails and arch their backs when first mounted, but others will take a few crow hops until warmed up.

cold-blooded (kohld bluhd-ehd)–(1) Refers to a horse of unknown breeding or one that does not have an ancestor or pure blood. (2) Refers to certain animals whose blood temperature depends on the environment. (3) Refers to certain horse breeds that carry slightly fewer red blood cells.

cold-jawed (kohld jawd)–A horse that is difficult to control with the bit and bridle. Also called hard-mouthed.

cold-storage eggs (kohld stohr-ihj ehgz)–Eggs held in storage at a temperature of 45°F (7.2°C) or less for more than thirty days.

colectomy (kō-lehck-tō-mē)–Surgical removal of the colon.

colic (kohl-ihck)–Severe abdominal pain; technically means pertaining to the colon.

coliform bacteria (kō-lih-fohrm bahck-tēr-ē-ah)–A group of bacteria predominantly inhabiting the intestines.

colitis (kō-lī-tihs)–Inflammation of the colon.

collagen (kohl-ah-jehn)–A flexible, fibrous protein; "glue."

collar (kohl-ehr)–(1) The padded and reinforced part of a harness that girdles the neck. Straps for pulling are attached to it by means of hames. (2) A restraint device around the neck of an animal.

collateral relative (kō-lah-tehr-ahl rehl-ah-tihv)–Animals related not by being ancestors or descendents, but by having one or more common ancestors.

collected (keh-lehck-tihd)–A horse that has full control over its limbs at all gaits and is ready and able to respond to the signals or aids of its rider.

collecting tubules (kō-lehck-tihng too-byoolz)–Hollow tubes that carry urine from the cortex to the renal pelvis.

collection (keh-lehk-shuhn)–(1) The balance of a horse and rider which makes for instinctive coordinated movement. (2) The accumulation of sperm for use in artificial insemination.

collop (kohl-ehp)–A unit of grazing area that can support a full-grown horse or cow for one year.

coloboma (kō-lah-bō-mah)–A defect of some ocular tissue; used with other ocular term to describe location.

colon (kō-lihn)–The part of the large intestine located between the cecum and rectum.

colonoscopy (kō-luhn-ohs-kō-pē)–Endoscopic examination of the inner surface of the colon; the scope passes from rectum into the colon.

colony (kohl-ō-nē)–A cluster of bacteria or fungi grown on a culture medium, usually originating from a single bacterium, spore, or inoculation transfer.

color defect (kuhl-ər dē-fehckt)–Any color that is not removable in wool scouring, such as urine stain, dung stain, canary yellow stain, and black fibers.

colostomy (kō-lahs-tō-mē)–The surgical production of an artificial opening between the colon and the body surface.

colostrum (kō-lah-struhm)–The first milklike substance produced by the female after parturition; it is thick, yellow, and high in protein and antibodies.

colotomy (kō-loht-ō-mē)–Surgical incision into the colon.

colpindach (kohl-pihn-dahch)–A young heifer.

colt (kōlt)–An intact male horse or donkey usually under four years of age.

columbiformes (kō-luhm-bih-fōrmz)–A group of dovelike birds that includes doves and partridges.

coma (kō-mah)–A deep state of unconsciousness.

comatose (kō-mah-tōz)–An abnormal condition of a deep state of unconsciousness.

comb (kōm)–(1) A fleshy crest on the head of fowls. (2) An instrument consisting principally of metal, bone, plastic, etc., in which long, narrow, thin teeth are cut for the purpose of arranging and cleaning hair on an animal. Also to clean and arrange the hair of an animal with such an instrument.

combing (kōm-ihng)–An operation in the manufacture of worsted yarn by which the long fibers are separated from the short fibers and arranged parallel to each other.

combining ability (kohm-bīn-ihng ah-bihl-ih-tē)–The level of hybrid vigor produced by a breed of animal when crossed. General combining ability may refer to the ability of a breed to produce hybrid vigor when crossed with many other breeds. Specific combining ability may refer to the ability of a breed to produce hybrid vigor when crossed with another specific breed.

comedo (kōm-eh-dō)–A blackhead or the buildup of sebum and keratin in a pore; plural is comedones (kōm-eh-dō-nehz).

comfort zone (kohm-fōrt zōn)–The range of temperature in which an animal feels more comfortable and that makes no demand upon the animal's temperature-regulating mechanism.

commensal (koh-mehn-sahl)–A parasitic organism that lives attached to a host and shares its food.

commercial (koh-muhr-shehl)–(1) A low-grade meat market. (2) Any commodity produced for the market. (3) A nonpurebred animal.

commercial herd (ko-muhr-shehl hərd)–A herd of animals that will eventually be slaughtered for meat.

common (kohm-ehn)–A low, market grade of meat animals.

common-use range (kohm-ehn yoos rānj)–A range containing grass, forbs, and browse that allows two or more kinds of stock to graze together during the entire season or separately during part of a season.

communicable (koh-mū-nih-kah-buhl)–Readily transferred from one individual to another.

communicable disease (kuh-mū-nih-kah-buhl dih-zēz)–A disorder that is transmitted from animal to animal or by contact with contaminated objects.

communis (koh-mū-nihs)–Gregarious; flocking together, as sheep.

compact bone (kohm-pahckt bōn)–The hard, dense, outer shell of bone; also called cortical bone (kōr-tih-kahl).

compactness (kohm-pahckt-nehs)–In animal judging, a closely knit, firmly united animal.

compensatory gain (kohm-pehn-sah-tohr-ē gān)–Weight gain in livestock at an above-normal rate following a period of little or no gain.

complementability (kohm-pleh-mehn-tah-bihl-ih-tē)–In crossbreeding animals, the degree to which two or more breeds match so that the strengths of one breed cover the weaknesses of the other.

complemental feed (kohm-pleh-mehn-tahl fēd)–Fodder given to livestock in place of, or as a supplement to, range feed.

complementary livestock (kohm-pleh-mehn-tahr-ē līv-stohck)–Any livestock yielding a product that contributes to the success of another.

complete blood count (kohm-plēt bluhd cownt)–A diagnostic evaluation of blood to determine the number of erythrocytes, leukocytes, and thrombocytes per cubic millimeter of blood; abbreviated CBC.

complete carcinogen (kohm-plēt kahr-sihn-ō-jehn)–An agent that can act as both initiator and promoter of cancer.

complete ration (kohm-plēt rah-shuhn)–A single feed (usually a commercially formulated ration) that fulfills all of the nutritional requirements of an animal except for water.

compress (kohm-prehs)–A pad of gauze or other material applied to put pressure on any part of the body to reduce swelling or control hemorrhage.

computed tomography (kohm-puh-tehd tō-moh-grah-fē)–A procedure in which ionizing radiation with computer assistance passes through the patient and shows the internal body structures in cross-sectional views; also called CT scan or CAT (computed axial tomography) scan.

conalbumin (kohn-ahl-bū-mihn)–A protein found in egg white; has the property of binding iron in an iron-protein complex that turns pink in the presence of oxygen.

conceive (kohn-sēv)–To become pregnant.

concentrate ratio (kohn-sehn-trāt rā-shō)–The amount of concentrates in comparison to the amount of other feeds fed.

concentrate (kohn-sehn-trāt)–A type of feed that is high in total digestible nutrients and low in fiber. High energy feed like grain.

concentrated feed (kohn-sehn-trā-tehd fēd)–Feeds that have a high caloric density and are high in relatively completely digestible substances such as fats, proteins, starches, and sugars. The definition of concentrates as being those feeds below 18% crude fiber is arbitrary.

conception (kohn-sehp-shuhn)–The beginning of a new individual resulting from fertilization.

conception rate (kohn-sehp-shuhn rāt)–In cattle breeding, the percentage of first services that conceive.

concolorous (kohn-kuhl-ər-uhs)–Of a single uniform color.

concussion (kohn-kuhsh-uhn)–Shaking of the brain caused by injury.

condemned (kohn-dehmd)–Describing an animal, carcass, or food that has been declared unfit for human consumption.

condition (kohn-dihsh-uhn)–(1) The state of wool regarding the amount of yolk and other foreign matter it contains. (2) The general appearance and/or state of health of an animal, seed, fruit, or flower at a show. (3) To get an animal in good health and appearance by proper feeding and grooming. (4) The degree or amount of fat on a breeding animal.

conductivity (kohn-duhck-tihv-ih-tē)–The property of transmitting waves or impulses.

condyle (kohn-dīl)–A rounded projection on a bone.

cones (kōnz)–Specialized cells of the retina that react to color and fine-detail.

confinement operation (kohn-fīn-mehnt)–A production system where animals are raised in a relatively small area. Usually, the environment is controlled for the animal's comfort.

conformation (kohn-fōr-mā-shuhn)–The shape and body type of an animal.

conformation score (kohn-fōr-mā-shuhn skōr)–In the selection of replacement bulls and heifers, a numerical value indicating the overall appearance of the animals. It is figured using numerical values assigned to such physical characteristics as frame size, bone size, degree of muscling, sex, breed and sex character, and structural soundness.

congenial (kohn-jēn-yehl)–Able to cross-fertilize readily; to unite easily.

congenital (kohn-jēn-ih-tahl)–Present at birth.

congestion (kohn-jehs-zhuhn)–An abnormal accumulation of fluid.

congestive heart failure (kohn-jehs-tihv hahrt fāl-yər)–A syndrome that reflects insufficient cardiac output to meet the body's needs; abbreviated CHF.

conjugate (kohn-jū-gāt)–Coupled or in pairs.

conjugation (kohn-jū-gā-shuhn)–Side-by-side association (synapsis) of homologous chromosomes as in meiosis.

conjuctiva (kohn-juhnck-tī-vah)–A mucous membrane that covers the underside of each eyelid; covers much of the exposed surface of the eyeball when the eyelid is closed.

conjunctivitis (kohn-juhnck-tih-vī-tihs)–Inflammation of the conjunctiva.

conjunctivoplasty (kohn-juhnck-tih-vō-plahs-tē)–Surgical repair of the conjunctiva.

connective tissue (koh-nehck-tihv tihs-yoo)–Fibrous tissue that binds and supports various structures; includes bones, muscles, tendons, and ligaments.

conscious (kohn-shuhs)–Awake, aware, and responsive.

conscious proprioceptive deficit (kohn-shuhs prō-prē-ō-sehp-tihv dehf-ih-siht)–A neurologic defect in which the animal appears not to know where its limbs are.

conservative grazing (kohn-sehr-veh-tihv grā-zihng)–(1) The practice of limiting the number of livestock on a grazing range in such proportion as would not exhaust the range in successive seasons. (2) A degree of grazing that causes little or no soil disturbance.

constipation (kohn-stah-pā-shuhn)–A condition of prolonged gastrointestinal transit time, making the stool hard, dry, and difficult to pass.

constitution (kohn-stah-too-shuhn)–The physical makeup of an animal.

constriction (kohn-strihckt-shuhn)–To narrow.

contact bedding (kohn-tahct behd-ihng)–A substrate with which an animal comes into direct contact; also called direct bedding.

contact surface (kohn-tahckt suhr-fahs)–The tooth surface that touches adjacent teeth.

contagion (kohn-tā-juhn)–The spread of disease by direct or indirect contact.

contagious (kohn-tā-juhs)–A disease that can be spread from one animal to another by direct or indirect contact; infectious.

contaminated (kohn-tah-mih-nā-tehd)–The presence of microorganisms.

contemporary group (kohn-tehm-poh-rehr-ē groop)–A group of animals that are of the same breed and sex and have been raised in the same management group (same location on the same feed and pasture).

contentment (kohn-tehnt-mehnt)–In animal judging, a characteristic of a stable, as contrasted to a nervous, animal.

continuous grazing (kohn-tihn-yoo-uhs grā-zihng)–Grazing for the entire grazing season; synonymous with set stocking.

contour feather (kohn-tər feh-thər)–The body or flight feathers; arranged in rows.

contract (kohn-trahckt)–A common term for catching a disease.

contraction (kohn-trahck-shuhn)–Tightening.

contraindication (kohn-trah-ihn-dih-kā-shuhn)–Recommended not to be used.

contralateral (kohn-trah-lah-tər-ahl)–The opposite side.

control (kohn-trōl)–(1) The standard "normal" against which experimental results are compared; also called experimental control. (2) The prevention of losses from animal diseases, insect pests, weeds, etc., by any method.

controlled environment (kohn-trōld ehn-vī-ərn-mehnt)–An environment for animals that is kept at the correct temperature and other conditions to maximize animal comfort.

contusion (kohn-too-zhuhn)–An injury that does not break the skin; characterized by pain, swelling, and discoloration; also called a bruise.

convalescence (kohn-vah-lehs-ehns)–The period of recovery following an illness.

convergence (kohn-vər-jehns)–Simultaneous inward movement of both eyes.

convergent strabismus (kohn-vər-gehnt strah-bihz-muhs)–Deviation of the eyes toward each other; crossed eyes; also called esotropia (ehs-sō-trō-pē-ah).

convulsion (kohn-vuhl-shuhn)–A clinical sign in an animal in which the muscles contract and relax violently, producing aimless body movements.

cool out (kool owt)–(1) The reduction of grain in the ration of show animals after the show season; usually using corn and barley with oats and bran to lighten feed. (2) The process of walking a post-exercise, sweating horse to aid cooling.

cooler shrinkage (kool-ər shrihnk-ihj)–The loss in weight of an animal's carcass during storage in a cooler, which usually amounts to 2.5% of the carcass weight.

cooling basket (kool-ihng bahs-keht)–A wire basket used for protecting young rabbits during periods of high temperatures.

coon footed (koon foot-ehd)–A horse having long pasterns and shallow heels.

coop (koop)–(1) A small light enclosure, usually made of wire netting, used to confine fowls. (2) A heap, usually of manure.

coprodeum (kōp-rō-dē-uhm)–A rectal opening into the cloaca.

coprophagia (kō-prō-fā-jē-ah or kohp-rohf-ah-jē)–The ingestion of fecal material; eating stool.

coprophagy (kōp-rō-fah-jē)–The ingestion of feces. See pica.

copulation (kōp-ū-lā-shuhn)–The sexual union of animals, which results in the deposition of the male gametes (sperm) in close proximity to the female gametes (eggs).

cor (kōr)–The heart.

cord (kōrd)–Any long, cylindrical, flexible structure like the spermatic cord or umbilical cord.

corium (kōr-ē-uhm)–(1) Specialized, highly vascular cells that nourish the hoof and horn. (2) True skin; also called dermis.

cornea (kōr-nē-ah)–The transparent anterior portion of the sclera.

corneal ulceration (kōr-nē-ahl uhl-sihr-ā-shuhn)–The surface depression on the cornea.

corner (kor-nehr)–The outer pair of incisor teeth in the upper and lower jaws of a horse.

corn-fed (korn fehd)–Designating an animal or fowl fattened on corn prior to marketing; hence, well finished and of high grade.

cornicle (kor-nih-kehl)–An abortive spur on the leg of a hen that hardens with age. It does not develop into a regular spur, as in the cock.

corona radiata (kō-rō-nah rād-ih-ah-tah)–The granular material adhering to an ovum after ovulation that gives it a sunburst appearance; cumulous oophorus.

coronary artery (kōh-oh-nār-ē ahr-tər-ē)–A blood vessel that carries blood from the aorta to the myocardium.

coronary band (kōr-oh-nār-ē bahnd)–The junction that produces the hoof wall located between the skin and the horn of the hoof; also called the coronet (kōr-ō-neht).

coronet (kōr-oh-neht)–The part of a horse's hoof where it joins the skin. Also called coronamen, coronary band.

corpora nigra (kōr-pōr-ah nī-grah)–The black pigmentation at the edge of the iris in equine and ruminants.

corpus (kōr-puhs)–The body or middle portion.

corpus albicans (kōr-puhs ahl-bih-kahnz)–The small bit of scar tissue that remains on the surface of the ovary after the regression of the corpus luteum.

corpus cavernosum (kōr-puhs kahv-ər-nō-suhm)–The highly vascular, sponge-like tissue of the penis in some species.

corpus hemorrhagicum (kōr-puhs hehm-ō-rahj-ih-kuhm)–The temporary blood clot that forms in the crater formed by the follicle after ovulation and prior to the development of the corpus luteum.

corpus luteum (kōr-puhs loo-tē-uhm)–The ruptured graafian follicle that takes on a yellow fatty substance after ovulation. If conception does not occur, the tissue gradually disappears. If conception does occur the tissue becomes functional, producing progesterone.

corral (koh-rahl)–A small enclosure for handling livestock at close quarters.

corrected weight (coh-rehck-tehd wāt)–A means of comparing the growth of animals that are of a different weight. For example to compare lambs in a flock that are weaned at around ninety days, divide the weaning weight of each lamb by its age in days and multiply by 90. This puts all of the lambs on a ninety-day basis.

cortex (kōr-tehckz)–The outer layer or region.

cosset (kohs-iht)–A lamb raised without the help of its dam.

costa (kohs-tah)–A rib (when there is only one), the midvein of a simple leaf or other organ.

costal (kohs-tahl)–A rib.

costochondral (kohs-tō-kohn-drahl)–The rib and its cartilage.

cottonseed (koht-ehn-sēd)–The seed of cotton after the lint has been removed. It is a source of cottonseed oil, the residue being used as a stock feed.

cotty wool (koht-ē wool)–Tangled or matted wool on a sheep's back. This condition is caused by insufficient wool grease being produced by the sheep, usually due to breeding, injury, or sickness. This type of defective wool is more common in the medium and coarse wools. The fibers cannot be separated without excessive breakage in manufacturing.

cotyledon (koht-eh-lē-dohn)–Fetal placental elevation that adheres to the maternal caruncles; part of the placentome; also called button.

cough (kohf)–The sudden noisy expulsion of air from the lungs.

counterfeit (kown-tehr-fiht)–Referring to deceptive characteristics of good breeding or physique in an animal which it does not actually possess.

coupling (kuhp-lihng)–(1) The loin of an animal. (2) Mating; breeding; copulation.

cover (kuhv-ər)–(1) The flesh, hide, and fat on a fattened animal. (2) To copulate with a female, as a bull covers a cow. (3) To incubate, to hatch eggs, as a hen. (4) Shelter and protection for animals and birds.

covert (kuhv-ehrt or kō-vehrt)–One of the small feathers covering the bases of large feathers in fowls.

covey (kuhv-ē)–A flock of birds; quail, partridge.

cow (kow)–An intact female bovine or llama that has given birth.

cow calf operation (kow kahf ohp-ər-ā-shuhn)–A system of raising cattle whose main purpose is the production of calves that are sold at weaning.

cow chips (kow chihpz)–Dried cow manure.

cow heifer (kow hehf-ər)–A cow that has calved only once.

cow horse (kow hōrs)–A horse or pony trained in rounding up cattle, etc., usually on the western range in the United States.

cow index (CI) (kow ihn-dehcks)–A method of evaluating the genetic value of dairy cows. In beef cattle selection, a score indication of a cow's calf-producing ability. The formula is (adjusted weaning weight of the calf ÷ the weight of the cow at weaning) × 100.

cow month (kow mohnth)–The tenure on a range or artificial pasture of a cow for one month. The quantity of feed or forage required for the maintenance of a mature cow in good condition for one month.

cow sense (kow sehns)–The ability or intelligence in some horses which slow unusual dexterity in working with cattle.

cowhide (kow-hīd)–(1) The hide from a cow, steer, or bull. (2) A coarse, leather whip. (3) To whip an animal with cowhide.

cow-hocked (kow-hohckd)–Standing with the joints of the hocks bending inward, with the toes pointing outward.

Cowper's gland (kow-pehrz glahnd)–An accessory gland in the male reproductive system that produces a fluid that moves ahead of the seminal fluid, cleansing and neutralizing the urethra.

cowy (kow-ē)–(1) A bull that lacks masculinity. (2) Milk with an off odor or flavor.

coxofemoral (kohcks-ō-fehm-ō-rahl)–A connection between the hip and femur.

cracks (krahckz)–Hoof wall defects that form because the hoof is too long and not trimmed frequently enough.

cradle (krā-duhl)–A device used to prevent an animal from licking or biting an injured area.

cranial (krā-nē-ahl)–Toward the head.

cranial cavity (krā-nē-ahl kah-vih-tē)–A hollow space that contains the brain within the skull.

craniocaudal projection (krā-nē-ō-kaw-dahl prō-jehck-shuhn)–X-ray beam passes from cranial to caudal; used to describe extremity radiographs; also called anterioposterior projection (ahn-ter-e-ō-poh-ster-e-ər prō-jehck-shuhn).

craniotomy (krā-nē-oht-ō-mē)–A surgical incision or opening into the skull.

cranium (krā-nē-uhm)–The portion of the skull that encloses the brain.

craw (kroh)–The crop of a chicken or bird.

creatinine (krē-aht-ih-nihn)–A chemical compound containing nitrogen, carbon, hydrogen, and oxygen, which is present in the urine and results from the metabolism of protein.

creep (krēp)–An area where young piglets spend most of their time, and which has openings too small to allow the sow to enter.

creep feed (krēp fēd)–A high energy feed that is fed to young animals in special feed devices so that adult animals cannot gain access to the feed.

creep feeding (krēp fēd-ihng)–A system of feeding young domestic animals by placing a special fence around special feed for the young. The fence excludes mature animals but permits the young to enter.

cremation (krē-mā-shuhn)–The burning or incineration of dead bodies.

crepitation (krehp-ih-tā-shuhn)–A cracking sensation that is felt and heard when broken bones move together.

crepuscular (krē-puhs-kuh-lahr)–Becoming active at twilight or prior to sunrise.

crest (krehst)–(1) A high ridge. (2) The upper margin of the neck. (3) The root of the mane in equine. (4) The tuft of feathers on the heads of certain fowls. (5) A cock's comb.

cretinism (krē-tihn-ihzm)–A severe deficiency of thyroid secretion dating from birth or early life, which results in arrested physical and mental development, dwarfism, sluggishness, and a thickened tongue. (2) Used to describe the physical characteristics of the deficiency. See hypothyroidism.

cria (krē-ah)–A young llama.

cribber (krihb-ər)–A horse that has the nervous habit of biting some stationary object, usually the crib, while holding its neck muscles rigid and noisily sucking air into its stomach. The condition is referred to as wind sucking, cribbing, crib biting.

cribbing (krihb-ihng)–A vice of equine in which an object is grasped between the teeth and pressure is applied.

crimp (krihmp)–The characteristic form of succession of waves in wool fibers; a desirable crimp is close and distinct; the amount of wave in a wool fiber.

crimping (krihmp-ihng)–Passing a feed crop through a set of corrugated rollers that are set close together.

crissum (krihs-ehm)–(1) The area surrounding the cloaca on a bird. (2) Feathers around the cloaca.

critical temperature (krih-tih-kahl tehm-pər-chər)–The environmental temperature below which the heat production in a warm-blooded animal increases to prevent a lowering of body temperature.

critter (kriht-ər)–Any animal or fowl.

crop (krohp)–(1) Any product of the soil. In a narrow sense, the product of a harvest obtained by labor, as distinguished from natural production or wild growth. (2) The young fowls or animals bred on a farm. (3) Craw; a saclike enlargement in the esophagus of many birds to store food. (4) Part of a cow's body just behind the shoulder blades. (5) A whole tanned hide. (6) An identification notch or earmark on an animal. (7) To cut. (a) To clip the ears or tail of an animal, especially a dog. (b) To bite off, as a cow crops grass. (c) To cut off the wattles of a bird. (d) To clip the hair of an animal; to shear.

crop pasture (krohp pahs-chər)–A sown crop that is normally harvested but may be used for pasture if necessary, as oats, wheat, soybeans, etc.

crop-bound (krohp bownd)–Referring to domestic fowl with a distended crop from which food does not pass out in the regular manner. Also called impacted crop.

crop-eared (krohp ērd)–(1) Describing an animal with its ears cut. (2) Designating a small-eared horse.

cross (krohs)–(1) An animal that results from mating individuals belonging to different species, races, breeds, varieties, etc. See crossbred, hybrid. (2) To mate individuals belonging to different species, varieties, races, breeds, etc.

cross mating (krohs mā-tihng)–The crossing of two or more strains of animals within the same breed.

cross tying (krohs tī-ihng)–A method of using two ropes to secure a horse so that the head is level.

crossbred (krohs-brehd)–Offspring resulting from mating two different breeds within the same species.

crossing-over (krohs-ihng ō-vər)–The exchange of parts between homologous chromosomes that occurs during the synapsis of the first division of meiosis.

crosstie (krohs-tī)–To tie together three of the feet of an animal as it lies on the ground.

crotch (krohtch)–The area between the hind legs of an animal.

croup (kroop)–The muscular area around and above the tail base in equine.

crown (krown)–The top of an animal's head.

crude fiber (krūd fī-behr)–The part of feeds containing the cellulose, lignin, and other structural carbohydrates as determined by the proximate analysis.

crude protein (krūd prō-tēn)–A measure or estimate of the total protein in a feed determined by multiplying the total nitrogen content by 6.25.

crumbles (kruhm-behlz)–Crushed pellets of feed.

crupper (kruhp-ər)–A leather strap with a padded semicircular loop. The loop end goes under the tail of a horse and the strap end is affixed to the center of the back band of a harness or to the cantle of a saddle to prevent the saddle from slipping over the withers.

crus (kruhs)–The portion of the hindlimb between the stifle and tarsal joints.

crust (kruhst)–(1) A collection of dried serum and cell debris. (2) The outer wall of an animal's hoof. (3) Rough, tanned hide of a sheep or a goat.

crutch (kruhtch)–(1) The area between the hind legs of a sheep. Sometimes called crotch. (2) To clip or shear wool from around the crutch of a sheep, removing stained or dingy wool.

crutching (kruhtch-ihng)–The process of clipping wool from dock, udder, and vulva of sheep prior to lambing; also called tagging.

cryosurgery (krī-ō-sihr-jər-ē)–The destruction of tissue using extreme cold.

crypt (krihpt)–A blind sac.

cryptorchid (kriph-tōr-kihd)–A male animal in which one or both testicles remained in the body cavity and did not descend into the scrotum during embryonic development.

cryptorchidism (krihp-tōr-kih-dihzm)–A developmental defect in which one or both testis fail to descend into the scrotum; also termed undescended testicle(s).

crystalluria (krihs-tahl-yoo-rē-ah)–The excretion of naturally produced angular solid of definitive form (crystals).

crystals (krihs-tahlz)–A naturally produced angular solid of definitive form.

cubing (kū-bihng)–Processing a feed by grinding and then forming it into a hard form called a cube. Cubes are larger than pellets.

Cuboni test (kū-bō-nē tehst)–A chemical urine test to determine pregnancy in a mare.

cuckoo lamb (koo-koo lahm)–A lamb born after the usual lambing season.

cud (kuhd)–Regurgitated food particles, fiber, rumen fluid, and rumen microorganisms by ruminants.

cud chewer (kuhd choo-ər)–A ruminant.

cudding (kuhd-ihng)–The act of chewing cud.

cull (kuhl)–The removal of an animal from the rest because it does not meet a specific standard or is unproductive.

culling chute (kuhl-ihng shoot)–A chute through which sheep pass in single file so that their fleeces and other qualities can be judged so that the poorest animals can be culled out.

culture (kuhl-chehr)–(1) (a) The growing of microorganisms in a special medium. (b) The microorganisms that are so grown. (2) Bacteria used in making dairy and other products.

culturing (kuhl-chehr-ihng)–The artificial propagation of pathogenic or nonpathogenic organisms on nutrient media or living plants.

cup waterer (kuhp wah-tər-ər)–A bowl-like watering device. To receive water, the animal pushes against a valve in the bottom of the cup or bowl.

cups (kuhps)–Deep indentations of the incisors in the center of the occlusal surface in young permanent teeth.

curative (kūr-ah-tihv)–(1) A remedy used in the cure of diseases. (2) Relating to the cure of diseases.

curb (kuhrb)–An enlargement on the caudal aspect of the hind leg below the hock.

curb bit (kuhrb biht)–A bridle bit so designed that a slight pull on the reins will exert great pressure on the horse's tongue or jaw.

curd (kuhrd)–(1) When milk becomes sour naturally or by the addition of an acid, it flocculates and settles as soft curd of casein and fat with the surrounding liquid consisting of dissolved substances such as sugars and salts, known as whey. The curd is sold as cottage cheese. (2) The coagulated part of milk that results when milk is clotted by rennet, by natural souring, or by the addition of a starter.

cure (kyoor)–To heal; to restore health.

cured forage (kyoord fohr-ihj)–(1) Dry range grasses, harvested or standing, slightly weathered, nutritious, and palatable stock feed. (2) Any forage preserved by drying.

curettage (kyoo-reh-tahzh)–The removal of material or growth from the surface of a cavity.

curling comb (kuhr-lihng kōm)–A comb used to untangle and curl an animal's hair coat for a stock show.

current cross (kuhr-ehnt krohs)–An animal offspring developed in the first season of cross-mating.

currette (kwoor-reht)–An instrument with cupped head to scrape material from cavity walls.

currycomb (kuhr-ē-kōm)–A comb with rows of metallic teeth used in dressing the coats of horses and cattle.

curry (kuhr-ē)–To comb and dress the coat of an animal to improve its appearance.

cushing (kuhsh-ihng)–A common term for copulation of llamas.

cushion (koosh-ehn)–(1) The inside face of a ham. (2) The fleshy pads on a horse's hips. (3) The frog of a horse's hoof. (4) The mass of feathers on the rear or back of a bird, especially in the Cochin hen.

cutability (kuht-ah-bihl-ih-tē)–An estimate of the percentage of salable meat (muscle) from a carcass versus the percentage of waste fat.

cutaneous (kū-tā-nē-uhs)–Pertaining to the skin.

cuticle (kū-teh-kl)–One of the four major parts of an eggshell.

cut (kuht)–(1) A slash wound. (2) The action of a horse's hooves striking its legs or other hooves in walking or running, interfering with its gait. (3) (a) An animal separated from the main herd. (b) To separate an animal from the main herd. (c) A severing of the stem or a part of a plant. (4) (a) A reduction in numbers, amount, size, etc. (b) To reduce in numbers, amount, size, etc. (5) To castrate; emasculate. (6) To sever the jugular vein of an animal or fowl for slaughter.

cuts (kuhtz)–Small groups of animals that have been separated from the main herd.

cutting (kuht-ihng)–(1) Removing a horse or cow from the main herd. (2) The striking of a horse's feet against its joints while running.

cutting chute (kuht-ihng shoot)–A narrow passageway into which animals are driven to remove certain ones from the main herd.

cutting fat (kuht-ihng faht)–The fat removed during the trimming or cutting of animal carcasses, as contrasted with offal fat.

cutting horse (kuht-ihng hōrs)–A trained horse used to separate animals from the main herd.

cutting needle (kuht-ihng nē-dl)–A needle that has two or three opposing cutting edges.

cut-up (kuht-uhp)–A high-flanked animal usually slight in body capacity; also referred to as wasp-waisted.

cyanosis (sī-ah-nō-sihs)–An abnormal condition of blue discoloration; implies insufficient oxygen levels.

cyclic (sihck-lihck or sī-klihck)–In animals, coming in estrus (heat) at regular intervals.

cyclopia (sī-klō-pē-ah)–A congenital anomaly characterized by a single orbit.

cycloplegia (sī-klō-plē-jē-ah)–Paralysis of the ciliary muscle caused by injury.

cyesis (sī-eh-sihs)–Pregnancy.

cyst (sihst)–A closed sac containing fluid or semisolid material.

cystalgia (sihs-tahl-jē-ah)–Urinary bladder pain.

cystectomy (sihs-tehck-tō-me)–Surgical removal of all or part of the urinary bladder.

cystitis (sihs-tī-tihs)–Inflammation of the urinary bladder.

cystocele (sihs-tō-sēl)–Displacement of the urinary bladder through the vaginal wall.

cystocentesis (sihs-tō-sehn-tē-sihs)–Surgical puncture of the urinary bladder (to remove fluid).

cystodynia (sihs-tō-dihn-ē-ah)–Urinary bladder pain.

cystogram (sihs-tō-grahm)–A radiographic film of the urinary bladder after contrast material has been placed into the urinary bladder via a urinary catheter.

cystography (sihs-tohg-rah-fē)–A radiographic study of the urinary bladder after contrast material has been placed into the urinary bladder via a urinary catheter.

cystopexy (sihs-tō-pehck-sē)–Surgical fixation of the urinary bladder to the abdominal wall.

cystoplasty (sihs-tō-plahs-tē)–Surgical repair of the urinary bladder.

cystoscope (sihs-toh-skōp)–A fiberoptic instrument used to access the interior of the urinary bladder.

cystoscopy (sihs-tohs-kō-pē)–Endoscopic examination of the urinary bladder.

cystostomy (sihs-tohs-tō-mē)–The surgical creation of a new opening between the skin and urinary bladder.

cystotomy (sihs-tah-tō-mē)–A surgical incision into the urinary bladder.

cytology (sī-tohl-ō-jē)–The study of cells.

cytophagy (sī-tohf-ah-jē)–The engulfing of cells by other cells.

cytoplasm (sī-tō-plahzm)–Cellular material that is not part of the nucleus.

cytotoxic agent (sī-tō-tohcks-sihck ā-jehnt)–A substance that kills or damages cells.

D

dacryoadenitis (dahck-rē-ō-ahd-ehn-ī-tihs)–Inflammation of the lacrimal gland.

dacryocystectomy (dahck-rē-ō-sihs-tehck-tō-mē)–The surgical removal of the lacrimal sac.

dacryocystitis (dahck-rē-ō-sihs-tī-tihs)–Inflammation of the lacrimal sac and abnormal tear drainage.

dacryocystotomy (dahck-rē-ō-sihs-toh-tō-mē)–An incision into the lacrimal sac.

dairy (dahr-ē)–(1) (a) A plant in which milk is processed and where dairy products are manufactured and sold. (b) Pertaining to that which is related to the production, processing, or distribution of milk and its products. (2) A place where milk is kept.

dairy herd improvement association (DHIA) (dahr-ē hard ihm-proov-mehnt ah-sō-sē-ā-shuhn)–A cooperative organization whose purpose is the testing of dairy cows for milk and fat production and the recording of feed consumed.

dairy type (dahr-ē tīp)–A cow that indicates an ideal conformation for dairy production. The cow is angular; carries no surplus flesh, but shows evidence of good feeding; has a good development of the udder and milk veins; and shows a marked development of the barrel in proportion to its size.

daisy cutter (dā-sē kuh-tər)–A horse that seems to skim the surface of the ground at the trot. Such horses are often predisposed to stumbling.

dally (dahl-ē)–To loop the end of a lariat around the horn of a saddle so that the horse may serve as an anchor in roping animals.

dam (dahm)–A female animal, especially bovine or rodent species, that has produced offspring.

dapple (dahp-ehl)–A circular pattern in an animal's coat color in which the outer portion is darker than the center.

dapple bay (daph-ehl bāy)–A term used to describe a horse that has a black mane and tail and light chestnut-colored body, or a light chestnut-colored body covered in part by small rings of darker color.

dapple gray (dahp-ehl grā)–A coat-color pattern of some animals, in which the gray color is overlaid with spots of lighter or darker tones.

dark-chestnut (dahrck chehst-nuht)–A term used to describe a brownish-black, mahogany, or liver-colored horse.

dark-dun (dahrk duhn)–A term used to describe the mouse color of an animal's coat.

data (dah-tah)–A mass of accumulated information or results of an experiment.

daughter (daw-tehr)–(1) The female offspring. (2) The primary division or first generation offspring of any plant, regardless of sex.

daughter cell (daw-tehr sehl)–A newly formed cell resulting from the division of another cell.

deacon (dē-kehn)–A veal calf that is marketed before it is a week old. Also called bob, bob veal.

dead mouth (dehd mowth)–Describing a horse whose mouth is no longer sensitive to direction by rein and bit.

dead weight (dehd wāt)–The weight of an animal after slaughter and when all the offal has been removed (virtually the weight of all the salable meat).

deaf ear (dehf ēr)–The fold in the skin of a fowl just below the ear.

deafness (dehf-nehs)–Complete or partial hearing loss.

death loss (dehth lohs)–The reduction of the number of animals through death as a result of plant poisoning, accident, or disease; is different from reduction by causes such as straying, theft, or sales.

debarking (dē-bahrck-ihng)–A surgical procedure that cuts the vocal folds to soften a dog's bark; also called devocalization (dē-vō-kahl-ih-zā-shuhn).

debeaker (dē-bēck-ər)–A device for cutting off the tip of the beak of young chickens and turkeys to reduce pecking, cannibalism, and egg eating.

debeaking (dē-bēck-ihng)–The removal of about one-half of the upper beak and a small portion of the lower beak in poultry to prevent feather picking, cannibalism, and fighting; also called beak trimming in poultry.

debilitated (dē-bihl-ih-tāt-ehd)–Weakened or loss of strength.

debilitation (dih-bihl-ih-tā-shuhn)–Loss of strength, or a weakened condition.

debridement (dē-brīd-mehnt)–The removal of foreign material and devitalized or contaminated tissue.

decalcification (deh-kahl-sih-fih-kā-shuhn)–The removal of calcium from bones of animals.

decay (dē-kā)–(1) The decomposition of organic matter by anaerobic bacteria or fungi in which the products are completely oxidized. See putrefaction. (2) Any chemical or physical process that causes deterioration or disintegration.

deciduous teeth (deh-sihd-yoo-uhs tēth)–The temporary set of teeth that erupt in young animals and are replaced at or near maturity.

decubital ulcer (dē-kyū-bih-tahl uhl-sihr)–Focal loss of skin or mucous membrane as a result of prolonged pressure; also called bedsore.

deep (dēp)–Positioned below the surface.

deep-litter system (dēp-lih-tər sihs-tehm)–The placing of wood shavings, sawdust, straw, etc., mixed with hydrated lime on the floor of a chicken house to about the depth of 6 inches (15 centimeters). The droppings are not removed and the litter is changed only when the birds are sold.

defecation (dehf-ih-kā-shuhn)–Voiding of excrement; movement of the bowels.

defect (deh-fehckt or dē-fehkt)–(1) Any blemish, fault, irregularity, imperfection in an animal, fowl, or farm product that reduces its usability or impairs its value. (2) In aminals or fowls, a departure from breed or variety specification. Because of a tendency to be inheritable, a serious defect is also a disqualification for registration as a purebred.

deferred grazing (deh-fehrd grāz-ihng)–The keeping of livestock from a pasture until there is enough vegetation to support the animals, or in the western range of the United States, until the seeds of the herbage have matured.

defibrillation (dē-fihb-rih-lā-shuhn)–The use of electrical shock to restore the normal heart rhythm.

deficiency (dē-fihsh-ehn-sē)–(1) An insufficiency in reference to the amount, volume, proportion, etc.; a lack; a state of incompleteness. (2) Absence, deletion, or inactivation of a segment of a chromosome.

deflea (dē-flē)–To remove fleas from an animal, frequently by using an insecticide.

deformity (dē-fōr-mih-tē)–Any physical deviation from the normal in animals or plants caused by injury or disease.

degrease (dē-grēs)–To remove the wool oil from wool fiber and to obtain the fat from which lanolin is made.

degree of grazing (deh-grē ohf grāz-ihng)–A term used to define the closeness of grazing. The degrees are: ungrazed, lightly grazed, moderately grazed, closely grazed, and severely grazed.

dehiscence (dē-hihs-ehns)–A disruption or opening of the surgical wound.

dehorn (dē-hōrn)–Mechanical, heat, or chemical removal of horns or horn buds.

dehorning clippers (dē-hōrn-ihng klihp-ərz)–A device with long handles attached to sharp blades that are used to remove the horns of a beef animal through a scissorlike action.

dehydrate (dē-hī-drāt)–To remove moisture from a substance.

dehydration (dē-hī-drā-shuhn)–A condition of excessive loss of body water or fluid.

delactation (dē-lahck-tā-shuhn)–(1) Cessation of giving milk. (2) Cessation of suckling young; weaning.

deleterious (dehl-eh-tē-rē-uhs)–Injurious.

deletion (dē-lē-shuhn)–The absence of a segment of a chromosome involving one or more genes.

deliming (dē-līm-ihng)–The soaking of an animal hide in a sulfuric acid bath to remove the lime used in dehairing the hide.

delivery (deh-lihv-ər-ē)–The expulsion of the fetus.

delousing (dē-lows-ihng)–The extermination of lice by insecticides.

demiluster (dehm-eh-luhs-tehr)–Wool that has some luster but not enough to be classified as luster wool.

demyelination (dē-mī-lih-nā-shuhn)–The destruction or loss of myelin.

denature (dē-nā-tuhr)–(1) To make a product unfit for human consumption without destroying its value for other purposes, e.g., denatured alcohol. (2) To

change the properties of a protein, as to coagulate egg white.

dendrites (dehn-drīts)–Rootlike structures of the neuron that receive impulses and conduct them toward the cell body.

dens (dehnz)–A toothlike structure that protrudes from the cranial part of the axis (cervical vertebrae II).

density (dehn-sih-tē)–(1) Mass per unit volume. (2) The number of wild animals per unit of area. (3) The degree of closeness with which wool fibers are packed together.

dental calculus (dehn-tahl kahl-kyoo-luhs)–An abnormal mineralized deposit (along with bacteria and food particles) that forms on the teeth; also called tartar (tahr-tahr).

dental caries (dehn-tahl kār-ēz)–The decay and decalcification of teeth; producing a hole in the tooth.

dental cup (dehn-tahl kuhp)–A small dark-colored depression in the wearing surface of the incisor teeth of horses, having a rim of hard, glistening, white enamel, which disappears as the teeth wear down. Also known as dental star.

dental formula (dehn-tahl fōrm-yoo-lah)–A guide to the types and numbers of teeth found in the mouth.

dental pad (dehn-tahl pahd)–The hard surface of the upper mouth of cattle that serves in place of upper teeth.

dental plaque (dehn-tahl plahck)–A soft deposit on the teeth of bacteria and bacterial by-products.

dental star (dehn-tahl stahr)–Marks on the occlusal surface of the incisor teeth appearing first as narrow, yellow lines, then as dark circles near the center of the tooth.

denticulate (dehn-tihck-ū-lāt)–Diminutive of dentate; with small teeth of the dentate type.

dentin (dehn-tihn)–The connective tissue surrounding the tooth pulp and cementum.

dentition (dehn-tihsh-shuhn)–The arrangement of the teeth.

deoxyribonucleic acid (DNA) (dē-ohck-sē-rī-bō-nū-klē-ihck ah-sihd)–A nucleic acid that is the carrrier of genetic information. DNA is composed of adenine, guanine, cytosine, thymine, deoxyribose, and phosphoric acid.

depauperatum (deh-poh-pehr-iht-uhm)–Stunted.

depot (dē-pō)–The body area in which a substance can be accumulated or stored and from which it is distributed.

depot fat (dē-pō faht)–Fat that is accumulated or stored in the body.

depraved appetite (dē-prāvd ahp-ih-tīt)–A craving in animals and fowl for items not normally eaten by; may be caused by a diet deficiency. May include eating clay soil. See pica.

depression (dē-prehs-shuhn)–Lowering.

dermal (dər-mahl)–Of the skin; through or by the skin.

dermatitis (dər-mah-tī-tihs)–Inflammation of the skin (dermis).

dermatologist (dər-mah-tohl-ō-jihst)–A specialist who studies the skin.

dermatology (dər-mah-tohl-ō-jē)—The study of the skin.

dermatomycosis (dər-mah-tō-mī-kō-sihs)–An abnormal skin condition due to growth of a superficial fungus.

dermatophyte (dər-mah-tō-fīt)–Superficial fungi that are found on the skin.

dermatosis (dər-mah-tō-sihs)–An abnormal skin condition; plural is dermatoses (der-mah-toh-sēz).

dermis (dər-mihs)–The true skin; also called the corium.

Descemet's membrane (dehs-eh-māz mehm-brān)–The innermost or deepest layer of the cornea.

descending (dē-sehnd-ihng)–Progressing downward or caudally.

determinant growth (dē-tər-mih-nahnt grōth)–The type of growth that stops in an animal after it reaches a certain age.

deutectomy (doo-tehck-tō-mē)–The removal of the yolk sac from newly hatched chicks.

deutoplasm (doo-tō-plahzm)–The nutritive material or yolk in the cytoplasm or vitellus of an ovum or egg.

dewclaw (doo-klaw)–The functionless, rudimentary first digit of dogs and cats, found on the medial side of the forelimbs; also the accessory claw of the ruminant foot.

dewlap (doo-lahp)–The pendulous skin fold hanging from the throat of any animal, particularly a member of the ox (cattle) tribe and certain fowl.

deworm (dē-wehrm)–To rid an animal of internal parasitic worms.

DHIA records (dē-aych-ī-ā reh-kōrdz)–Production records of cows tested under the supervision of the tester of a Dairy Herd Improvement Association.

diagnosis (dī-ahg-nō-sihs)–Determination of the cause of disease; to know completely; plural is diagnoses (dī-ahg-nō-sēz).

diagnostic antigens (dī-ahg-noh-stihck ahn-tih-jehnz)–Biological agents that are prepared for use in the diagnosis of a specific disease, e.g., tuberculin, which is injected under the skin of cattle to determine if they have tuberculosis.

diakinesis (dī-ah-kī-nē-sihs)–A stage of meiosis just before metaphase of the first division: the homologous chromosomes are associated in pairs near the nucleus and have undergone most of the decreases in length.

diallel crossing (dī-ah-lēl krohs-ihng)–In animals, a method of progeny testing in which two males, bred at different times to the same females, are compared with the two sets of progeny.

dialysis (dī-ahl-ih-sihs)–A procedure to remove blood waste products when the kidney(s) are no longer functioning.

diaphragm (dī-ah-frahm)–The layer or sheet of muscle and connective tissue that forms the wall between the thoracic and abdominal cavities of mammals and aids in the process of breathing.

diaphysis (dī-ahf-ih-sihs)–The shaft of a long bone, which is mainly composed of compact bone.

diarrhea (dī-ah-rē-ah)–Abnormal frequency and liquidity of fecal material.

diastole (dī-ahs-stohl-ē)–Dilation (of the heart ventricles).

dickey (dihck-ē)–A slang term for a donkey or a small bird.

dicoumarol (dī-koo-mehr-ahl)–An anticoagulant produced by molds on spoiled clovers.

diecious (dī-ē-sihs)–Animals that are either male or female, i.e., each individual animal has either male or female reproductive organs but not both.

diencephalon (dī-ehn-cehf-ah-lohn)–The interbrain that provides connections between the cerebral hemispheres and the brain stem; contains the thalamus, epithalamus, and hypothalamus.

diestrus (dī-ehs-truhs)–The period of the estrous cycle that occurs between metestrus and proestrus.

diet (dī-eht)–The type and amount of food and drink habitually ingested by a person or an animal. See ration.

dietary fiber (dī-eh-tah-rē fī-bər)–The generic name for plant materials that are resistant to the action of normal digestive enzymes.

diethylstilbestrol (DES) (dī-ehth-ihl-stihl-behs-trohl)–A synthetic estrogenic hormone that has been used to stimulate faster growth and the deposition of additional fat in steers on feed.

differential (dihf-ər-ehn-shahl)–Diagnostic evaluation of the number of blood cell types per cubic millimeter of blood.

differential diagnosis (dihf-ər-ehn-shahl dī-ahg-nō-sihs)–Determination of possible causes of diseases; a list of possible disease causes.

differentiation (dihf-ehr-ehn-shē-ā-shuhn)–(1) The development of different kinds of organisms in the course of evolution. (2) The development or growth of a cell, organ, or immature organism into a mature organism.

diffuse (dih-fūs)—Widespread.

diffusion (dih-fū-shuhn)–The movement of solutes from an area of high concentration to low concentration.

digest (dī-jehst)–(1) To convert food within the body into a form that can be assimilated. (2) To soften, dissolve, or alter any substance by heat and chemical action.

digestibility (dī-jehs-tih-bihl-ih-tē)–(1) The readiness with which a substance may be converted into an absorbable form within the body. (2) The rate or amount of a nutrient digested, i.e., the difference between the amount of a nutrient fed and the amount found in the feces.

digestible (dī-jehs-tih-buhl)–(1) That feed consumed and digested by an animal as opposed to that which is evacuated by the animal. (2) Feed that can be converted into an absorbable form within the body.

digestible crude protein (dī-jehs-tih-buhl krood prō-tēn)–The total nitrogenous compound protein fed to livestock less the nitrogenous compounds eliminated in the feces.

digestible energy (DE) (dī-jehs-tih-buhl ehn-ər-jē)–The proportion of energy in a feed that can be digested and absorbed by an animal.

digestible nonnitrogenous nutrient (dī-jehs-tih-buhl nohn-nī-trohj-eh-nuhs nū-trē-ehnt)–The total digestible nutrients less the digestible protein in a foodstuff.

digestible nutrient (dī-jehs-tih-buhl nū-trē-ehnt)–That portion of a nutrient that can be digested and absorbed into a human or an animal body.

digestible protein (dī-jehs-tih-buhl prō-tēn)–The proportion of protein in a feed that can be digested and absorbed by an animal; usually 50 to 80 percent of crude protein.

digestion (dī-jehst-chuhn)–The process of breaking down foods into nutrients that the body can use.

digestion coefficient (dī-jehst-chuhn kō-eh-fihsh-ehnt)–The amount of a given feed that is digested by an animal expressed as a percentage of the gross amount eaten.

digestive tract (dī-jehs-tihv trähck)–The mouth, esophagus, digestive organs; stomach or stomachs, crop, gizzard, and the small and large intestines, and anus; all of the organs of an animal or fowl through which food passes.

digestion trial (dī-jehst-chuhn trī-uhl)–An experiment with any feed by an animal or fowl to determine the amount of any substance that can be digested.

digit (dihj-iht)–The distal extremity made up of phalanges (i.e., toes, claws, hooves, etc.).

digitigrade (dihg-iht-ih-grād)–Walking on toe surfaces.

di hybrid (dī hī-brihd)–An individual who is heterozygous with respect to two pairs of genes.

dilatation (dihl-ah-tā-shuhn)–Stretching beyond normal.

dilate (dī-lāt)–To widen.

dilation (dī-lā-shuhn)–Widening.

diluent (dihl-ū-ehnt)–Any gaseous, liquid, or solid inert material that serves to dilute or carry an active ingredient, as in an insecticide or fungicide. Also called carrier.

dilute (dī-loot)–(1) To make more liquid by mixing with water, alcohol, etc. (2) To weaken by mixing with another element.

diluted color (dī-loot-ehd kuhl-ər)–A term applied to feather color in poultry. It refers to soft colors such as light tan, cream or buff, light yellow, and light blue.

diluted feed (dī-loot-ehd fēd)–Feed that has a high concentration of roughage or fiber.

diluters (dī-loo-tərs)–A type of fluid that is used to increase the volume of semen (thus diluting the sample).

dingy (dihn-jē)–Wool that is dark or grayish in color and generally heavy in shrinkage.

diopter (dī-ohp-tər)—A unit of measurement for the refractive power of a lens.

dip (dihp)–Any chemical preparation into which livestock or poultry are submerged briefly to rid them of insects, mites, ticks, etc.

diploid (dihp-loyd)–(1) Having one genome comprising two sets of chromosomes. Somatic tissues of higher plants and animals are ordinarily diploid in chromosome constitution in contrast with the haploid (monoploid) gametes.

diplopia (dih-plō-pē-ah)–Double vision.

dipping (dihp-ihng)–A method of preserving seasoned wood by submersion in an open tank of creosote or similar preservative.

dipping vat (dihp-ihng vaht)–A pit filled with a liquid containing insecticides, ovicides, repellents, etc., through which animals are forced to pass for disinfestation.

disbud (dihs-buhd)–The removal of horn growth in kids or calves by use of a hot iron or caustic substance; also called debudding.

disc–See disk.

disc fenestration (dihsk fehn-ih-strā-shuhn)–The removal of an intervertebral disc by perforation and scraping out its contents. Also spelled disk fenestration

discharge (dihs-chahrj)–An exudate or abnormal material coming from a wound or from any of the body openings, e.g., a bloody discharge from the nose.

discospondylitis (dihs-kō-spohn-dih-lī-tihs)–A destructive, inflammatory disorder that involves the intervertebral discs, the vertebral end plates, and vertebral bodies (seen primarily in dogs and swine).

discriminate breeder (dih-skrihm-eh-nāt brēd-ər)–An animal that will only breed with a certain mate.

disease (dih-zēz)–Deviation from normal health or production.

disease control (dih-zēz kohn-trōl)–Any procedure that tends to inhibit the activity or effect of disease-causing organisms, or which modifies conditions favorable to disease.

disease resistant (dih-zēz rē-zihs-tehnt)–Designating plants, animals, or people not readily susceptible to, or able to withstand, a particular disease.

disinfect (dihs-ihn-fehckt)–To destroy or render inert disease-producing microorganisms on inanimate objects.

disinfectant (dihs-ehn-fehck-tahnt)–A chemical agent that kills or prevents the growth of microorganisms on inanimate objects.

disk (dihsck)–Intervertebral disk. Also spelled disc.

dismount (dihs-mownt)–To get down, alight, as from a horse's back.

disorientation (dihs-ōr-ē-ehn-tā-shuhn)–A condition in which the animal appears mentally confused.

displaced abomasum (dihs-plāsd ahb-ō-mā-suhm)–A disease of ruminants in which the abomasum dilates and migrates either to the left or right of its normal position; abbreviated DA and is denoted LDA (left displaced abomasum) or RDA (right displaced abomasum) depending on its location.

disposition (dish-peh-zihsh-ehn)–The temperament or spirit of an animal.

disqualification (dihs-kwohl-ih-fih-kā-shuhn)–In animal husbandry, a defect in characters or form for breed types that disqualifies an animal from an exhibition or from breed registration.

dissect (dī-sehckt)–To cut or divide a plant or animal into pieces for examination.

distal (dihs-tahl)–Farthest from the midline or from the beginning of a body structure.

distal contact surface (dihs-tahl kohn-tahckt sər-fihs)–The contact surface of a tooth that is farthest from midline.

distal convoluted tubules (dihs-tahl kohn-vō-lūd-ehd too-boolz)–The hollow tubes located between the loops of Henle and the collecting tubules.

distal spots (dihs-tahl spohts)–Dark circles on a white coronet band.

distention (dihs-tehn-shuhn)–Enlargement via stretching or dilation.

distichia (dihs-tēck-ē-ah)–Double row of eyelashes, usually resulting in conjunctival injury.

distichiasis (dihs-tē-kē-ī-ah-sihs)–An abnormal condition of a double row of eyelashes that usually results in conjunctival injury.

distributor (dih-strihb-yeh-tehr)–A device in some bulk milk tanks that spreads the milk over a cooling surface.

diuresis (dī-yoo-rē-sihs)–Increased excretion of urine.

diuretic (dī-yoo-reht-ihck)–A substance that increases urine secretion.

diverticulitis (dī-vər-tihck-yoo-lī-tihs)–Inflammation of a diverticulum.

diverticulum (dī-vər-tihck-yoo-luhm)–A pouch occurring in the wall of a tubular organ; plural is diverticula (dī-vər-tihck-yoo-lah).

dizygotic twins (dī-zī-goht-ihck twihnz)–Twins that develop from two separate fertilized ova.

dobbin (dohb-ihn)–An affectionate designation for a gentle horse.

docile (dohs-ihl)–Refers to an animal that is gentle in nature.

dock (dohck)–(1) To cut short the tail of an animal (most commonly a lamb); usually for sanitary reasons and to facilitate breeding in females. (2) The area around the tail of sheep or other animals. (3) A leather case to cover a horse's tail when clipped or cut.

dockage (dohck-ihj)–The weight deducted from stags and pregnant sows to compensate for the unmerchantable parts of an animal.

docking (dohck-ihng)–The removal of the distal portion of the tail; also means to reduce in value.

doddie (dohd-ē)–A polled cow.

doe (dō)–An adult intact female goat, rabbit, or deer.

dog (dohg)–(1) *Canis familiaris,* family Canidae; a domesticated, carnivorous animal which is used as a household pet, a watchdog, a herder for sheep, cows, etc. (2) A low-quality beef animal.

dogie (dogy, dogey, doge)–(dō-gē)–A term used on western ranges in the United States for a motherless calf.

dogtrot (dohg-troht)–A slow, gentle trot.

dolichocephalic (dō-lih-kō-seh-fahl-ihck)–A narrow, long head; examples include collie and greyhound.

domesticate (deh-mehs-tih-kāt)–To bring wild animals under the control of humans over a long period of time for the purpose of providing useful products and services; the process involves careful handling, breeding, and care.

dominance (dohm-eh-nehns)–(1) The tendency for one gene to exert its influence over its partner. There are varying degrees of dominance, from partial to complete to overdominance. (2) The tendency of one animal in a group to exert its social influence or presence over others in the group. Also referred to as social order or pecking order (called social dominance).

dominant (dohm-eh-nahnt)–In genetics, a gene or trait that expresses itself over the other. See recessive.

dominant gene (dohm-eh-nahnt jēn)–The gene that prevents its allele from having a phenotypic effect. See recessive.

donkey (dohng-kē)–An ass.

dorsal (dōr-sahl)–Refers to the back, or toward the back; the opposite of ventral.

dorsal plane (dōr-sahl plān)–An imaginary line dividing the body into dorsal and ventral portions; also called the frontal plane (frohn-tahl).

dorsal punctum (dōr-sahl puhnck-tuhm)–A small spot near the upper medial canthus where the nasolacrimal duct begins.

dorsal recumbency (dōr-sahl rē-kuhm-behn-sē)–Lying on the back; also called supine (soo-pīn).

dorsal root (dōr-sahl root)–The portion of the spinal nerve that enters the uppermost or dorsal portion of the spinal cord that carries afferent sensory impulses.

dorsoventral projection (dōr-sō-vehn-trahl prō-jehck-shuhn)–An x-ray beam that passes from the back to the belly; abbreviated D/V.

dosage (dō-sihj)–The amount of medication based on units per weight of the animal.

dosage interval (dō-sihj ihn-tər-vahl)–The time between separate administrations of a drug.

dose (dōs)–The amount of medication measured in milligrams, milliliters, units, or grams.

dose syringe (dōs sər-ihnj)–A syringe with a long pipe or nozzle that is used to force medicine down the throats of animals and poultry and to wash out the throats or crops of fowls.

double (duhb-uhl)–Designating a team of horses, e.g., to work horses double.

double cover (duhb-uhl kuhv-ər)–Coitus of a mare, cow, or other female animal with a male on successive days to ensure conception. See double mating.

double cross (duhb-uhl krohs)–The crossing of two single cross hybrids of plants or animals.

double immune (duhb-uhl ihm-yoon)–Designating a hog that has been vaccinated with both live virus and antihog cholera serum for protection against hog cholera.

double mating (duhb-uhl mā-tihng)–(1) A rarely used breeding practice with Barred Plymouth Rocks to produce birds of exhibition type. (2) The mating of female livestock twice during a given estrous period to the same or different males to ensure conception. See double cover.

double reins (duhb-uhl rānz)–Two sets of reins attached to a horse's double bit.

double-rigged saddle (duhb-uhl rihgd sah-duhl)–A saddle that has two cinches.

down (down)–The soft, furry or feathery covering of young animals and birds that occurs on certain adult birds under the outer feathers.

downer (down-ər)–(1) An animal which, in transit in a truck or railroad car, has fallen or lain down. (2) A diseased animal unable to get up.

downy (dow-nē)–Covered with very short, weak, soft hairs.

draft (drahft)–Any feedstuff obtained as a by-product of the distillation of grains.

draft animal (drahft ahn-ih-mahl)–An animal used for work stock and, in some countries, is still used for plowing and pulling heavy loads.

drain (drān)–A device by which a channel can be established for the exit of fluids from a wound.

draining pen (drān-ihng pehn)–A pen with a sloping concrete floor adjoining a dipping vat where dipped animals stand while excess fluid drains back into the vat.

drake (drāk)–A mature male duck.

drape (drāp)–(1) Cloth arranged over a patient's body to provide a sterile field around the area to be examined, treated, or incised. (2) A term that is applied to a cow or ewe incapable of bearing offspring, especially to such an animal selected for slaughter.

drawing (droh-ihng)–The eviscerating of poultry.

drawing blood (droh-ihng bluhd)–A common term for collecting a blood sample.

drench (drehnch)–To give medicines in liquid form by mouth and forcing the animal to drink.

dress (drehs)–(1) To curry and brush an animal. (2). To remove the feathers and blood from a bird that has been killed.

dress out (drehs owt)–(1) To remove the feathers or skin and to cut up and trim the carcass of an animal after slaughter. (2) The percentage of carcass weight to live weight.

dressage (druh-sahzh)–A method of riding in which a rider guides (rather than using hands, feet, or legs) a trained horse through natural maneuvers. Also called high school, haute ecole.

dressed weight (drehsd wāt)–The weight of a sheep, cow, or hog carcass. (2) The weight of a bird that had had its blood and feathers removed but not its intestines, head, or feet.

dressing (drehs-sihng)–(1) An application of medicine or bandage to a wound. (2) The removal of feathers and blood from a bird. (3) The trimming of excess fat and bone from a meat carcass.

dressing comb (drehs-ihng kōm)–A special comb used to prepare an animal's coat for a show. Also called currycomb.

dressing defect (drehs-ihng dē-fehckt)–A defect of dressed poultry characterized by one or more of the following: pin feathers left on the carcass; incomplete bleeding; a cut, tear, or abrasion in the skin longer than two inches; a broken bone; feed left in the crop; dirty feet; dirty body; dirty vent.

dressing loss (drehs-ihng lohs)–The loss in weight between a live animal and its dressed carcass.

dressing percent (drehs-ihng pər-sehnt)–The carcass weight divided by live weight and multiplied by 100.

dressing weight (drehs-sihng wāt)–The weight of a dressed animal or bird in contrast to its live weight.

drinking cup (drihnck-ihng kuhp)–A metal or porcelain cup just large enough for an animal's muzzle. It contains a valve or float that is activated when the animal pushes its muzzle against it.

drive (drīv)–(1) The moving of livestock under human direction. (2) To herd livestock.

driveway (drĭv-wā)–(1) In the range country, that land which is set aside for the movement of livestock from place to place, e.g., from the home range to the shipping point. Also called stock driveway, stock route. (2) A path or passage, sometimes paved, for the movement of vehicles and/or livestock. (3) The farm road from the building site to the gate at the highway. (4) Vehicle passage into or through a barn.

drive-ins (drĭv ihnz)–Designating those cattle that are herded in to market, as contrasted to those transported there by truck, etc.

driving (drī-vihng)–Horses harnessed and controlled from behind.

drop (drohp)–(1) To give birth to young, as to drop a calf. (2) To shoot an animal and to cause it to fall.

drop band (droph bahnd)–A group of ewes or nanny goats that is managed separately at the season of bearing young. Also called drop herd.

droppings (drohp-ihngz)–The excrement of animals and birds (in birds it is a composite of feces and urine). See manure.

dropsy (drohp-sē)–Widespread edema (swelling) throughout the body of an animal.

drove (drōv)–A collection or mass of animals of one species, as a drove of cattle.

drug (druhg)–An agent used to diagnose, prevent, or treat a disease.

drug residue (druhg rehz-ih-doo)–Any amount of drug that is left in an animal's body.

dry (drī)–(1) To cause a pregnant cow to stop giving milk shortly before she drops her calf. (2) Designating a cow who has ceased to give milk shortly after she drops her calf.

dry band (drī bahnd)–A band of sheep without lambs.

dry basis (drī bā-sihs)–Designating a product, e.g., soil, fertilizer, or feed, which is analyzed for its constituents calculated on the basis of oven-dried material.

dry cow (drī kow)–A cow that has ceased to give milk.

dry lot (drī loht)–A bare, fenced-in area used as a place to keep livestock for feeding and fattening.

dry matter (drī mah-tər)–The total amount of matter, as in a feed, less the moisture it contains.

dry off (drī ohf)–To bring the lactation period of a cow to an end.

dry period (drī pehr-ē-ohd)–That time just before parturition when a cow or other lactating animal ceases to give milk.

dry picked (drī pihckd)–To dress a bird without scalding it. It may be bled by cutting the jugular vein, and its medulla oblongata punctured to relax the feather follicles to facilitate plucking.

dry rendered (drī rehn-dehrd)–The residues of animal tissues cooked in open, steam-jacketed vessels until the water has evaporated. The fat is removed by draining and pressing the solid residue.

dry weight (drī wāt)–The weight of a product or material less the weight of the moisture it contains; the weight of the residue of a substance that remains after virtually all the moisture has been removed from it. Also called dry matter.

drying off (drī-ihng ohf)–Ending the production of milk when milk yield is low or prior to freshening.

dual purpose (dool pər-puhs)–Animals that are bred and used for both meat and milk production.

dual-use range (dool ūs rānj)–A range containing a forage combination of grass, forbs, and browse that allows two or more kinds of stock (such as cattle and sheep) to graze the area to advantage at the same time through the entire season, or separately, during a part of the season.

dub (duhb)–To cut the combs and wattles from cockerels when they are twelve to sixteen weeks old, to prevent injury from freezing or from fighting.

duck (duhck)–(1) A bird, family Anatidae, especially the domesticated duck raised for meat and eiderdown. (2) The female duck. See drake. (3) A heavy cotton or linen cloth.

duckling (dihck-lihng)–A young duck that still has downy plumage.

ductless glands (duhckt-lehs glahndz)–Those glands of the body whose secretions pass directly into the bloodstream or into the lymph, as one of the endocrine glands.

ducts (duhcks)–A passage with well defined walls.

ductus deferens (duhck-tuhs dehf-ər-ehnz)–The excretory duct at the end of the vas deferens.

dull (duhl)–A lack of shine to haircoat; also used to describe behavior that is more lethargic than normal.

dummy (duhm-ē)–A sleepy or stupid horse, especially one that has suffered inflammation of the brain. (2) A strongly built frame suggesting the shape of a female animal, often covered with the hide of such a female, which is used to excite a male animal sexually.

dumpy (duhm-pē)–(1) Designating an animal or fowl that does not feel very well, but that is not seriously ill. (2) An animal that is short in length and height and is fat.

dun (duhn)–An animal's color characterized by a dark dorsal stripe over the withers and shoulders. Recognized shades are mouse dun, buckskin dun, and claybark dun.

dung (duhng)–Manure; the feces or excrement of animals and birds. See manure.

dung locks (duhng lohcks)–In sheep, britch wool locks that are encrusted in hardened dung.

dunging pattern (duhn-gihng pah-tərn)–The tendency for animals to eliminate wastes in a particular location.

duodenum (doo-ō-dē-nuhm)–The proximal or oral portion of the small intestine; located between the pylorus and jejunum.

duplicate genes (doo-plih-kiht jēnz)–Factors with the same recessive phenotypic expression.

dura matter (doo-rah mah-tər)–The thick, tough, outermost layer of the meninges.

dust (duhst)–An insecticide, fungicide, etc., which is applied in a dry state, or the application of these to plants, animals, or fowls.

dusting (duhs-tihng)–A cleaning method in which chinchillas roll in dust.

dwarf (dwahrf)–An abnormally undersized animal. Dwarfing may be caused by disease, lack of water, or mineral deficiency.

dyad (dī-ahd)–The univalent chromosome, composed of two chromatids, at meiosis. The pair of cells formed at the end of the first meiotic division.

dyschezia (dihs-kē-zē-ah)–Difficulty defecating.

dyscrasia (dihs-krā-zē-ah)–Any abnormal condition of the blood.

dysecdysis (dihs-ehck-dī-sihs)–Difficult or abnormal shedding.

dysentery (dihs-ehn-tər ē)–A number of disorders marked by inflammation of the intestine.

dysesthesia (dihs-ehs-thē-zē-ah)–Impaired sensation.

dysfunction (dihs-fuhnckt-shuhn)–Not working normally.

dyspepsia (dihs-pehp-sē-ah)–An impairment of digestion.

dysphagia (dihs-fā-jē-ah)–Difficulty swallowing or eating.

dysplasia (dihs-plā-zē-ah)–Abnormal development or growth.

dyspnea (dihsp-nē-ah)–Difficult or labored breathing.

dysrhythmia (dihs-rihth-mē-ah)–The loss of normal heart rhythm; also called arrhythmia (ā-rihth-mē-ah).

dystocia (dihs-tōs-ē-ah)–A difficult birth.

dystrophy (dihs-trō-fē)–A defective growth.

dysuria (dihs-yoo-rē-ah)–Difficult or painful urination.

E

E. coli (ē cō-lī)–Abbreviated name for the bacteria *Escherichia coli* (ehsh-ehr-ih-kē-ah cō-lī)

ear covert (ēr kō-vuhrt)–The very small feathers of a bird that cover the ear.

ear implant (ēr ihm-plahnt)–A small pellet containing a growth regulator or other medication that is placed beneath the skin of the ear of a beef animal.

ear marking (ēr mahr-kihng)–The process of removing parts of the ears of livestock so as to leave a distinctive pattern; done for the purpose of designating ownership.

ear notcher (ēr nohch-ər)–A punch used to cut notches in the ears of animals for identification.

ear notching (ēr nohch-ihng)–An identification method used in swine in which notches of various patterns are cut in the ear.

ear tag (ēr tahg)–A tag fastened in an animal's ear for the purpose of identification.

ear tagging (ēr tahg-ihng)–The placement of identification device in the ear.

eared (ērd)–(1) Pertaining to an animal restrained by its ears. (2) Designating the presence of earlike tufts of feathers in some birds.

early speed (ər-lē spēd)–The ability of a horse to obtain its maximum speed in one or two strides.

easy (ē-zē)–(1) A command to a horse to move slower. (2) Designating an animal that has a low or weak back or pasterns, as easy in the back.

easy keeper (ē-zē kē-pər)–An animal that thrives with a minimum of feed, attention, and care.

ecchymosis (ehck-ih-mō-sihs)–A purplish patch of bleeding into the skin; also called a bruise.

ecdysis (ehck-dī-sihs)–Shedding or molting.

echocardiography (ehck-ō-kahr-dē-ohg-rah-fē)–A diagnostic procedure that uses sound waves to evaluate the heart structures.

echoic (eh-kō-ihck)–The ultrasound property of producing adequate levels of reflections (echoes) when sound waves are returned to the transducer and displayed.

eclosion (eh-klō-zhehn)–The process of hatching from an egg.

ecraseur (ē-krah-zehr)–A surgical instrument consisting of a fine chain or cord that is looped around the scrotum or a diseased part and gradually tightened, thus severing the enclosed part.

ectoderm (ehck-tō-dərm)–The outer of the three basic layers of the embryo which gives rise to the skin, hair, and nervous system.

ectopic pregnancy (ehck-toph-ihck prehg-nahn-sē)–A fertilized ovum implanted outside the uterus.

ectothermic (ehck-tō-thehr-mihck)–Deriving heat from without the body; cold-blooded, as lizards and snakes.

ectropion (ehck-trō-pē-ohn)–Eversion or turning outward of the eyelid.

eczema (ehck-zeh-mah)–A nonparasitic skin disease characterized by a red inflammation of the skin, development of papules, vesicles, and pustules, serous discharge, formation of crusts, severe itching, and alopecia.

edema (eh-dē-mah)–An abnormally large amount of fluid in the intercellular tissue spaces.

effective progeny number (EPN) (eh-fehck-tihv proh-jehn-ē nuhm-bər)–An indication of the amount of information available for estimation of expected progeny differences in sire evaluation.

efferent (ē-fər-ahnt)–To carry away; in the nervous system efferent nerves carry motor impulses away from the CNS; also called descending tracts.

efferent ducts (ē-fər-ahnt duhckz)–Very small ducts located in the testes that connect the rete tests to the epididymis.

effete (eh-fēt)–Designating an exhaustion of the ability in animals to produce young.

efficacy (ehf-ih-kah-sē)–The extent to which a drug causes the intended effects; "effectiveness."

effluent (ehf-floo-ehnt)–Discharge.

effusion (eh-fū-zhuhn)–Fluid escaping from blood or lymphatic vessels into tissues or spaces.

egestion (ē-jehs-chuhn)–Indigestible waste material that is excreted from the digestive tract.

egg (ehg)–(1) The reproductive body produced by a female organism: the ovum. (2) The oval reproductive body produced by females of birds, reptiles, and certain other animal species, enclosed in a calcerous shell or strong membrane within which the young develop.

egg cell (ehg sehl)–Female germ cell.

egg tooth (ehg tooth)–A scale on the tip of the upper mandible of the embryo chick, used as a reinforcement for the beak for breaking open the shell at hatching. Also called pip.

ejaculate (ē-jahck-ū-lāt)–The discharge of semen from the reproductive tract of the male.

elastin (ē-lahs-tihn)–A protein substance that is found in tendons, cartilage, connective tissue, and bone. Elastin is not softened as much as collagen by heat in the presence of water.

elastration (ē-lahs-trā-shuhn)–Bloodless castration in which rubber bands constrict the spermatic cord, causing it and the testicles to wither away. It is, however, not always successful.

elastrator (ē-lahs-trā-tər)–A bloodless castration device using small elastic bands.

elatus (ē-lah-tuhs)–Tall.

elbow (ehl-bō)–The joint in a front leg of an animal that corresponds to the elbow joint of a person's arm.

electric prod(der) (ē-lehck-trihck prohd-ər)–A portable, battery-powered, canelike device used to control or drive livestock by giving them a slight electrical shock.

electrical stimulation (ē-lehck-trih-kahl stihm-yoo-lā-shuhn)–A method of collecting semen by means of an electric current from a probe placed in the rectum of the male.

electrocardiogram (ē-lehck-trō-kahr-dē-ō-grahm)–A record of the electrical activity of the myocardium; also known as ECG or EKG.

electrocardiograph (ē-lehck-trō-kahr-dē-ō-grahf)–An instrument used to produce a record of the electrical activity of the myocardium.

electrocardiography (ē-lehck-trō-kahr-dē-ohg-rah-fē)–The process of recording the electrical activity of the myocardium.

electroejaculation (ē-lehck-trō-ē-jahck-yoo-lā-shuhn)–A method of collecting semen for artificial insemination or examination in which electrical stimulation is applied to the nerves to promote ejaculation.

electroejaculator (ē-lehck-trō-ē-jahck-yoo-lā-tər)–A rectal probe and power source to apply current to nerves to promote ejaculation.

electroencephalogram (ē-lehck-trō-ehn-sehf-ah-lō-grahm)–A record of the electrical activity of the brain.

electroencephalograph (ē-lehck-trō-ehn-sehf-ah-lō-grahf)–An instrument to record the electrical activity of the brain.

electroencephalography (ē-lehck-trō-ehn-sehf-ah-lohg-rah-fē)–The process of recording electrical activity of the brain; abbreviated EEG.

electrolyte (ē-lehck-trō-līt)–A substance that dissociates into ions when placed in solution; "charged substances"; examples include Na⁺, Cl⁻, and K⁺.

electromyogram (ē-lehck-trō-mī-ō-grahm)–A record or printout of the strength of muscle contractions due to electrical stimulation.

electromyography (ē-lehck-trō-mī-ohg-rah-fē)–The process of recording the strength of muscle contractions due to electrical stimulation; abbreviated EMG.

electroretinogram (ē-lehck-trō-reh-tihn-ō-grahm)–A record of the electrical activity of the retina; abbreviated ERG.

electroretinography (ē-lehck-trō-reh-tihn-ohg-rah-fē)–The procedure of recording the electrical activity of the retina.

elevation (ehl-ah-vā-shuhn)–Lifting.

elevator (ehl-eh-vā-tər)–An instrument used to reflect tissue from bone.

eligible to registry (ehl-ih-jeh-behl too reh-jihs-trē)–Designating an animal whose sire and dam are registered in the same breed registry and that meets other rules as to age, color, etc., of that breed.

emaciation (ē-mā-sē-ā-shuhn)–Marked wasting or excessive leanness.

emasculate (ē-mahs-kyū-lāt)–To castrate.

emasculatome (ē-mahs-kyū-lah-tohm)–An instrument used to crush and sever the spermatic cord through intact skin.

emasculator (ē-mahs-kyū-lā-tər)–An instrument used in castrations to crush and sever the spermatic cord.

embolism (ehm-bō-lihzm)–Blockage of a vessel due to a foreign object (for example, fat or air).

embolus (ehm-bō-luhs)–Mass, aggregation of blood, or collection of air forced from one vessel to a smaller one resulting in obstruction; plural is emboli (ehm-bō-lī).

embryo (ehm-brē-ō)–The developing zygote after implantation.

embryo sac (ehm-brē-ō sahck)–A sac that contains the embryo in its very early life in animals. Also called blastodermic vesicle.

embryo transfer (ehm-brē-ō trahnz-fər)–The removal of an embryo from a female of superior genetics and placing it in the reproductive tract of another female.

embryology (ehm-brē-ohl-ō-jē)–The science that deals with the study of the embryo.

embryonic (ehm-brē-ohn-ihck)–(1) Pertaining to the embryo or its development. (2) Underdeveloped; immature. See rudimentary.

embryonic vesicle (ehm-brē-ohn-ihck vehs-ih-kuhl)–The sac containing the developing embryo.

embryotomy (ehm-brē-oh-tō-mē)–The cutting up of fetus to enable removal from uterus.

emesis (ehm-eh-sihs)–Vomiting.

emetic (ē-meh-tihck)–A substance that induces vomiting.

emission (ē-mihsh-uhn)–A discharge.

emphysema (ehm-fih-zē-mah)–Chronic lung disease caused by enlargement of the alveoli or changes in the alveolar walls.

emulsification (ē-muhl-sih-fih-kā-shuhn)–The breakdown of large fat globules into smaller parts.

emulsify (ē-muhl-sih-fī)–To suspend small globules of one liquid in a second liquid in which it will not mix.

emulsion (ē-muhl-shuhn)–A mixture in which one liquid is suspended as tiny drops in another liquid, such as oil in water.

enamel (ē-nahm-ahl)–The hard, white substance covering the dentin of the crown of the tooth.

encapsulation (ehn-kahp-sū-lā-shuhn)–Enclosure in a capsule or sheath.

encephalitis (ehn-sehf-ah-lī-tihs)–Inflammation of the brain.

encephalocele (ehn-sehf-ah-lō-sēl)–Herniation of brain through a gap in the skull.

encephalomalacia (ehn-sehf-ah-lō-mah-lā-shē-ah)–Abnormal softening of the brain.

encephalomyelitis (ehn-cehf-ah-lō-mī-ih-lī-tihs)–Inflammation of the brain and spinal cord; sleeping sickness.

encephalopathy (ehn-sehf-ah-lohp-ah-thē)–Any disease of the brain.

enclosure (enh-klō-zhehr)–A fenced area that confines animals.

encrustation (ehn-kruhs-tā-shuhn)–A crust or hard coating on or in a body.

encyst (ehn-sihst)–To become enclosed in a sac, bladder, or cyst.

endectocide (chnd-chck-tō-sīd) An agent that kills both internal and external parasites.

endemic (ehn-dehm-ihck)–The ongoing presence of disease within a group or geographic region.

endocarditis (ehn-dō-kahr-dī-tihs)–Inflammation of the endocardium or epithelial lining membrane of the heart.

endocardium (ehn-dō-kahr-dē-uhm)–The inner layer of the heart.

endocrine (ehn-dō-krihn)–Secreting internally (into the bloodstream).

endocrine gland (ehn-dō-krihn glahnd)–A gland that secretes its chemical substances directly into the bloodstream, which then transports them throughout the body, e.g., the pituitary, pineal, thyroid, parathyroid, thymus, adrenal, pancreas.

endocrinologist (ehn-dō-krihn-ohl-ō-jihst)–A specialist who studies the hormonal system.

endocrinology (ehn-dō-krihn-ohl-ō-jē)–The study of the hormonal system.

endocrinopathy (ehn-dō-krih-nohp-ah-thē)–A disease of the hormone-producing system.

endoderm (ehn-dō-dərm)–The inner layer of embryo.

endogenous (ehn-doh-jehn-uhs)–Originates from within the body.

endolymph (ehn-dō-lihmf)–The fluid that fills one of the semicircular canals (cochlear canal).

endometrium (ehn-dō-mē-trē-uhm)–The inner layer of the uterus.

endoparasite (ehn-dō-pahr-ah-sīt)–Any of the various parasites which lives within the body of its host.

endorphin (ehn-dōr-fihn)–A brain-produced, natural, opioid like chemical that raises the pain threshold.

endoscope (ehn-dō-skōp)–A tubelike instrument with lights and refracting mirrors that is used to internally examine the body or organs.

endoscopy (en-dohs-kō-pē)–A visual examination of the interior of any body cavity using an endoscope.

endosteum (ehn-dohs-tē-uhm)–The tissue lining the medullary cavity of bone.

endothelium (ehn-dō-thē-lē-uhm)–A cellular covering that forms the lining of the internal organs and blood vessels.

endotoxin (ehn-dō-tohck-sihn)–A toxin produced within an organism and liberated only when the organism disintegrates or is destroyed.

endotracheal intubation (ehn-dō-trā-kē-ahl ihn-too-bā-shuhn)–The passage of a tube through the oral cavity, pharynx, and larynx into the windpipe.

enema (ehn-ah-mah)–The introduction of fluid into the rectum.

enrich (ehn-rihch)–To add a substance or vitamin to food products.

ensilage (ehn-sī-lahj)–Any green crop preserved for livestock feed by fermentation in a silo, pit, or stack, usually in chopped form. Also called silage.

ensile (ehn-sīl)–To place green plant material, such as green crops of grain, grasses, cornstalks, etc., in a silo in such a manner as to bring about proper fermentation for preservation and storage. See silage.

ensiling (ehn-sī-lihng)–The process in which a forage is chopped, placed in a storage unit that excludes oxygen, and allowed to ferment to a sufficient level of acetic acid.

enteric (ehn-tār-ihck)–The intestines.

enteritis (ehn-tər-ī-tihs)–Inflammation of the small intestine.

enterocolitis (ehn-tehr-ō-kō-lī-tihs)–Inflammation of the small intestines and the large intestine.

enterostomy (ehn-tər-ohs-tō-mē)–Surgically creating an opening into the small intestines through the abdominal wall.

entozoon (ehn-tō-zō-ohn)–Internal animal parasites, e.g., stomach worms.

entrails (ehn-trālz)–The visceral organs of the body, particularly the intestines.

entropion (ehn-trō-pē-ohn)–Inversion or turning inward of the eyelid.

enucleation (ē-nū-klē-ā-shuhn)–The removal of an organ in whole; usually used for removal of the eyeball.

enurination (ehn-yoo-rihn-ā-shuhn)–Behavior in rabbits where the buck jumps up in some fashion and squirts another rabbit with a stream of urine.

enzymatic (ehn-zih-maht-ihck)–Referring to a reaction or process that is catalyzed by an enzyme or group of enzymes.

enzyme (ehn-zīm)–A substance that chemically changes another substance.

eosinophil (ē-ō-sihn-ō-fihl)–A class of granulocytic leukocyte that detoxify allergens or parasitic infections.

eosinophilic (ē-ō-sihn-ō-fihl-ihck)–Things that stain readily with acidic, or pink, dyes in many commonly used stains such as H & E, Giemsa's, and Wright's.

epaxial (ehp-ahcks-ē-ahl)–The region along the dorsal vertebral column.

ependyma (eh-pehn-dih-mah)–The lining of the ventricles of the brain and central canal of the spinal cord.

epicardium (ehp-ih-kahr-dē-uhm)–The external layer of the heart; also part of the inner layer of the pericardial sac.

epidemic (eph-ih-dehm-ihck)–The sudden and widespread outbreak of disease within a group; also called epizootic (ehp-ih-zō-oh-tihck).

epidemiology (ehp-ih-dē-mē-ohl-ō-jē)–The study of relationships determining the frequency and distribution of diseases.

epidermis (ehp-ih-dər-mihs)–The outermost layer of skin.

epididymis (ehp-ih-dihd-ih-mihs)–The tube at the upper part of each testis that secrete parts of semen, stores sperm before ejaculation, and provides a passageway for sperm out of the testis.

epididymitis (ehp-ih-dihd-ih-mī-tihs)–Inflammation of the epididymis.

epidural (ehp-ih-doo-rahl)–Located above the dura matter.

epidural anesthesia (ehp-ih-doo-rahl ahn-ehs-thē-zē-ah)–The absence of sensation to a region following injection of a chemical into the space above the dura mater of the spinal cord.

epidural hematoma (ehp-ih-doo-rahl hē-mah-toh-mah)–A mass or collection of blood above or superficial to the dura matter.

epigenetic (ehp-ih-jeh-neht-ihck)–As used in reference to cancer, an effect that does not directly involve a change in the sequence of bases in DNA. See deoxyribonucleic acid.

epiglottis (ehp-ih-gloht-ihs)–The lidlike cartilage covering the larynx.

epilepsy (ehp-ih-lehp-sē)–Recurrent seizures of nonsystemic origin. Also called fits, falling sickness.

epinephrine (ehp-ih-nehf-rihn)–A hormone, also known as adrenaline, that is produced by the medulla of the adrenal glands in mammals.

epiphora (ē-pihf-ōr-ah)–Excessive tear production.

epiphysis (eh-pihf-ih-sihs)—-The wide end of a long bone, which is covered with articular cartilage and is composed of cancellous bone.

episcleritis (ehp-ih-sklehr-ī-tihs)–Inflammation of the tissue of the cornea.

episiotomy (eh-pihz-ē-oht-o-mē)–Surgical incision of the perineum and vagina to facilitate delivery of the fetus and prevent damage to maternal structures.

epispadias (ehp-ih-spā-dē-uhs)–An abnormal condition where the urethra opens on the dorsum of the penis.

epistasis (eh-pihs-tah-sihs)–The type of gene action where genes at one locus affect or control the expression of genes at a different locus.

epistaxis (ehp-ih-stahck-sihs)–A nosebleed.

epistatic (ehp-ih-stah-tihck)–Designating a condition of genetics in which one factor prevents a factor other than its allelomorph from exhibiting its normal effect on the development of the individual.

epithalamus (eph-ē-thahl-ah-muhs)–The portion of the diencephalon that includes the olfactory centers and the pineal gland.

epithelialization (ehp-ih-thē-lē-ahl-ih-zā-shuhn)–Healing by growth of epithelium over an incomplete surface.

epithelium (ehp-ih-thē-lē-uhm)–A cellular covering that forms the outer layer of the skin and covers the external surfaces of the body.

epizootic (ehp-ih-zō-oh-tihck)–Designating a disease of animals that spreads rapidly and affects many individuals of a kind at the same time, thus corresponding to an epidemic in people.

epulis (ehp-uhl-uhs)–A benign tumor arising from periodontal mucous membranes.

equilibrium (ē-kwihl-ih-brē-uhm)–A state of balance.

equine (ē-kwīn)–Species name for horses (includes donkeys and mules).

equitation (ehck-wih-tā-shuhn)–Horsemanship, horsewomanship.

erection (ē-rehck-shuhn)–The process whereby a body part is made to stand upright; a process where the penis becomes engorged with blood causing it to be firm, turgid, and ready for breeding.

erosion (ē-rō-shuhn)–Focal loss of epithelium, only to the level of the basement membrane.

eructation (ē-ruhck-tā-shuhn)–Belching or raising gas orally from the stomach.

erythema (ehr-ih-thē-mah)–Skin redness.

erythematous (ehr-ih-thehm-ah-tuhs)–Redness.

erythrocyte (eh-rihth-rō-sīt)–A mature red blood cell (oxygen carrying cell).

erythrocytosis (eh-rihth-rō-sī-tō-sihs)–An abnormal increase in red blood cells.

erythroderma (eh-rihth-rō-dər-mah)–An abnormal redness of the skin.

erythropoietin (ē-rihth-rō-poy-ē-tihn)–A hormone produced by the kidney that stimulates red blood cell production in the bone marrow.

escape (eh-skāp)–A fowl or animal that has gotten out of its enclosure.

Escherichia coli (ehsh-ehr-ih-kē-ah kō-lī)–One of the bacterial species in the fecal coliform group. Commonly referred to as *E. coli*.

escutcheon (eh-skuhch-ehn)–That part of a cow that extends upward just above and back of the udder where the hair turns upward in contrast to the normal downward direction of the hair. Also called milk shield, milk mirror.

esophageal reflux (ē-sohf-ah-jē-ahl rē-fluhcks)–The return of stomach contents into the esophagus; also called gastroesophageal reflux.

esophagoplasty (ē-sohf-ah-gō-plahs-tē)–Surgical repair of the esophagus.

esophagoscopy (ē-sohf-ah-gohs-kō-pē)–Endoscopic examination of the esophagus; the scope is passed from the oral cavity through the esophagus.

esophagus (ē-sohf-ah-guhs)–A collapsible tube that leads from the oral cavity to the stomach.

essential amino acid (eh-sehn-shahl ah-mē-nō ah-sihd)–Any of the amino acids that cannot be synthesized in the body from other amino acids or substances or that cannot be made in sufficient quantities for the body's use.

essential host (eh-sehn-shahl hōst)–A host for one stage in the development of a parasite without which the parasite cannot develop to maturity.

estimated breeding value (EBV) (ehs-tih-mā-tehd brē-dihng vahl-ū)–In beef cattle, an estimate of the value of an animal as a parent, expressed as a ratio with 100 being average.

estimated relative producing ability (ERPA) (ehs-tih-mā-tehd rehl-ah-tihv prō-doo-sihng ah-bihl-ih-tē)–In dairy cattle, a prediction of 305-day, two times per day milking, mature equivalent production compared with the production of other cows in the herd.

estivate (ehs-tih-vāt)–A dormant state in which an animal reduces its body temperature, heart and respiration rates, and metabolism in summer.

estray (eh-strā)–A wandering domesticated animal of unknown ownership.

estriol (ehs-trē-ohl)–An estrogenic substance derived from the urine of pregnant animals which, when administered to female animals, causes them to come into heat. Also called theelol.

estrogen (ehs-trō-jehn)–A hormone or group of hormones produced by the developing ovarian follicle; it stimulates female sex drive and controls the development of feminine characteristics.

estrone (ehs-trōn)–A product derived from the urine of pregnant animals or from the ovary, which may be administered to female animals to cause them to come into heat. Also called theelin, folliculin, follicular hormone.

estrous cycle (ehs-truhs sī-kuhl)–Reproductive phases beginning at puberty that vary at regular intervals to prepare the uterus to receive a fertilized ovum; it is measured from the beginning of estrus or heat period to the beginning of the next.

estrus (ehs-truhs)–The period of the reproductive cycle in which the female is receptive to the male.

estrus synchronization (ehs-truhs sihn-krō-nih-zā-shuhn)–Using synthetic hormones to make a group of females come into heat at the same time.

ethology (ē-thohl-ō-jē)–The science of animal behavior related to the environment.

etiological (ē-tē-ō-lohj-ih-kahl)–Pertaining to the causes of diseases.

etiology (ē-tē-ohl-ō-jē)–The study of disease causes.

eupeptic (ū-pehp-tihck)–Having normal digestion.

euploid (ū-ployd)–An organism or cell in which the chromosome number is the exact multiple of the monoploid or haploid number. Terms used for euploid series are haploid, diploid, triploid, tetraploid, etc.

eustachian tube (yoo-stā-shuhn toob)–A narrow duct that leads from the middle ear to the nasopharynx; helps equalize air pressure in the middle ear with that of the atmosphere; also called auditory tube.

euthanasia (yoo-thehn-ā-zhah)–Inducing the death of an animal quickly and painlessly; "putting an animal to sleep."

euthyroidism (yoo-thī-royd-ihzm)–A condition of normal thyroid function.

evaginated (ē-vahj-ih-nā-tehd)–Turned inside out.

eversion (ē-vər-shuhn)–Turning outward.

everted (ē-vehrt-ehd)–Turned inside out.

eviscerate (ē-vihs-ər-āt)–The removal or exposure of internal organs.

ewe (ū)–An intact female sheep.

excessive (ehck-sehs-ihv)–More than normal.

excise (ehck-sīz)–To surgically remove.

excisional biopsy (ehcks-sih-shuhn-ahl bī-ohp-sē)–Removing an entire mass, tissue, or organ to examine.

exclosure (ehcks-klō-zhuhr)–An area of land fenced in to prevent all or certain kinds of animals from entering it and used for ecological experiments involving biotic factors such as the grazing pressure of livestock.

excreta (ehcks-krē-tah)–Matter that is excreted; waste matter discharged from the body; materials cast out of the body. Excrement; waste matter discharged from the bowels; manure.

excretion (ehcks-krē-shuhn)–The act of eliminating or the material that is eliminated.

exfoliative (ehcks-fōl-ē-ah-tihv)–Falling off or sloughing.

exhalation (ehcks-hah-lā-shuhn)–The act of breathing out; also called expiration.

exocrine gland (ehck-soh-krihn glahnd)–A gland that secretes its chemical substances into ducts that lead out of the body or to another organ.

exogenous (ehcks-ah-jehn-uhs)–Originates from outside the body.

exomphalos (ehcks-ohm-fah-lohs)–A hernia that has escaped through the umbilicus.

exophthalmos (ehcks-ohp-thahl-mōs)–An abnormal protrusion of the eyeball.

exoskeleton (ehck-sō-skehl-eh-tohn)–Collectively, the external plates of the body wall.

exosmosis (ehck-sohs-mō-sihs)–The slow diffusion of the more dense fluid through a membrane to mingle with the less dense fluid.

exostosis (ehck-sohs-tō-sihs)–A benign growth on the bone surface, e.g., splints, bone spavins, etc.

exotic (ehg-zoht-ihck)–Foreign, unfamiliar, new, imported.

exotoxin (ehck-sō-tohck-sihn)–A soluble toxin excreted by specific bacteria and absorbed into the tissues of the host.

exotropia (ehck-sō-trō-pē-ah)–Deviation of the eyes away from each other.

expected progeny difference (EPD) (ehcks-pehk-tehd proh-jehn-ē dihf-ər-ehns)–In beef cattle selection, a score used to measure the amount of performance difference in the offspring of a certain sire as compared to the performance level of the herd average.

expectorant (ehck-spehck-tō-rahnt)–A drug that causes expulsion of mucus from the respiratory tract.

exsanguination (ehcks-sahn-gwih-nā-shuhn)–Removal or blood loss from the body.

extenders (ehcks-tehn-dərs)–Additive used to extend the lifespan of sperm cells.

extension (ehcks-tehn-shuhn)–To straighten a joint or to increase the angle between two bones.

extensor (ehcks-tehn-sər)–Muscle that straightens a limb at a joint.

exteriorize (ehcks-tēr-ē-ōr-īz)–To move an internal organ to the outside of the body.

external auditory canal (ehcks-tər-nahl aw-dih-tōr-ē kah-nahl)–Passageway or tube that transmits sound from the pinna to the tympanic membrane.

extracapsular extraction (ehcks-trah-kahp-soo-lahr ehcks-trahck-shuhn)–Cataract removal that leaves the posterior lens capsule intact.

extraction (ehcks-trahck-shuhn)–To remove; is used to describe surgical removal of a tooth.

extraocular (ehcks-trah-ohck-yoo-lahr)–Outside the eyeball.

extrinsic muscles (ehcks-trihn-sihck muhs-uhlz)–Six muscles that attach the outside of the eyeball to the bones of the orbit.

exudate (ehcks-yoo-dāt)–Material that has escaped from blood vessels and is high in protein, cells, or solid materials derived from cells.

eyeing (ī-ihng)–Clipping the wool from around the face of closed-faced sheep to prevent wool blindness.

eyetooth (ī-tooth)–Either of the two canine teeth in the upper jaw. Also called dog tooth.

F

F₁–The first filial generation; the first generation of a given mating.

F₂–The second generation progeny produced by crossing two F₁ individuals.

F₁ females (ehf-wuhn fē-mā-uhlz)–Female cattle from the first cross of mating of two different purebreeds of cattle.

F₁ generation (ehf-wuhn jehn-ər-ā-shuhn)–The first offspring from purebred parents of different breeds or varieties; F₁ stands for first filial.

face (fās)–(1) The bare skin of a fowl's head around and below its eyes. (2) The front part of the head of an animal, including the eyes, nose, and mouth.

facet (fahs-eht)–A small, smooth area.

factor (fahck-tōr)–A unit of inheritance occupying a definite locus on one or both members of a definite chromosome pair whose presence is responsible for the development or modification of a certain character of the individual who possesses that genotype; a determiner or gene.

facultative (fahck-uhl-tā-tihv)–Designating an organism that is capable of living under more than one condition, e.g., as a saprophyte and as a parasite; as an aerobic or anaerobic organism.

Fahrenheit scale (fahr-ehn-hīt scāl)–A temperature scale in which the freezing point of water is 32°F and the boiling point is 212°F. Named after Gabriel Daniel Fahrenheit (1686–1736), a German physicist.

fall farrow (fahl fahr-ō)–A pig born in the autumn.

fall lamb (fahl lahm)–A lamb that is usually born in the spring and sold in the fall.

fallopian tube (fahl-lō-pē-ahn toob)–One of the two tubes or ducts connected to the uterine horn of mammals and leading to the ovary; functions in transporting the ovum from the ovary to the uterus; more commonly called oviduct.

false heat (fahlz hēt)–The displaying of estrus by any female animal when she is pregnant or is out of season. It can occur in a healthy animal or may be caused by diseased ovaries.

false molt (fahlz mōlt)–The shedding of feathers by a bird due to unnatural causes. See molt.

false pregnancy (fahlz prehg-nahn-sē)–Not carrying a fetus but demonstrating maternal behavior and characteristics; also called pseudocyesis (soo-dō-sī-ē-sihs) or pseudopregnancy.

fancier (fahn-sē-ər)–A breeder who shows particular interest in, and the development of, a particular breed or type of animals or plants, e.g., a rose fancier, dog fancier, Hereford fancier.

fancy (fahn-sē)–A top-quality grade for many vegetables, fruits, flowers, poultry, and livestock.

fancy points (fahn-sē poyntz)–Indications of purity of breeding, e.g., the set of the horns, the carriage of the ear, the color of the hair on certain parts of the body, and the general style of an animal.

far side (fahr sīd)–The right side of a horse.

farrier (fār-ē-ər)–(1) A person who cares for equine feet including trimming and shoeing. (2) An obsolete term for a horse and cattle veterinarian.

farrowing (fār-ō-ihng)–Giving birth to pigs or guinea pigs.

farrow-to-finish (fār-ō too fihn-ihsh)–A full-service swine operation that houses breeders, newborns, weanlings, and feeder stock.

farrow-to-wean (fār-ō too wēn)–A swine operation that houses breeding sows and offspring until they reach weaning age or weight.

farrowing crate (fār-ō-ihng krāt)–A holding pen that limits sow movement prior to and during parturition.

farrowing house (fār-ō-ihng hows)–Any of several types of structures especially designed for a sow and her litter of pigs.

farrowing pen (fār-ō-ihng pehn)–A sow holding area that has guardrails and floor junctures that allow young pigs to escape; used prior to and during parturition; the pen is larger than a crate.

fascia (fahsh-ē-ah)–A sheet of fibrous connective tissue that covers, supports, and separates muscle.

fasciitis (fahs-ē-ī-tihs)–Inflammation of the fascia.

fast-stepping trot (fahst stehp-ihng troht)–One form of the trot, a horse's gait, characterized by long rapid steps.

fat (faht)–(1) (a) The tissues of an animal that bear an oily or greasy substance. (b) Designating any animal or fowl that abounds in fat. (2) The oily substance in milk.

fatigue (fah-tēg)–Loss of energy.

fatling (faht-lihng)–A young farm animal that is fattened for slaughter.

fatten (faht-tehn)–(1) To add or put on fat, as an animal. (2) To cause an animal or fowl to put on fat.

fattening (faht-tehn-ihng)–(1) The feeding of animals or fowls so that they put on fat.

(2) Designating a feed that, when fed to livestock, causes them to add fat, as fattening grain.

fattening range (faht-tehn-ihng rānj)–A productive range devoted primarily to fattening livestock for market.

favor (fā-vehr)–To protect; to use carefully; as a horse favors a lame leg.

fawn (fawn)–(1) A coat color for an animal that is a soft, grayish-tan. (2) A young deer.

feather (fehth-ər)–(1) The epidermal structure partly embedded in the skin follicles of birds. Feathers vary greatly in color, size, and shape, and generally cover the body of a bird. (2) Plumage (plural). (3) The long hair of certain breeds of horses that grows below the knees and hocks.

feather follicle (fehth-ər fohl-ih-kl)–The depression in the skin of a bird from which the feather grows.

feather-legged (fehth-ər lehgd)–Designating the breeds of chickens that have feathers on the shanks and toes.

feathering (feh-thər-ihng)–(1) The fringe of hair around an equine foot just above the hoof; also used to describe the fringe of hair on the caudal aspects of canine limbs. (2) Digital stroking of the vagina, performed in whelping bitches to stimulate uterine contractions.

febrile (fē-brī-ahl)–Having a fever.

fecal (fē-kahl)–Pertaining to excrement, manure, and waste material passed from the bowels.

fecal incontinence (fē-kahl ihn-kohn-tihn-ehns)–The inability to control bowel movements.

fecalith (fēk-ah-lihth)–A stonelike fecal mass.

feces (fē-sēz)–Solid body wastes expelled through the rectum; also called stool, bowel movements, dung (duhng), excreta (ehcks-krē-tah), manure (mahn-oo-ər), or defecation (dehf-eh-kā-shuhn).

fecund (fē-kuhnd or feh-kuhnd)–Fruitful; fertile; prolific.

fecundity (fē-kuhn-dih-tē)–The ability to produce offspring.

feed (fēd)–(1) Harvested forage, such as hay, silage, fodder, grain, or other processed feed for livestock. See forage. (2) The quantity of feed in one portion. (3) To furnish with essential nutrients.

feed additives (fēd ahd-ih-tihvz)–Material added to livestock feed that is not a nutrient but enhances the growth efficiency of the animal.

feed analysis (fēd ah-nahl-ih-sihs)–The chemical or material analysis of a commercially prepared feed, printed on a tag and fastened to the bag in which the feed is to be sold.

feed bunk (fēd buhnk)–A forage and grain feeding station.

feed conversion ratio (fēd kohn-vehr-shuhn rā-shō)–The rate at which an animal converts feed to meat. If an animal requires four pounds of feed to gain one pound, it is said to have a four to one (4:1) feed conversion ratio.

feed efficiency (fēd eh-fih-shehn-sē)–A term for the number of pounds of feed required for an animal to gain one pound of weight.

feed energy utilization (fēd ehn-ər-jē ū-tihl-ih-zā-shuhn)–The percentage of energy obtained from a feed that is used for an animal's bodily functions.

feed hopper (fēd hohp-ər)–A container that allows feed to drop down gradually as more is consumed.

feed out (fēd owt)–To prepare animals for market by fattening.

feeder (fē-dər)–Beef cattle that are placed in a feedlot based on age and weight for finishing to a determined grade.

feeder calf (fē-dər kahf)–A weaned calf that is under one year of age and is sold to be fed for more growth.

feeder lamb (fē-dər lahm)–In marketing or stock judging, a lamb that is a feeder.

feeder pig (fē-dər pihg)–A barrow or gilt carrying enough age and flesh so as to be ready to place in a feedlot for finishing; usually a pig weighing less than 120 pounds.

feeder-to-finish (fē-dər too fihn-ihsh)–A swine operation that raises weanling pigs to market weight.

feedgrain (fēd-grān)–Any of several grains, and most commonly used for livestock or poultry feed, such as corn, grain sorghum, oats, and barley.

feeding (fēd-ihng)–Providing animals with the desirable quality and quantity of feeds; making feed available to animals. (2) The ingestion of feed by an animal or plant.

feeding ratio (fēd-ihng rā-shō)–The weight of feed consumed divided by the increase in weight of an animal, during a given time interval.

feeding standard (fēd-ihng stahnd-ahrd)–Established standards that state the amounts of nutrients that should be provided in rations for animals of various ages and classes in an attempt to attain the optimum economy of growth, gain, or production.

feeding value (fēd-ihng vahl-ū)–A term referring to the nutritive value of different feeds, i.e., expressing the amount of nutrients furnished by each feed and the degree of their digestibility.

feedlot (fēd-loht)–A confined area where an animal is fed until it is slaughtered.

feedstuff (fēd-stuhf)–One of, or a mixture of, the substances that form the nutrients; namely, proteins, carbohydrates, fats, vitamins, minerals, and water. A feedstuff is different from a feed in that a feedstuff is not normally fed by itself but is mixed with other feedstuffs to formulate a feed.

feline (fē-līn)–Species name for cat.

fell (fehl)–The elastic tissue just under the hide of an animal attached to its flesh; fascia.

felting (fehl-tihng)–(1) A property of wool fibers to interlock when rubbed together under heat, moisture, or pressure. (2) The manufacturing of felts from the furs of the rabbit and of other animals.

femoral artery (fehm-ōr-ahl ahr-tēr-ē)–The main artery carrying blood to an animal's hind legs.

femoral pores (fehm-ōr-ahl poorz)–Sexually dimorphic glands that are prominent in mature lizards, located on the ventral surface of lizard thighs; also called femoral glands.

femur (fē-muhr)–The proximal hindlimb long bone that articulates proximally with the pelvis and distally with the tibia and fibula; also called the thigh bone.

fence (fehns)–(1) A hedge or barrier of wood, metal, stone, or plants erected to enclose an area to prevent trespassing or the straying of animals. (2) To enclose an area with a fence.

fenestration (fehn-ih-strā-shuhn)–A perforation.

feral (fehr-ehl)–Wild or untamed.

fertile (fǝr-tihl)–(1) Productive; producing plants in abundance, as fertile soil. (2) Capable of growing or of development, as a fertile egg. (3) Capable of reproducing viable offspring.

fertile egg (fǝr-tihl ehg)–A fertilized, avian egg capable of embryonic development.

fertility (fǝr-tihl-ih-tē)–The ability of an animal or fowl to produce offspring.

fertilization (fǝr-tihl-ih-zā-shuhn)–The union of ovum and sperm.

fertilize (fǝr-tih-līz)–To fecundate the egg of an animal.

fetid (feh-tihd or fē-tihd)–Having a disagreeable odor.

fetlock joint (feht-lohck joynt)–The joint between the metacarpals or metatarsals and phalanx I in livestock; also called the metacarpo/metatarsophalangeal joint.

fetotomy (fē-toh-tō-mē)–The cutting apart of a fetus to enable removal from the uterus.

fetter (feht-ehr)–A shackle for the feet of horses.

fetus (fē-tuhs)–An unborn animal.

fever (fē-vehr)–A higher than normal temperature in animals.

feverish (fē-vehr-ihsh)–Pertaining to animals that have a temperature higher than normal because of disease, etc.

fibers (fī-bǝrz)–Threads or filaments usually arranged in bundles.

fibrillation (fih-brih-lā-shuhn)–Rapid, random, and ineffective heart contractions.

fibrinogen (fih-brihn-ō-jehn)–One of the clotting proteins in plasma.

fibroblast (fī-brō-blahst)–Fiber-producing cells of the connective tissue.

fibroid (fī-broyd)–A benign tumor arising from the smooth muscle of the uterus; also called leiomyoma (lī-ō-mī-ō-mah).

fibroma (fī-brō-mah)–A tumor composed of fully developed connective tissue.

fibula (fihb-ū-lah)–The smaller of the two hindlimb long bones that articulates proximally with the femur and distally with the tarsus.

fighting teeth (fī-tihng tēth)–A set of six teeth in llamas that include an upper vestigial incisor and an upper and lower canine on each side.

filial (fihl-ē-ahl)–Refers to offspring; the meaning of the F in F_1 and F_2.

filiform (fihl-ih-fōrm)–Threadlike.

fill (fihl)–(1) The increase in weight and form of livestock that have been watered and fed after arriving at their destination. (2) To feed and water livestock at the end of the shipment to make up for the loss of weight en route.

filler (fihl-ǝr)–The nonessential matter in a manufactured or mixed feed, such as high-fiber materials, oat hulls, screenings, etc.

filly (fihl-ē)–An intact female horse or donkey usually under four years of age.

filterable virus (fihl-tehr-eh-buhl vī-ruhs)–A virus that is capable of passing through the pores of a filter that does not allow passage of the ordinary bacteria.

fimbria (fihm-brē-ah)–A fingerlike extension; plural is fimbriae (fihm-brē-ā).

find (fihnd)–To give birth to young, e.g., a cow finds a calf.

finish (fihn-ihsh)–The degree of fatness. This term is often used interchangeably with condition, but as finish, the fat should lay smoothly over the body in a proper degree to suit the market.

finisher pig (fihn-ihsh-ər pihg)–Swine over 100 pounds that are sent to slaughter.

finishing (fihn-ihsh-ihng)–The increased feeding of an animal just prior to butchering, with results in rapid gains and increased carcass quality.

fire (fīr)–To treat a spavin or ringbone on a horse with a strong liniment in an attempt to cure or alleviate lameness.

firing (fīr-ihng)–Marking a series of skin blisters with a hot needle over an area of lameness.

fission (fihsh-uhn)–A form of reproduction, common among bacteria and protozoa, in which a unit or organism splits into two or more whole units.

fissure (fihs-shər)–A cracklike sore.

fistula (fihs-tyoo-lah)–An abnormal passage from an internal organ to the body surface or between two internal organs; plural is fistulae (fihs-tyoo-lā).

fit (fiht)–To condition livestock for use, sale, or exhibit.

fitting cow ration (fiht-ihng kow rā-shuhn)–Ration fed to cows that are being prepared for exhibition, sale, or calving.

fitty (fiht-ē)–Referring to a horse that has fits when overheated.

flaccid (flah-sihd or flahs-ihd)–Without rigidity; lax and weak.

flagella (flah-jehl-ah)–Whiplike appendages of certain single-celled aquatic animals and plants, including some bacteria, the rapid movement of which produces motion.

flank (flānck)–The side of the body between the ribs and ilium.

flank cinch (flahnck sihnch)–A cinch that is separate from the saddle and pulled tight in front of the hips and under the flanks of a horse. It is used in rodeos to cause the horse to buck.

flank incision (flahnck ihn-sihz-shuhn)–A surgical cut perpendicular to the long axis of the body, caudal to the last rib.

flap (flahp)–A mass of tissue for grafting in which part of the tissue is still adhered to the body.

flat foot (flaht foot)–In a horse, a foot of which the angle is less than 45°, or one in which the sole is not concave, or one with a low, weak heel.

flat boned (flaht bōnd)–Reference made to the cannon bone region of a horse's leg, which is constituted of the bone, ligaments, and tendons.

flatulence (flaht-yoo-lehns)–Excessive gas formation in the gastrointestinal tract.

flavus (flah-vuhs)–Yellow.

flay (flā)–To remove the skin from a carcass.

flea (flē)–Small, wingless insects that feed upon the blood of warmblooded animals.

flea-bitten gray (flē biht-tehn grā)–A coat color of horses that is gray with darker hairs throughout.

flea-bitten white (flē biht-tehn whīt)–A coat color of horses that is white with numerous, tiny, dark spots.

fledge (flehj)–To acquire the feathers necessary for flight, for example, a young bird just fledged.

fledgling (flehdj-lihng)–A young bird that has recently acquired its flight feathers.

fleece (flēs)–(1) The wool from all parts of a single sheep, which consists of the crinkly hair up to 12 inches in length. This waviness enables the wool to be matted together into felt or spun into yarn, twine, or thread. (2) To shear sheep.

flehmen (fleh-mehn)–An action by a bull, boar, or ram associated with courtship and sexual activity. The lip curls upward and the animal inhales in the vicinity of urine or the female vulva.

flesh (flehsh)–(1) The portion of an animal body that consists mainly of muscle. (2) Plumpness or corpulence, especially in such phrases as good flesh, etc.

flesh-colored (flehsh kuhl-ərd)–A coat color of animals in which the hair color is the same as the skin.

fleshed (flehshd)–(1) Designating muscles or lean meat. (2) Designating a pelt with the flesh or fatty pieces removed from the inner surface.

fleshy (flehsh-ē)–(1) Fat or corpulent. (2) Designating the soft or edible portions of meat.

flexion (flehck-shuhn)–To bend a joint or to reduce the angle between two bones.

flexor (flehck-sər)–A muscle that bends a limb at a joint.

flexor tendon (flehck-sər tehn-dohn)–A tendon that causes the fetlock joint to bend.

flexure (flehck-shər)–A bent or curved part of a structure.

flight coverts (flīt cuh-vərts)–The stiff feathers located at the base forward of the flight feathers (primaries) and covering their base.

flights (flītz)–The primary feathers of the wing. The term is sometimes used to denote both primaries and secondaries.

flighty (flī-tē)–Nervous.

float (flōt)–An instrument used in filing or rasping an equine's premolar and/or molar teeth.

floaters (flō-tərz)–Particles that cast shadows on the retina suspended in the vitreous fluid; also called vitreous floaters.

floating (flō-tihng)–Filing off the sharp edges of equine teeth.

flock (flohck)–(1) Several birds or domestic mammals, such as sheep, which are tended as a unit. Also called herd, band. (2) In stock judging, one ram, of any age, and four ewes of varying ages as designated by the show.

flock mating (flohck mā-tihng)–The indiscriminate breeding of fowls.

flocking tendency (flohck-ihng tehn-dehn-sē)–The habit of congregating in large flocks, inherent in sheep.

flow (flō)–The amount of milk produced per cow herd, etc., at a specified time.

flow rate (flō rāt)–The rate at which the cow lets down milk.

fluff (fluhf)–(1) The downy part of a feather. (2) The soft feathers on the thighs and posteriors of birds.

flukes (flookz)–Flatworms of the class Trematoda which, at maturity, are internal parasites of vertebrate animals and humans, but, in snails, usually have intermediate stages.

fluoroscopy (floor-ohs-kō-pē)–A procedure to visually examine internal body structures in motion using radiation to project images on a fluorescent screen.

flush (fluhsh)–(1) To increase the feed allowance to ewes or sows with a protein-rich supplement feed a short time before and during the breeding season. (2) A vigorous or abundant, sudden new growth. (3) To introduce and shortly afterwards withdraw an irrigating solution of mild antiseptic of medicinal value, as to flush the vagina or uterus.

flushed (fluhshd)–An animal that receives extra feed and care prior to breeding.

flushing (fluhsh-ihng)–The act of increasing feed prior to breeding or embryo transfer to increase the number of ova released.

flutter (fluht-tər)–A cardiac arrhythmia in which atrial contractions are rapid but regular.

fly (flī)–Any winged insect, such as a moth, bee, gnat, etc. Specifically a two-winged insect of the family Muscidae.

fly strike (flī strīk)–An infestation with maggots.

flying stall (flī-ihng stahl)–A portable stall for a horse in which the animal remains while being loaded or unloaded from a vessel or while in transit.

foal (fōl)–(1) The unweaned young of the horse or mule. (2) In stock shows, a horse foaled on or after January 1 of the year shown.

foal heat (fōl hēt)–Estrus that occurs directly after parturition (usually not fertile).

foaling (fō-lihng)–Giving birth to horses or donkeys.

focus (fō-kuhs)–A localized region.

fodder (fohd-ər)–Feed for livestock, specifically the dry, cured stalks and leaves of corn and the sorghums. In the case of corn, the ears may be removed from the stalk leaving the stover. See forage, roughage.

fold (fohld)–A pen enclosure for sheep or cattle.

fold unit (fohld yoo-niht)–A small house, complete with covered-in run, for the controlled grazing of poultry.

folic acid (fō-lihck ah-sihd)–A B vitamin found in the leaves of leguminous and other plants, in yeast, liver meal, and wheat. Folic acid is needed in hemoglobin formation and for growth. Also called pteroylglutamic acid.

follicle (fohl-lih-kl)–(1) A small anatomical cavity; particularly, a small blisterlike development on the surface of the ovary that contains the developing ovum. (2) A small sac, gland, or pit for secretion or excretion. The hairs of an animal grow out of pits called follicles. (3) The growth that appears on the surface of the ovary late in the estrous cycle containing the developing ovum.

follicle-stimulating hormone (fohl-lih-kl stihm-yoo-lā-tihng hōr-mōn)–A hormone, produced by the pituitary gland, that promotes growth of ovarian follicles in the female and sperm in the male.

fomentation (fō-mehn-tā-shuhn)–A poultice; the external application of warm moist cloths, or other objects to ease pain.

fomite (fō-mīt)–Inanimate carriers of disease.

fontanelle (fohn-tah-nehl)–The soft spot remaining at the junction of the sutures; usually this closes after birth.

food (food)–Anything that, when taken into the body, nourishes the tissues and supplies body heat. Also known as aliment and nutriment.

foot (foot)–In quadrupeds, the terminal portion of the leg that rests upon the ground.

foot pad (foot pahd)–The cushions on the bottom of the feet of such animals as cats and dogs.

forage (fōr-ahj)–(1) That portion of the feed for animals that is secured largely from leaves and stalks of plants, such as the grasses and legumes used as hays. (2) To search for, spread out, or seek for food.

forage crops (fōr-ahj krohpz)–Those plants or parts of plants that are used for feed before maturing or developing seeds (field crops).

forage feeds (fōr-ahj fēdz)–(1) Bulky type feeds composed largely of pasture grasses, hays, silage, etc. (2) A mixture of ground or processed feeds that is composed largely of forages.

forage legumes (fōr-ahj lehg-ūmz)–Any of the legume plants that are grown or used largely as forage for livestock, such as alfalfa, clover, etc.

foramen (fō-rā-mehn)–An opening in a bone through which blood vessels, nerves, and ligaments pass.

foramen magnum (fōr-ā-mehn mahg-nuhm)–An opening located in the occipital bone for passage of the spinal cord.

forceps (fōr-sehpz)–In veterinary medicine, a plierslike instrument used for grasping, pulling, and compressing.

forearm (fōr-ahrm)–The part of the foreleg supported by the radius and ulna; between the elbow and carpus.

foreflank (fōr-flahnck)–That section of the body of a hog just behind the lower shoulder or foreleg.

forefoot (fōr-foot)–To rope the front legs of a running animal usually from horseback.

forehand (fōr-hahnd)–The front of the horse, including head, neck, shoulders, and forelegs; that portion of the horse in front of the center of gravity.

forehobble (fōr-hohb-uhl)–A strap or rope that is tied around the forelegs of an animal to prevent straying.

foreign matter (fōr-ihn mah-tər)–Any material, substance, etc., that is unnatural to, or not commonly developed in, a product.

foreleg (fōr-lehg)–The lower portion, above the pastern, of either front leg or four-legged animals. It is commonly used in reference to horses.

forelock (fōr-lohck)–In maned animals the most cranial part of the mane, hanging down between the ears and on to the forehead.

foremilk (fōr-mihlk)–The first 25 to 50 millimeter of milk to be withdrawn from the udder at the beginning of milking in contrast to middle milk and strippings. Foremilk is of poor quality chemically and bacteriologically and should be rejected.

foremilk cup (fōr-mihlk kuhp)–A metal milking cup fitted with a dark shelf. The first milk at each milking is drawn into the cup. If there is mastitis, clots of milk will be seen on the dark shelf.

forepastern (fōr-pahs-tərn)–A term commonly used to describe the portion of the front leg next to the hoof.

forepunch (fōr-puhnch)–A kind of counterpunch used to make a place for the nailheads to be set into a horseshoe.

forequarters (fōr-kwahr-tərz)–The front two quarters of an animal.

forestomachs (fōr-stoh-mahcks)–The three nonglandular stomachs found in ruminants that are actually outpouchings of the esophagus; specifically, the rumen, reticulum, and omasum.

forestripping (fōr-strihp-ihng)–The removal of a small amount of milk by hand from each teat prior to the milking operation; forestrippings are usually discarded because of their high bacterial and low fat content.

forging (fōr-jihng)–The noisy striking of the foreshoe with the toe of the hind shoe by a horse when walking, trotting, or running. Also called clicking, striking.

formed elements (fōrmd ehl-eh-mehnts)–The portion of blood that includes red blood cells, white blood cells, and platelets.

fortified (fōr-teh-fīd)–Designating a product to which amounts of a vitamin, as vitamin A or vitamin D have been added.

fossa (fohs-ah)–A hollow or depressed area.

foundation herd (fown-da-shuhn hərd)–Breeding stock; cows, bulls, and heifers or calves retained for replacement; ewes, rams, and lambs for replacement; and nannies, billies, and kids for replacement.

founder (fown-dər)–An inflammation of the tissue that attaches the hoof to the foot. Also called laminitis.

four footing (fōr foo-tihng)–Throwing an animal by means of a rope around the feet.

four-cornered gait (fōr kōr-nehrd gāt)–A horse's gait in which each foot is placed on the ground individually. Also called single-foot, rack.

fourth ventricle (fōrth vehn-trih-kuhl)–The most posterior cavity in the brain.

four-tooth sheep (fōr tooth shēp)–A two-year-old sheep.

four-way hybrid (fōr wā hī-brihd)–The hybrid that results from mating two single crosses. Also called double cross. See single cross.

fovea centralis (fō-vē-ah sehn-trah-lihs)–A pit located in the macula where vision is the sharpest.

fowl (fowl)–Refers to a bird, usually poultry.

fox trot (fohcks troht)–A slow, short, broken type of gait in which the head usually nods.

fracture (frahck-shər)–Breaking of a part, especially a bone.

free-choice feeding (frē choys fē-dihng)–A type of feeding routine whereby feed, water, salt, etc., are provided in unlimited quantities and an animal is left to regulate its own intake.

freemartin (frē-mahr-tihn)–A sexually imperfect; usually a sterile female calf twinborn with a male calf.

free-ranging (frē rānj-ihng)–Allowing animals, especially poultry, to roam freely and eat as they wish without any sort of confinement.

freeze branding (frēz brahn-dihng)–An identification method done by clipping hair from the brand area, wetting skin with alcohol, then applying a branding iron cooled in liquid nitrogen or dry ice and alcohol.

Friedman test (frēd-mahn tehst)–A test for pregnancy in which a small amount of urine from the tested animal is injected into the bloodstream of a virgin female rabbit. Pregnancy is indicated by certain changes in the ovaries of the rabbit.

frenulum (frehn-yoo-luhm)–A band of connective tissue that connects the tongue to the ventral surface of the oral cavity.

frequency (frē-kwehn-sē)–The number of cycles per unit of time.

fresh (frehsh)–(1) Designating a cow that has recently dropped a calf. (2) Designating an egg of good quality.

fresh manure (frehsh mahn-oo-ər)–A recently excreted animal dung whose direct contact can be harmful to plant tissues because of rapid chemical and fermentive changes that take place.

freshen (frehsh-ehn)–Giving birth in dairy animals.

friable (frī-ah-buhl)–Easily crumbled.

fribs (frihbz)–Short second cuts of wool resulting from faulty shearing; also small-sized dirty or dungy locks.

frizzle feather (frihz-uhl fehth-ər)–A term used to denote feathers that are curled and that curve outward and forward, a characteristic of Frizzle chickens.

frog (frohg)–A V-shaped pad of soft horn between the bars on the sole of the equine hoof.

frowsy (frow-sē)–A wasty, lifeless-appearing, dry and harsh wool, lacking in character. Also spelled frowzy.

fryer (frī-ər)–Any young chicken approximately eight to twenty weeks old, of either sex, weighing more than 2½ pounds but not more than 3½ pounds, which is sufficiently soft-meated to be cooked tender by frying.

fulguration (fuhl-gər-ā-shuhn)–The destruction of living tissue by electric sparks generated by a high-frequency current.

full bloods (fuhl bluhdz)–A term referring to purebred animals.

full feed (fuhl fēd)–Feed or ration being fed to the limit of an animal's appetite.

full mouth (fuhl mowth)–A state in sheep or goats when an animal has a full set of permanent teeth. This occurs at approximately the age of four.

full-mouthed (fuhl mowthd)–A horse having all the permanent teeth and cups present.

fulling (fuhl-ihng)–The operation of shrinking and felting a woolen fabric to make it thicker and denser. The individual yarns cannot be distinguished on a fulled fabric.

functional efficiency (fuhnck-shuhn-ahl ē-fihsh-ehn-sē)–In cattle, the production of such traits as red meat per unit, genetic excellence, libido, ability to copulate, estrus, ovulation, fertilization, embryo survival gestation, parturition, and mothering ability of the cow.

fundus (fuhn-duhs)–The bottom or base of an organ; in opthalmology refers to the caudal part of the interior of the eye.

fungistat (fuhnj-ih-staht)–An agent or chemical material that prevents the growth and reproduction of, but does not kill, fungi.

fungous (fuhng-guhs)–Pertaining to a fungus, as a fungous disease.

fungus (fuhn-guhs)–A lower order of plant organisms of the kingdom Thallophyta that contain no chlorophyll, has no vascular system, and is not differentiated into roots, stems, or leaves. Fungi are familiar as molds, rusts, smuts, rots, and mushrooms. Plural is fungi (fuhn-jī).

fur (fər)–Short, fine, soft hairs.

fur-slip (fər slihp)–The shedding of hair patches from rough handling in chinchillas.

furuncle (fyoo-ruhng-kuhl)–A localized skin infection in a gland or hair follicle; also called a boil.

G

gait (gāt)–The way an animal moves.

gaited (gāt-ehd)–A definite rhythmic movement of a horse such as trot, canter, pace, etc.; certain breeds are selected and bred on the basis of their ability to perform the various gaits.

galactagogue (gah-lahck-tah-gohg)–Any substance that promotes the flow of milk.

galactose (gah-lahck-tōs)–A white crystalline sugar obtained from lactose (milk sugar) by hydrolysis.

gallop (gahl-ohp)–A fast, three-beat gait of a horse, in which two diagonal legs are paired, their single beat falling between the successive beats of the other two legs, the hind one of which makes the first beat of the three.

Galvayne's groove (gahl-vānz groov)–A mark of the labial surface of the equine tooth; used to determine age; usually appears around eleven years of age.

gambrel (gahm-brehl)–A wooden or metal rod whose ends are inserted in the hocks of hogs, and which is used to support animals while butchering. Also called gamble.

gamete (gahm-ēt)–A sex cell (sperm or egg); also called germ cell.

gametogenesis (gahm-ē-tō-jehn-eh-sihs)–The process in animals, male or female, involving the production of gametes; ovigenesis (females) or spermatogenesis (males).

gamma globulin (gahm-ah glohb-ū-lihn)–A specific protein fraction of the fluid part of the blood that includes the bodies that protect against certain infections (immune bodies).

gander (gahn-dehr)–A mature male goose.

ganglion (gahng-glē-ohn)–A knotlike mass of nerve cell bodies located outside the central nervous system; plural is ganglia (gahng-glē-ah) or ganglions.

gangrene (gahng-grēn)–Necrosis associated with loss of circulation.

garget (gahr-giht)–The abnormal changes in the udder and its secretion as a result of mastitis.

gaskin (gahs-kihn)–The muscular portion of the hindlimb between the stifle and hock.

gastrectomy (gahs-trehck-tō-mē)–The surgical removal of all or part of the stomach.

gastric (gahs-trihck)–Refers to the stomach.

gastric dilatation (gahs-trihck dihl-ah-tā-shuhn)–A condition usually seen in deep chested canines in which the stomach fills with air and expands.

gastritis (gahs-trī-tihs)–Inflammation of the stomach.

gastroduodenostomy (gahs-trō-doo-ō-deh-nohs-tō-mē)–The removal of part of the stomach and duodenun and making a new connection between them.

gastroenteritis (gahs-trō-ehn-tehr-ī-tihs)–Inflammation of the stomach and small intestine.

gastroenterologist (gahs-trō-ehn-tər-ohl-ō-jihst)–A specialist who studies the stomach and small intestine.

gastroenterology (gahs-trō-ehn-tər-ohl-ō-jē)–The study of the stomach and small intestine.

gastrointestinal (gahs-trō-ihn-tehst-ihn-ahl)–The digestive tract (literally means pertaining to the stomach and intestines).

gastropexy (gahs-trō-pehcks-ē)–Surgical fixation of the stomach to the abdominal wall.

gastroscopy (gahs-trohs-kō-pē)–Endoscopic examination of the stomach; the scope is passed from the oral cavity through the stomach.

gastrostomy (gahs-trohs-tō-mē)–Surgical production of an artificial opening between the stomach and abdominal wall.

gastrotomy (gahs-troht-ō-mē)–A surgical incision into the stomach.

gavage (gah-vahzh)–Feeding or administering medication through a tube passed into the stomach.

gee (jē)–A command that directs a horse, mule, or ox to turn to the right; whereas haw means to turn to the left.

geese (gēs)–The plural of goose.

geld (gehld)–(1) Designating an animal that is sterile. (2) To render sterile, as in castration.

gelding (gehld-ihng)–A castrated male horse, donkey, or llama.

gene (jēn)–The simplest unit of inheritance. Physically, each gene is apparently a nucleic acid with a unique structure. It influences certain traits. Sometimes called a trait determiner.

gene pool (jēn pool)–The genetic base available to animal breeders for stock improvement.

gene splicing (jēn splīs-ihng)–The technique of inserting new genetic information in a plasmid.

gene transfer (jēn trahnz-fər)–The process of moving a gene from one organism to another.

genera (jehn-ehr-ah)–The plural of genus.

general anesthesia (jehn-rahl ahn-ehs-thē-zē-ah)–The absence of sensation and consciousness.

generation (jehn-ehr-rā-shuhn)–The group of individuals of a given species that have been reproduced at approximately the same time; the group of individuals of the same genealogical rank.

generation interval (jehnr-eh-rā-shuhn ihn-tehr-vahl)–The period of time between the birth of one generation and the birth of the next.

generic drug (jehn-ār-ihck druhg)–A medication not protected by a brand name or trademark; also called nonproprietary drug.

genesis (jehn-eh-sihs)–The origin, or evolutionary development, as of a soil, plant, or animal.

genetic (jehn-eh-tihck)–Genes (heredity units that contain DNA); inherited.

genetic base (jehn-eh-tihck bās)–The breeding animals available for a producer to use.

genetic drift (jehn-eh-tihck drihft)–The gradual change in a plant or animal species because of rearrangement of the genes due to the environment or unknown causes.

genetic engineering (jehn-eh-tihck ehn-jih-nēr-ihng)–Alteration of the genetic components of organisms by human intervention. Also known as biogenetics.

genetic index (jehn-eh-tihck ihn-dehcks)–An estimate of the future predicted difference of a young bull. See predicted difference.

genetic trait summary (GTS) (jehn-eh-tihck trāt suhm-ahr-ē)–The comparative ranking of beef sires derived from evaluating the conformation of the daughters.

genetics (jeh-neh-tihcks)–(1) The science that deals with the laws and processes of inheritance in plants and animals. (2) The study of the ancestry of some special organism or variety of animal. Also called breeding.

genital (jehn-ih-tahl)–Pertaining to the organs of reproduction.

genital eminence (jehn-ih-tahl ehm-ih-nehns)–In sexing chicks, a very small, shiny or glistening projection that is the rudimentary male copulatory organ. Also called male process.

genitalia (jehn-ih-tā-lē-ah)–The reproductive organs.

genome (jē-nōm)–A complete set of chromosomes (hence of genes) inherited as a unit from one parent.

genotoxicity (jē-nō-tohck-sihs-ih-tē)–The quality of being damaging to genetic material.

genotype (jē-nō-tīp)–The genetic makeup of an individual for a particular trait.

genus (jē-nuhs)–A group of species of plants or animals believed to have descended from a common direct ancestor that are similar enough to constitute a useful unit at this level of taxonomy.

germ (jərm) A common term for any pathogenic microorganism, but especially bacterial or viral organisms.

germ cell (jərm sehl)–A cell capable of reproduction or of sharing in the reproduction of an organism, which may divide to produce new cells in the same organism, as contrasted with the somatic or body cells.

germ plasma (jərm plahz-mah)–A term for the reproductive and hereditary substance of individuals that is passed on from the germ cell in which an individual originates in direct continuity to the germ cells of succeeding generations.

germ spot (jərm spoht)–The germinal disc on the surface of the yolk of an egg. The blastoderm of a fertilized egg and the point at which embryonic development starts in the making of a chick.

germicide (jehr-mih sīd)–Any agent that kills germs.

gestation (jehs-tā-shuhn)–The period of development of the fetus in the uterus (conception to birth).

gestation period (jehs-tā-shuhn pehr-ē-ohd)–The length of time from conception to birth of young in particular species. Usual gestation periods for farm animals are: mare, 330 to 340 days; cow, 230 to 285 days; ewe, 145 to 150 days; sow, 112 to 115 days; goat, 148 to 152 days; jennet (female ass), 360 to 365 days.

get (geht)–Offspring.

gib (gihb)–A neutered male ferret.

gigantism (jī-gahn-tihzm)–In animals, abnormal overgrowth of a part or all of the body. Also called giantism.

gilt (gihlt)–A young female pig that has not yet farrowed.

gingiva (jihn-jih-vah)–The mucous membrane that surrounds the teeth and forms the mouth lining; also called the gums.

gingivectomy (jihn-jih-vehck-tō-mē)–The surgical removal of gum tissue.

gingivitis (jihn-jih-vī-tihs)–Inflammation of the gums.

girth (gihrth)–(1) The circumference of the body of an animal behind the shoulders. (2) A band or strip of heavy leather or webbing that encircles a pack animal's body; used to fasten a saddle or pack on its back. Also called cinch.

gizzard (gihz-ahrd)–The muscular posterior stomach of birds, which has muscular walls and a thick, horny lining; its principal function is the grinding or crushing of coarse feed particles. Also called ventricules

glabrous (glā-bruhs)–Smooth, devoid of hair or surface glands.

gland (glahnd)–A group of specialized cells that secrete material used outside itself.

gland cistern (glahnd sihs-tərn)–The area on an udder where milk collects prior to entering the teat cistern.

glans penis (glahnz pē-nihs)–The bulge at distal end of penis.

glaucoma (glaw-kō-mah)–A group of disorders resulting from increased intraocular pressure.

globe (glōb)–A sphere; a term used to describe the eyeball.

globule (glohb-ūl)–A collection of several molecules of fat that takes on a spherelike appearance, and is insoluble in water.

glomerulonephritis (glō-mər-yoo-lō-neh-frī-tihs)–Inflammation of the glomeruli.

glomerulus (glō-mər-yoo-luhs)–A cluster of capillaries; plural is glomeruli (glō-mər-yoo-lī).

glossitis (glohs-ī-tihs)–Inflammation of the tongue.

glottis (gloh-tihs)–The vocal apparatus; consists of vocal folds and the opening between them.

gloves (gluhvz)–White paws.

glucagon (gloo-kah-gohn)–A pancreatic hormone that increases blood glucose.

gluconeogenesis (gloo-kō-nē-ō-jehn-ih-sihs)–The production of new glucose.

glutinous (gloo-tih-nuhs)–Sticky.

glycerol (glihs-ehr-ahl)–One of the components of a fat molecule; a fat molecule is composed of three fatty acids attached chemically to glycerol.

glycogen (glī-kō-jehn)–$C_6H_{10}O_5$; A carbohydrate similar to starch, found abundantly in the liver and stored in lesser amounts in other tissues and organs. Also called animal starch.

glycolytic (glī-kō-liht-ihck)–Pertaining to the chemical breakdown of sugars to lactic acid.

glycosuria (glī-kohs-yoo-rē-ah)–Glucose (blood sugar) in the urine.

glucosuria (gloo-kōs-yoo-rē-ah)–Glucose (blood sugar) in the urine.

gnotobiotic (nōt-ō-bī-oh-tihck)–Germ free animals that have been introduced to one or two known nonpathogenic microorganisms.

go off feed (gō ohf fēd)–(1) To cease feeding with a normal appetite. (2) To refuse feed in the amounts and kinds previously eaten.

go stale (gō stā-uhl)–(1) To suffer sperm deterioration, as a bull. (2) As an animal, not to work at normal standards of production. (3) As an animal, to go off feed.

goat (gōt)–Any horned ruminant of the genus *Capra*, family Bovidae, especially the domestic goat, *C. hircus,* which is bred as a source of milk, meat, and wool or hair.

goat month (gōt muhnth)—The tenure of a mature goat on range or pasture for one month.

goatling (gōt-lihng)–A female goat between one and two years of age that has not borne a kid.

gobby (gō-bē)–A lumpy, unattractive condition of the fat covering the body of an animal, such as a sheep or beef animal.

goiter (goy-tehr)—-The enlargement of the thyroid gland that results from a deficiency of iodine.

golden bay (gōl-dehn bā)–The rich, yellowish-red of an animal's coat.

gomer bull (gō-mər buhl)–A bull used to detect female bovines in heat; the bull may have its penis surgically deviated to the side, may be treated with androgens, or may be vasectomized so as not to impregnate female; also called teaser bull.

gonad (gō-nahd)–The gamete producing gland (ovaries in females and testes in males).

gonadotropin (gohn-ah-dō-trō-pihn)–A hormone that stimulates the gonads.

goniometry (gō-nē-oh-meh-trē)–A procedure to measure the drainage angle of the eye.

goniotomy (gō-nē-oh-tō-mē)–An incision into the anterior chamber angle for treatment of glaucoma.

goose (goos)–(1) Any large, web-footed bird (intermediate in size between swans and ducks) of the subfamily Anserinae (family Anatidae) including the genus *Anser* and related genera. The domestic goose, *Anser domesticus,* includes a number of breeds that are kept for their flesh and feathers. (2) The female goose as distinguished from the male, or gander.

goose-rumped (goos ruhmpd)–An animal having a short, steep croup that narrows at the point of the buttocks.

goosestep (goos-stehp)–A peculiar walk or body action, locomotor incoordination (spastic gait) of swine which is caused by a nutritional deficiency of pantothenic acid, B_3 complex vitamin.

gore (gōr)–To pierce the body with an animal's horns.

gosling (gohs-lihng)–A very young or recently hatched goose.

gossypol (gohs-ih-pohl)–A material found in cottonseed that is toxic to swine and certain other simple-stomached animals.

graafian follicles (grahf-ē-ahn fohl-lihck-kuhlz)–Small sacs on ovaries; each sac contains a single ovum and secretes estrogen.

grade (grād)–(1) Any animal that has one purebred parent and one of unknown or mixed breeding. (2) Designating a herd, flock, brand, etc., of such animals. (3) The classification of a product, animal, etc., by standards of uniformity, size, trueness to type, freedom from blemish or disease, fineness, quality, etc.

grade animal (grād ahn-ih-mahl)–An animal with nonpurebred ancestors.

graded eggs (grād-ehd ehgz)–Eggs that have been sorted and labeled according to size and quality.

grading (grā-dihng)–(1) The classification of products, animals, etc., into grades. (2) The mating of a purebred animal with one of mixed or unknown breeding. (3) The smoothing of the land surface.

grading up (grā-dihng uhp)–The practice of improving a flock whereby purebred sires are mated to grade animals and their offspring. In three generations the offspring will be seven-eighths purebred and in some cases eligible for registration. Upgrading.

graft (grahft)–A tissue or organ for transplantation and implantation.

grain-fed (grān fehd)–Designating animals, such as cattle, which are being or have been fattened for market largely by the use of grain feeds.

gram (grahm)–A metric unit of weight, equal to 0.035 ounce.

Gram stain (grahm stān)–A staining method devised by a Danish physician, Hans Christian Joachim Gram, to aid in the identification of bacteria. Bacteria either resist discoloration with alcohol and retain the initial deep violet stain (gram-positive) or can be decolorized by alcohol and are stained with a contrast stain (gram-negative). Also called Gram's stain.

grand mal seizure (grahnd mahl sē-zhǝr)–Convulsions characterized by loss of consciousness and muscle contractions; also called tonic-clonic seizures.

granular (grahn-ū-lahr)–In the form of granules or small particles.

granulocyte (grahn-yoo-lō-sīt)–A cell that contains prominent grainlike structures in its cytoplasm.

granuloma (grahn-yoo-lō-mah)–A type of circumscribed inflammatory reaction, resulting in a nodule.

grass (grahs)–Cattle marketed directly off grass pastures and not grain-fed.

grass lamb (grahs lahm)–A lamb that is dropped in the springtime and is raised on pasture in the summer months and butchered in the fall, when pasture is less productive.

grass-fattened (grahs faht-tehnd)–Designating an animal that has been fattened on pasture or range, in contrast to one fattened on grain or other feed concentrate.

gray (grā)–(1) A color of an animal's coat that has white hairs mixed with black. (2) A cotton-lint color designation that is the darkest in chroma.

gray roan (grā rōn)–A coat color for a horse that is roan in combination with gray. See steel gray.

gray wool (grā wool)–Fleeces with a few dark fibers, a rather common occurrence in the medium wools produced by Down or black-faced breeds.

graze (grāz)–Eating grasses and plants that grow close to the ground.

graze off (grāz ohf)–To cause animals to feed on and almost consume the top growth of herbaceous vegetation.

graze out (grāz owt)–To allow domestic animals to feed abusively on certain palatable grasses or forbs, alone or in combination, until the vegetation ceases to exist on a particular pasture or range.

grazing (grā-zihng)–(1) Feeding available to animals on ranges and pastures. (2) The process of feeding by livestock on live or standing plants other than browse.

grazing capacity (grā-zihng kah-pah-siht-ē)– In range or pasture management, the ability of a grassed unit to give adequate support to a constant number of livestock for a stated period each year without deteriorating.

grazing land (grā-zihng lahnd)–Land used regularly for grazing; not necessarily restricted to land suitable only for grazing, excluding pasture and cropland used as part of farm crop rotation system.

grazing pressure (grā-zihng preh-shər)–The actual animal-to-forage ratio at a specific time. For example, three animal units per ton of standing forage.

grazing unit (grā-zihng yoo-niht)–(1) The quantity of pasturage used by an average, mature cow or its equivalent in other livestock in a grazing season in a given region. (2) Any division of the range that is used to facilitate range administration or the handling of livestock.

grazing value (grā-zihng vahl-ū)–The worth of a plant or cover for livestock and/or game that is determined by its palatability, nutritional rating, amount of forage produced, longevity, and area of distribution.

grease (grēs)–(1) See fat, lanolin. (2) Hog fat as distinguished from tallow, which is the fat of cattle and sheep. (3) To apply salve to a wound or irritation.

green broke (grēn brōk)–A term applied to a horse that has been hitched or ridden only once or twice.

green chop (grēn chohp)–Green forage that is cut with a field chopper and hauled to lots or barns for livestock feed in lieu of pasturing.

green geese (grēn gēs)–Geese full fed for fast growth and marketed at ten to thirteen weeks of age when they weigh 10 to 12 pounds (4.5 to 5.4 kilograms); also called junior geese.

green hay (grēn hā)–(1) Cured hay. (2) That hay which, on being cured, retains a green color.

green manure (grēn mahn-oo-ər)–Crops such as legumes or grasses that are grown to be plowed or spaded into the soil to increase humus content and improve soil structure.

green pellet (grēn pehl-eht)–A pellet made from alfalfa meal only, or a complete pellet that contains enough green roughage to color it.

gregariousness (greh-gahr-ē-ehs-nehs)–The tendency within a species population to flock or herd together.

gristle (grihs-ehl)–Cartilage.

groin (groyn)–The region between the lower abdomen and proximal thigh; also known as the inguinal area (ihng-gwih-nahl).

groom (groom)–(1) A person who curries, combs, washes, etc., an animal and cares for it generally. (2) To wash, curry, brush, and generally care for an animal.

grooming chute (groom-ihng shoot)–A portable chute in which cattle are held while they are being groomed for a show.

grow (grō)–(1) To live and to increase in stature and girth toward maturity. (2) To raise animals.

grow out (grō owt)–To feed cattle so that the cattle get a certain desired amount of growth without much, if any, fattening.

grower pig (grō-ər pihg)–Swine from about 40–100 pounds.

growth (grōth)–(1) The increase in size of a living organism. (2) A tumor, gall, etc. (3) The development of an organism from its earliest stage to maturity.

growth hormone (grōth hōr-mōn)–A hormone that promotes body growth and milk production.

growthy (grōth-ē)–A livestock judging term used to describe an animal that is large and well developed for its age.

grub hole (gruhb hōl)–A hole or wound in the hide of an animal caused by the larva of the common cattle grub.

grulla (gruhl-ah)–A coat color of some animals, especially the American quarter horse. It is a slate-blue bordering on a sooty black.

guaranteed analysis (gahr-ehn-tēd ah-nahl-ah-sihs)–In feed labels or tags, a listing of certain nutrients, usually crude protein, crude fiber, fat, and ash, guaranteeing a minimum or maximum percentage of each in the feed.

gullet (guhl-eht)–See esophagus.

gummer (guhm-ehr)–A sheep or goat having no teeth.

gummy-legged (guhm-ē lehgd)–A term applied to a horse having legs in which the tendons lack definition, or do not stand out clearly.

guttural pouch (guht-ər-ahl powch)–A large, air-filled ventral outpouching of the eustachian tube in equine.

gynandromorph (jih-nahn-drō-mohrph)–An individual of which one part of the body exhibits female characteristics and another part male characteristics. Also known as hermaphrodite.

gynecomastia (gī-neh-kō-mahs-tē-ah)–A condition of excessive mammary development in males.

gyp-rope (jihp-rōp)–The rope used by a trainer to rope or to exercise his horse.

gyrus (jī-ruhs)–An elevated portion; also called convolution (kohn-vō-loo-shuhn); plural is gyri (jī-rī) or convolutions.

H

hackamore (hahck-ah-mōr)–A type of head restraint for horses similar to a halter but provided with a loop or noose that may be placed about the nose of the horse to give additional restraint in handling unbroken horses.

hackles (hahck-ehlz)–(1) The long, narrow, neck plumage of male birds. (2) The erectile hairs on the backs of certain animals.

hair (hār)–The outgrowth of a cell in the epidermis of a plant or animal. In vast numbers it forms the coat of an animal and is frequently used as a fiber, such as wool. See pubescent.

hair ball (hār bahl)–The hair an animal has swallowed that has gathered in the stomach in the form of a ball; common in the stomach of cats and in the rumen of ruminants. See bezoar, trichobezoar.

hair slips (hār slihpz)–An animal hide that has been improperly salted and cured, allowing some decomposition to take place as indicated by slipping patches of hair.

hairlessness (hār-lehs-nehs)–In genetics, a lethal factor characterized in the Holstein breed by calves being born without hairs except around natural openings and at the end of the tail.

half-breed (hahf brēd)–Designating a horse that has a thoroughbred as one parent and a draft horse as the other. A popular, medium-heavyweight horse with considerable action.

half-brother (half-sister) (hahf brohth-ər, hahf sihs-tər)–Animals from the same mother but by different sires, or by the same sire from different mothers.

half-sib (hahf sihb)–A half-brother or half-sister.

half-stocking (hahlf stohck-ihng)–A white marking from the coronet to the middle of the cannon.

hallucination (hah-loo-sehn-ā-shuhn)–A false sensory perception.

halter (hahl-tər)–A head harness worn by animals for restraint that extends behind the head and over the nose; also called a head collar.

ham string (hahm strihng)–The large tendon above and behind the hock in the hind leg of quadrupeds.

hame (hām)–The wood or metal parts of the harness of a draft animal that fit about the collar to which the traces are attached for pulling.

hammerhead (hahm-ər-hehd)–A coarse-headed animal.

hamstrung (hahm-struhng)–Designating an injury to the tendon behind the cannon in the hind leg of an animal.

hand (hahnd)–(1) A unit of measurement equal to 4 inches (10 centimeters) that is used to measure the height of horses from the ground to a point at the shoulder. (2) The near horse in a team used for plowing.

hand breeding (hand mating) (hahnd brē-dihng)–A system of animal breeding in which the breeder controls the number of times coitus is performed; also called hand mating (hahnd mā-tihng).

hand feeding (hahnd fē-dihng)–A type of feeding routine whereby an animal is fed measured amounts of food, water, salt, etc., at fixed intervals.

hand gallop (hahnd gahl-ohp)–The restrained or slow gallop of horses. See canter.

hand milker (hahnd mihl-kər)–A person who milks a cow manually.

hand strip (hahnd strihp)–(1) To take the last bit of milk from a cow's udder, usually following machine milking. (2) To harvest a seed crop by hand.

handiness (hahn-dē-nehs)–A characteristic of a horse with good manners who has good coordination, and is not clumsy or awkward in movement or action.

handpick (hahnd-pihck)–(1) To harvest by hand, as contrasted to harvesting by machine. (2) To pluck the feathers of a fowl manually.

haploid (hahp-loyd)–An organism or cell with one set of chromosomes.

haploid number (hahp-loyd nuhm-bər)–In genetics, half (haploid) the number of chromosomes that are usually present in the nucleus; occurs during reduction division of meiosis.

hard breeder (hahrd brē-dər)–Designating a female animal that is difficult to breed or has difficulty conceiving

hard feeder (hahrd fē-dər)–An animal that stays in a thin condition even though well fed. Sometimes called hard keeper.

hard keeper (hahrd kē-pər)–An animal that is unthrifty and grows or fattens slowly regardless of the quantity and quality of feed. Also called hard feeder.

hard milker (hahrd mihl-kər)–A cow that milks slowly due to hardened or constricted sphincter muscles in the end of the teat, or a fleshy udder with limited space for the fast accumulation of milk.

hardiness (hahr-dē-nehs)–The state of being hardy.

hard-mouthed (hahrd-mowthd)–A term used when the membrane of the bars of a horse's mouth have become toughened and

the nerves deadened because of the continued pressure of the bit.

hardware disease (hahrd-wahr dih-zēz)–A condition found in ruminants in which metal objects, such as wire, nails, and screws, are swallowed with feed, and because of their weight, lodge in the reticulum which may lead to peritonitis or pericarditis. Also called traumatic reticuloperitonitis.

harsh (hahrsh)–Designating a fleece that lacks character, as rough hair.

hat racks (haht rahckz)–Thin cattle; canners.

hatch (hahch)–(1) To bring forth young from the egg by natural or artificial incubation. (2) The young produced from one incubation.

hatch out (hahch owt)–To emerge from an egg, as a fully developed chick comes forth from the shell.

hatchability (hahch-ah-bihl-ih-tē)–(1) In poultry farming, the quality of fertilized eggs that, when incubated, allows for normal embryonic development and the emergence of normal young. (2) In incubation practice, the percentage of fertile eggs that hatch.

hatchery (hahch-eh-rē)–A place, building, company, etc., where eggs are incubated; usually a commercial establishment where newly hatched young (chicks, poults, ducklings, etc.) are sold.

hatching egg (hahch-ihng ehg)–A fertile egg of good form and quality produced by a breeding flock that may be used for hatching.

haunch (hawnch)–(1) A pivot on the hind feet. It is commonly observed in stock horses used on ranches for culling or cutting. (2) A hindquarter of an animal.

haustration (haw-strā-shuhn)–The pouches of the cecum and rectum responsible for the sacculated appearance of the equine and porcine large intestine.

hay (hā)–Any leafy plant material, usually clover, fine-stemmed grasses and sedges, alfalfa, and other legumes, that has been cut and dried principally for livestock feeding. See fodder.

hay belly (hā behl-ē)–A term applied to animals having a distended barrel due to excessive feeding of bulky rations, such as hay, straw, or grass.

haylage (hā-lahj)–Forage that could have been cut for hay but is stored with a higher moisture content than hay, and with less moisture than silage.

head (hehd)–(1) Cows, asses, horses, collectively, as ten head of horses. (2) The part of the body that includes the face, ears, brain, etc. (3) To get in front of a band of sheep, herd of cattle, etc., so as to stop their forward movement (head them off). (4) A rounded projection on bone.

head gland (hehd glahnd)–The small secretory organ located on the head of snakes.

head shy (hehd shī)–Designating a horse on which it is difficult to put a bridle, to lead, or to work around its head.

headstall (hehd-stahl)–The part of a bridle that encircles a horse's head.

headstrong (hehd-strohng)–Designating an animal that tends to be stubborn.

health (hehlth)–The state wherein all body parts of plants, animals, and people are functioning normally.

heart (hahrt)–The organ of the body that, by its rhythmical contractions, circulates the blood. It is an edible by-product of slaughter animals and fowls.

heart girth (hahrt gərth)–The circumference around the thoracic cavity used to estimate an animal's weight and the capacity of the heart and lungs.

heartrate (hahrt-rāt)–The number of heartbeats per minute; abbreviated HR.

heat (hēt)–A common term for estrus.

heat detection (hēt dē-tehck-shuhn)–The process used in determining females that are in estrus. See gomer bull.

heat mount detector (hēt mownt dē-tehck-tōr)–A plastic device that is glued to the tailhead of a cow to determine when she comes into heat. Prolonged pressure from a mounting animal's brisket turns the detector a different color.

heat period (hēt pər-ē-ohd)–Estrus; the period during which a female is sexually receptive.

heat prostration (hēt proh-strā-shuhn)–Heat stroke; a condition of a person or an animal resulting from excessively hot weather; characterized by lethargy, inability to work, staggering gait, convulsions, and a high temperature. Death often occurs.

heat spot (hēt spoht)–A defect in a fertile egg that results from alternating high and low temperatures; characterized by the beginning of the development of the embryo without blood showing.

heat synchronization (hēt sihnck-rohn-ih-zā-shuhn)–Causing a group of cows or heifers to exhibit heat together at one time by artificial manipulation of the estrous cycle.

heat tolerant (hēt tohl-ər-ahnt)–Designating the ability of an animal to endure extreme heat conditions.

heaving (hē-vihng)–An extra contraction of the flank muscles during respiration; caused by loss of lung elasticity.

heavy (hehv-ē)–The late stages of pregnancy of a cow.

heavy breed (hehv-ē brēd)–A bird (usually referring to a chicken) that has a high meat-to-bone ratio and is therefore suitable for the table. See broiler.

heel (hēl)–(1) The end of the branches of a horseshoe. (2) The rear end of the foot.

heifer (hehf-ər)–A young female bovine that has not given birth.

held (hehld)–Designating the controlling of an animal in a small space or by other controlling devices.

helminths (hehl-mihnths)–Parasitic worms, such as the flatworms (flukes and tapeworms) and roundworms (hookworms and lungworms).

hemangioma (hē-mahn-jē-ō-mah)–A benign neoplasm composed of newly formed blood vessels.

hemangiosarcoma (hē-mahn-jē-ō-sahr-kō-mah)–A malignant tumor of vascular tissue.

hematemesis (hēm-ah-tehm-eh-sihs)–Vomiting blood.

hematochezia (hēm-aht-ō-kē-zē-ah)–The passage of bloody stools.

hematocrit (hē-maht-ō-kriht)–The percentage of erythrocytes in blood; "to separate blood"; also called crit, PVC, or packed cell volume.

hematology (hē-mah-toh-lō-jē)–The study of blood.

hematoma (hē-mah-tō-mah)–The collection of blood in tissue.

hematopoiesis (hē-mah-tō-poy-ē-sihs)–The formation of blood.

hematuria (hēm-ah-toor-ē-ah)–Blood in the urine.

heme (hēm)–A nonprotein, iron-containing portion of hemoglobin.

hemiparesis (hehm-ih-pahr-ē-sihs)–A weakness of one side of the body.

hemiplegia (hehm-ih-plē-jē-ah)–Paralysis of one side of the body.

hemisphere (hehm-ih-sfēr)–Half of a circle.

hemizygous (hehm-ē-zī-guhs)–The condition in which only one allele of a pair of genes is present in the cells of an individual plant or animal, the other one being absent.

hemoccult (hēm-ō-kuhlt)–A test for hidden blood in the stool.

hemoglobin (hē-mō-glō-bihn)–An erythrocyte protein that transports oxygen.

hemoglobinuria (hē-mō-glō-bihn-ū-rē-ah)–The presence of hemoglobin in the urine.

hemogram (hē-mō-grahm)–A record of the findings in examination of blood.

hemolysis (hē-mohl-eh-sihs)–The breaking down of blood cells.

hemolytic (hē-mō-liht-ihck)–Removing and destroying red blood cells.

hemoperitoneum (hēm-ō-pehr-ih-tō-nē-uhm)–Blood in the peritoneal cavity.

hemophilia (hē-mō-fihl-ē-ah)–A hereditary condition of deficient blood coagulation. Also called a bleeder.

hemoptysis (hē-mohp-tih-sihs)–The spitting of blood from the lower respiratory tract.

hemorrhage (hehm-ōr-ihdj)–A loss of blood (usually in a short period of time).

hemostasis (hē-mō-stā-sis)–To control or stop bleeding.

hemostat (hē-mō-staht)–An instrument to control bleeding.

hemostatic forceps (hē-mō-stah-tihck fōr-sehps)–A locking instrument used for grasping and ligating vessels and tissues to control or stop bleeding; also called hemostat.

hemostatics (hē-mō-stah-tihckz)–Substances that check internal hemorrhage.

hemothorax (hē-mō-thō-rahcks)–The accumulation of blood in the pleural cavity.

hen (hehn)–An intact female parrot, turkey, or chicken.

hen battery (hehn baht-ch-rē)–The number of individual hen-cage units arranged in a group, usually single or multiple decks, which have provision for watering, feeding, collecting eggs, and disposing of droppings.

hennery (hehn-eh-rē)–(1) A poultry farm, particularly one that specializes in the production of market eggs. (2) A building or enclosure where laying hens are kept.

heparin (hehp-ahr-ihn)–A type of anti-coagulant.

hepatic (heh-paht-ihck)–The liver.

hepatitis (hehp-ah-tī-tihs)–Inflammation of the liver.

hepatocytes (heh-paht-ō-sītz)–Liver cells.

hepatoma (heh-pah-tō-mah)–A liver neoplasm.

hepatomegaly (hehp-ah-tō-mehg-ah-lē)–Abnormal enlargement of the liver.

hepatotomy (hehp-ah-toht-ō-mē)–A surgical incision into the liver.

herbivore (hərb-ih-vōr)–An animal that is able to sustain life by only eating plants.

herd (hehrd)–A group of rabbits, pigs, horses, donkeys, mules, goats, or cattle; to tend animals in a herd.

herd book (hehrd book)–The recognized, official record of the ancestry of a purebred animal kept by the particular breed association.

herd improvement registry (hehrd ihm-proov-mehnt rehg-ihs-trē)–A type of registry maintained by certain purebred cattle breeder associations to record the production and yearly records of all producing cows of that breed in a given breeder's herd.

herd sire (hehrd sīr)–The male of the species kept for the sole purpose of reproduction. (The term is most commonly applied to cattle and horses but is also used with sheep and swine.) See bull, ram, stallion.

herd test (hehrd tehst)–A type of semi-official testing for milk production in which the whole herd of cows of milking age are included.

herding (hehr-dihng)–The control of animals on the range by guiding their direction and movements to procure grazing and water where and when desired.

heredity (hē-rehd-ih-tē or heh-rehd-ih-tē)–(1) The study or description of genes passed from one generation to the next through sperm and ova. (2) Genetic transmission of traits from parents to offspring. (3) The genetic constitution of an individual.

heritability (hehr-ih-tah-bihl-ih-tē)–The proportion of the differences among animals, measured or observed, that is transmitted to the offspring (values range from zero to one). The higher the heritability of a trait, the more accurately

the individual performance predicts breeding value and the more rapid the response due to selection for that trait.

heritability percent estimates (hehr-ih-tah-bihl-ih-tē pehr-sehnt ehs-tih-mahtz)–The percent of a trait that is inherited from an animal's parents and is not controlled by environment.

hermaphrodite (hər-mahf-rō-dīt)–A bisexual individual that possesses both male and female sex organs. In some species such individuals are capable of reproducing, in others they are sterile.

hermaphroditism (hər-mahf-rō-dih-tihzm)–The condition of an individual having both ovarian and testicular tissue.

hernia (hər-nē-ah)–A protrusion of a body part through tissues that normally contain it.

heterogametic (heht-ehr-ō-gah-meht-ihck)–Producing unlike gametes, particularly with regard to the sex chromosome. In species in which the male is of the X-Y type, the male is heterogametic, the female homogametic.

heterogen (heht-ehr-ō-jehn)–A variable group of plants or animals that arise as hybrids, sports, mutations, etc., certain types of which may or may not breed true.

heterogeneous (heht-ehr-ō-jē-nē-uhs)–Designating elements having unlike qualities.

heterologous serum (heht-ehr-ohl-ō-guhs sēr-uhm)–Serum derived from another species or disease.

heterosis (heht-ehr-ō-sihs)–The amount of superiority observed or measured in crossbred animals compared with the average of their purebred parents; hybrid vigor.

heterotrophic (heht-ehr-ō-trohf-ihck)–Referring to organisms that for their metabolism are dependent upon organic matter supplied from sources outside of their own bodies. See Autotrophic.

heteroxenous parasite (heht-ehr-ohck-sē-nuhs pahr-ah-sīt)–A parasite requiring several or different hosts for its complete development.

heterozygous (heht-ər-ō-zī-guhs)–The condition of having two different genes for a given genetic trait; usually one gene is dominant over the other.

hibernate (hī-bər-nāt)–A dormant state in which an animal reduces its body temperature, heart and respiration rates, and metabolism in winter.

hide (hīd)–The tanned or untanned skins of animals, especially those of cattle, horses, sheep, and goats.

hidebound (hīd-bownd)–Designating an animal whose skin is very tightly fastened to its body, often resulting from poor feeding and emaciation.

hidrosis (hī-drō-sihs)–Excessive sweating.

high blower (hī blō-ər)–A horse that has broken wind. (2) A horse that snorts at each exhalation while galloping.

high roller (hī rō-lər)–Designating a horse whose bucking action is higher than usual.

high school horse (hī skool hōrs)–Horse trained for performing certain relatively highly complicated routines.

hilum (hī-luhm)–A depression where vessels and nerves enter an organ.

hilus (hī-luhs)–(1) A concave depression on the medial aspect of the kidney where the ureters, blood vessels, and nerves enter. (2) The stalk of the ovary that serves as the attachment to the broad ligament.

hind cinch (hīnd sihnch)–The rear cinch strap that encircles a horse to keep the saddle from tipping up when roping. In normal use, saddles have only one cinch that is fastened just behind the forelegs.

hindgut (hīnd-guht)–The collective term for the cecum, small colon, and large colon.

hinny (hihn-ē)–The offspring of a stallion and a jenny (hinnies are sterile).

hip (hihp)–That region of one of the rear quarters of four-legged animals where the hind leg joins the pelvic region.

hip height (hihp hīt)–A measurement of cattle taken from the ground to the top of the hip. At a given age of the animal, hip height determines the frame score of a bull, steer, or heifer.

hippocampus (hihp-ō-kahm-puhs)–A portion of limbic system of the brain that involves memory.

hirsute (hehr-soot)–Covered with coarse hairs.

histiocyte (hihs-tē-ō-sīt)–Large macrophages found in loose connective tissue.

histology (hihs-tohl-ō-jē)–The study of the structure, composition, and function of tissues.

histopathologic (hihs-tō-pah-thō-lohj-ihck)–Designating abnormal changes in body structures, as observed by microscopic examination of sections of abnormal or diseased cells.

hitch (hihch)–(1) A horse or horses used to pull an implement or vehicle. (2) The stride of a horse when one of the hind legs is shorter than the other. (3) To fasten an animal to a post, rail, etc.

hives (hīvz)–In horses, small swellings under or within the skin similar to human hives. They appear suddenly over large portions of the body and can be caused by a change in feed. Also called urticaria.

hob (hohb)–An intact male ferret.

hobble (hohb-uhl)–(1) A type of restraint in which either the front feet or hind feet are placed in straps to keep them from moving. (2) To tie an animal's forelegs so as to prevent straying.

hock (hohck)–(1) The region of the tarsal joint in the hind leg of a horse or other quadruped.

hog bristles (hohg brihs-uhlz)–The coarse, stiff hairs on swine used in the manufacture of brushes; now synthetic bristles have largely replaced hog bristles.

hog down (hohg down)–To pasture hogs on a crop grown for stock feed, thus eliminating the harvesting process. Also called hog off.

hog holder (hohg hōl-dər)–A device consisting of a metal tube through which a cable is run. A loop on the end of the cable is placed on the snout of a hog to hold the hog for receiving vaccination, etc.

hog hurdle (hohg hər-duhl)–A portable partition used to move swine by blocking the area in which the pig should not go.

hog ring (hohg rihng)–A ring fastened in the end of a hog's nose to prevent the hog from rooting in the ground.

hot-tight (hohg-tīt)–Fencing that prevents an animal's escape.

hold (hōld)–(1) To restrain animals to a particular place. (2) To maintain condition, as a steer that holds flesh. (3) To store or retain in storage, as to hold eggs. (4) Not to market at harvest time or when an animal is fat, but to wait to sell for a better price.

hold up milk (hōld uhp mihlck)–In a cow, to cause a cessation in milk secretion as a result of undue excitement at milking time.

holding pen (hōld-dihng pehn)–A large pen in which sheep or other animals are held prior to being handled.

holosystolic (hō-lō-sihs-stohl-ihck)–The entire ventricular contraction phase; used to describe the length of a murmur; also called pansystolic (pahn-sihs-stohl-ihck).

Holter monitor (hōl-tər mohn-ih-tər)–An electrical recording of heart rates and rhythms for a 24-hour period via a recorder that is worn by the patient.

homeostasis (hō-mē-ō-stā-sihs)–Maintaining a constant internal environment.

homogametic (hō-mō-gah-meht-ihck)–Refers to the particular sex of the species that possesses two of the same kind of sex chromosome such that only one kind of gamete can be produced with respect to the kinds of sex chromosomes it contains; in mammals, the female is the homogametic sex (XX).

homogeneous (hō-mō-jē-nē-uhs)–Being of uniform character or nature throughout.

homogenized milk (hō-mohj-eh-nihzd mihlk)–Milk that has been treated in such a way as to break up the particles or globules of fat small enough that they will remain suspended and not rise to the top after standing.

homolog (hō-mō-lohg or hohm-ō-lohg)–One of a pair of structures having similar structure, shape, and function, as with two homologous chromosomes.

homologous (hō-mohl-ō-guhs)–Having a common origin but different function(s) in different species.

homologous chromosomes (hō-mohl-ō-guhs krō-mō-sōmz)–Pairs of chromosomes that are the same length that pair up to form a synapsis in meiosis.

homologous serum (hō-mohl-ō-guhs sēr-uhm)–Serum that is derived from the same species or like disease.

homothermic (hō-mō-thehr-mihck)–Refers to animals that are able to maintain a fairly constant body temperature; warm-blooded.

homozygote (hō-mō-zī-gōt)–An animal whose genotype, for a particular trait or pair of genes, consists of like genes.

homozygous (hō-mō-zī-guhs)–The condition of having two identical genes for a given trait.

homozygous recessive (hō-mō-zī-guhs rē-seh-sihv)–A recessive character that produces two kinds of gametes; one carries the dominant gene, and the other carries the recessive gene.

hooded (huhd-ehd)–Rats that have a white coat with a black "hood" over their head, shoulders, and pigmented eyes.

hoof (huhf)–The hard, horny, outer covering of the feet of horses, cattle, sheep, goats, and swine.

hoof pick (huhf pihck)–An instrument used to clean the sole, frog, and hoof wall.

hoof testers (huhf tehs-tərs)–An instrument used to test sensitivity in the equine foot.

hoof wall (huhf wahl)–A hard, horny substance made up of parallel fibers covering the digit of ungulates; wall should be dense, straight, and free from ridges and cracks.

hook bones (huhk bōnz)–The prominent bones on the dorsolateral area of ruminants formed by the cranial ends of the illia (plural of ilium); points of the hip. Also called hooks.

hopper feeding (hohp-ər fē-dihng)–Making available a sufficient quantity of feed to a rabbit, hog, or other animal for several days so the animal may eat as often as it wishes and not be limited to a certain amount.

hordeolum (hōr-dē-ō-luhm)–An infection of one or more glands of the eyelid; also called stye (stī).

horizontal silo (hōr-ih-zohn-tahl sī-lō)–A silo built with its long dimension parallel to the ground surface rather than perpendicular as in the case of an upright silo.

horizontal transmission (hōr-ih-zohn-tahl trahnz-mihs-shuhn)–Disease transfer from one animal to the other.

hormone (hōr-mōn)–An organ-produced chemical substance that is transported by the bloodstream to regulate the activity of another organ; literally means "to set in motion."

horn (hōrn)–(1) A natural, bonelike growth or projection on each side of the skull of most breeds of cattle, sheep, and goats that is a natural weapon of defense. (2) The pointed end of a blacksmith's anvil used for shaping hot metal. (3) The front, upraised projection of a riding saddle, the snubbing horn. (4) The outer hard covering of the hoofs of horses, cattle, sheep, and swine. (5) A broad term commonly used in describing various shadings of color in the beak of some breeds of fowl.

horned (hōrnd)–Designating an animal that has horns. See hornless, polled.

Horner's syndrome (hōr-nərz sihn-drōm)–A collection of signs that are characterized by sinking of the eyeball, ptosis of the upper eyelid, pupil constriction, and narrowing of the palpebral fissure relating to paralysis of the cervical sympathetic fibers.

hornless (hōrn-lehs)–Designating a well defined, polled condition of certain breeds of cattle, such as the Angus and red polled. See polled.

horny frog (hōrn-ē frohg)–The semisoft, elastic, V-shaped structure in the sole of a horse's foot. Also called foot pad.

horse (hōrs)–(1) *Equus caballus;* a quadruped of very ancient domestication used as a beast of burden, a draft animal, and a pleasure animal for riding, and in some areas as a meat animal. (2) A stallion. (3) Designating an implement that is drawn by a horse.

horse manure (hōrs mahn-oo-ər)–Dried horse excrement used as a medium for growing mushrooms, for making heat in hot beds, and as a soil amendment.

horsemanship (hōrs-mahn-shihp)–The ability to show a high degree of skill in handling horses.

hoss (hohs)–Horse (slang).

host (hōst)–Any organism, plant or animal, in or upon which another spends part or all of its existence, and from which it derives nourishment and/or protection.

host specific (hōst speh-sihf-ihck)–Designating a parasite that can live in or on only one host, to which it is therefore said to be specific.

hot (hoht)–(1) Designating a horse with a bad disposition. (2) Designating manure that heats upon decomposition. (3) An animal feed that contains a high percentage of concentrate or a feed containing a high level of salt.

hot-blooded (hoht bluhd-ehd)–(1) Designating a horse that has some thoroughbred or Arab blood. (2) Designating a horse that is nervous and at times even vicious. See cold-blooded.

hot carcass weight (hoht kahr-kuhs wāt)–The weight of a carcass immediately after slaughter before the carcass has had time to age and shrink. See cold carcass weight.

hot iron (hoht ī-ərn)–The heated iron rod or stamp with a handle used in branding cattle, etc.

hot manure (hoht mahn-oo-ər)–Fresh manure that is going through the process of heating due to fermentation. Horse manure is designated as one of the hot manures. Cow manure is one of the cold manures.

hothouse lamb (hoht-hows lahm)–A lamb that is dropped in the fall or early winter and is marketed at an age of six to twelve weeks.

hotis test (hoh-tihs tehst)–A test for the rapid detection of certain microorganisms in raw milk.

hover (huhv-ər)–(1) The sheet metal canopy surrounding a heat source under which incubator chicks are kept warm. (2) To cover chicks, as a hen covers her chicks.

humerus (hū-mər-uhs)–The proximal forelimb long bone that articulates proximally with the scapula and distally with the radius and ulna to form the elbow joint.

humor (hū-mər)–Any clear body liquid.

hump up (huhmp uhp)–The attitude an animal takes in order to expose less body surface to rain, wind, cold, etc.; to pull the feet together and push the back up.

hunters (huhn-tərz)–A riding horse of quality conformation and size, especially adapted for the chase and riding to hounds; a thoroughbred type of breeding about 16 hands high.

hurdle (hər-dl)–A broad made of plywood usually around 3 feet wide and 3 feet long that is used to herd pigs.

husbandry (huhz-bahn-drē)–In its earlier usage, the skill, or art, of tillage, crop production, and rearing of farm animals. Today the word is occasionally used as a synonym for farming.

hutch (huhch)–A boxlike cage or pen for a small animal, e.g., a rabbit hutch.

hybrid (hī-brihd)–Offspring resulting from mating of two different species.

hybrid chicks (hī-brihd chihcks)–Chicks that result from crossing two or more inbred lines of the same or different breeds, varieties, or strains.

hybrid vigor (hī-brihd vihg-ōr)–The increase of size, speed of growth, and vitality of a crossbreed over its parents. See heterosis.

hybridization (hī-brihd-ih-zā-shuhn)–The production of hybrids by natural crossing or by manipulated crossing.

hydrocephalus (hī-drō-sehf-ah-luhs)–An abnormal accumulation of cerebrospinal fluid within the brain; "water on the brain."

hydronephrosis (hī-drō-neh-frō-sihs)–Dilation of the renal pelvis as a result of an obstruction to urine flow.

hydrophilic (dī-drō-fihl-ihck)–Water loving; ionized form.

hydrophobia (hī-drō-fō-bē-ah)–Literally, fear of water; commonly refers to rabies, which is a misnomer.

hydrops (hī-drohps)–An abnormal accumulation of fluid in tissues or a body cavity; also called dropsy.

hydroureter (hī-drō-yoo-rē-tər)–Distention of the ureter with urine due to a blockage.

hygiene (hī-jēn)–The science of health; the rules or principles of maintaining health in people and animals; sanitation.

hymen (hī-mehn)–A membranous fold of tissue that partly or completely covers the external vaginal orifice.

hyperadrenocorticism (hī-pər-ahd-rēn-ō-kōr-tih-kihz-uhm or hī-par-ahd-rēn-ō-kōr-tih-sihzm)–A disorder caused by excessive adrenal cortex production of glucocorticoid; also called Cushing's disease.

hyperalbuminemia (hī-pər-ahl-byoo-mih-nē-mē-ah)–A blood condition of abnormally high albumin (type of protein) levels.

hypercalcemia (hī-pər-kahl-sē-mē-ah)–Abnormally high blood calcium levels.

hypercapnia (hī-pər-kahp-nē-ah)–Above normal levels of carbon dioxide in the blood.

hyperchromic (hī-pər-krō-mihck)–More than normal color.

hypercrinism (hī-pər-krī-nihzm)–A condition of excessive gland secretion.

hyperechoic (hī-pər-eh-kō-ihck)–A tissue that reflects more sound back to the transducer than the surrounding tissues; appears bright.

hyperemia (hī-pər-ē-mē-ah)–Excess blood in a part.

hyperesthesia (hī-pər-ehs-thē-zē-ah)–Excessive sensitivity.

hyperestrogenism (hī-pər-ehs-trō-jehn-ihz-uhm)–Elevated blood estrogen levels; seen in intact cycling female ferrets if not bred (ferrets are induced ovulators).

hyperextension (hī-pər-ehcks-tehn-shuhn)–Extreme straightening of a limb beyond its normal limit.

hyperglycemia (hī-pər-glī-sē-mē-ah)–Abnormally elevated blood glucose levels.

hypergonadism (hī-pər-gō-nahd-ihzm)–An abnormal condition of excessive hormone secretion from the gonads (ovaries in females and testes in males).

hyperimmunization (hī-pər-ihm-mūn-ih-zā-shuhn)–The process of increasing the immunity of an animal by increasing the injection of an antigen, subsequent to the establishment of an initial immunity; enhanced immunity.

hyperinsulinism (hī-pər-ihn-suh-lihn-ihzm)–A disorder of excessive insulin production.

hyperkalemia (hī-pər-kā-lē-mē-ah)–An excessive level of blood potassium.

hyperkinesis (hī-pər-kihn-ē-sihs)–Increased motor function or activity.

hyperlipidemia (hī-pər-lihp-ih-dē-mē-ah)–A blood condition of abnormally high fat levels.

hypernatremia (hī-pər-nā-trē-mē-ah)–Excessive levels of blood sodium.

hyperparasite (hī-pər-pahr-ah-sīt)–An organism that is parasitic on another parasite.

hyperparathyroidism (hī-pər-pahr-ah-thī-royd-ihzm)–An abnormal condition of excessive parathyroid hormone secretion resulting in hypercalcemia.

hyperphagia (hī-pər-fā-jē-ah)–Excessive hunger.

hyperpituitarism (hī-pər-pih-tū-ih-tah-rihzm)–A condition brought about by excessive production of one or more hormones by the pituitary gland, causing abnormal or excessive growth, as in gigantism.

hyperplasia (hī-pər-plā-zē-ah)–An abnormal increase in the number of normal cells that are in normal arrangement in a tissue.

hyperpnea (hī-pərp-nē-ah)–An abnormal increase in the rate and depth of respirations.

hypersensitivity (hī-pər-sehn-sih-tihv-ih-tē)–The violent reaction of an organism to attack by a pathogen; a condition in which the response to a stimulus is unusually prompt or excessive in degree.

hypersensitization (hī-pər-sehn-sih-tih-zā-shuhn)–An increased response to an allergen.

hypersusceptible (hī-pər-suh-sehp-tih-bl)–Designating a condition of abnormal susceptibility to infection, or to a poison, to which a normal individual is resistant.

hypertension (hī-pər-tehn-shuhn)–Abnormally high blood pressure.

hyperthermia (hī-pər-thər-mē-ah)–Increased body temperature.

hyperthyroidism (hī-pər-thī-royd-ihzm)–An abnormal condition of excessive thyroid secretion resulting in increased metabolic rate and autonomic nervous system disturbances.

hypertrophy (hī-pər-trō-fē)–Excessive size.

hypertropia (hī-pər-trō-pē-ah)–Deviation of one eye upward.

hyperventilation (hī-pər-vehn-tih-lā-shuhn)–Abnormally rapid deep breathing.

hyphema (hī-fē-mah)–Hemorrhage into the anterior chamber of the eye.

hypnosis (hihp-nō-sihs)–A condition of altered awareness; a "trancelike" state.

hypoadrenocorticism (hī-pō-ahd-rēn-ō-kōr-tih-kihz-uhm or hī-pō-ahd-rēn-ō-kōr-tih-sihzm)–A disorder caused by deficient adrenal cortex production of glucocorticoid; also called Addison's disease.

hypocalcemia (hī-pō-kahl-sē-mē-ah)–Abnormally low blood calcium levels.

hypocapnia (hī-pō-kahp-nē-ah)–Below normal levels of carbon dioxide in the blood.

hypochromic (hī-pō-krō-mihck)–Less than normal color.

hypocrinism (hī-pō-krī-nihzm)–A condition of deficient gland secretion.

hypodermis (hī-pō-dər-mihs)–The layer of tissue below the dermis of the skin.

hypoechoic (hī-pō-eh-kō-ihck)–A tissue that reflects less sound back to the transducer than the surrounding tissues; appears dark.

hypoglycemia (hī-pō-glī-sē-mē-ah)–Abnormally low blood glucose.

hypogonadism (hī-pō-gō-nahd-ihzm)–An abnormal condition of deficient hormone secretion from the gonads (ovaries in females and testes in males).

hypokalemia (hī-pō-kā-lē-mē-ah)–A deficient level of blood potassium.

hypomagnesemia (hī-pō-mahg-nē-sē-me-ah)–A deficiency of magnesium in the blood. See grass tetany.

hyponatremia (hī-pō-nā-trē-mē-ah)–A deficient level of blood sodium.

hypoparathyroidism (hī-pō-pahr-ah-thī-royd-ihzm)–An abnormal condition of deficient parathyroid hormone secretion resulting in hypocalcemia.

hypophysectomy (hī-pohf-ih-sehck-tō-mē)–Surgical removal of the pituitary gland.

hypophysis (hī-poh-fī-sihs)–The pituitary gland.

hypoplasia (hī-pō-plā-zē-ah)–An incomplete or decrease in the number of normal cells that are in normal arrangement in a tissue.

hypopnea (hī-pōp-nē-ah)–Abnormally slow or shallow respirations.

hypopyon (hī-pō-pē-ohn)–Pus in the anterior chamber of the eye.

hyposensitization (hī-pō-sehn-sih-tih-zā-shuhn)–A decreased response to an allergen.

hypospadias (hī-pō-spā-dē-uhs)–An abnormal condition where the urethra opens on the undersurface of the penis.

hypothalamus (hī-pō-thahl-ah-muhs)–A portion of the diencephalon located below the thalamus that regulates hormone release.

hypothermia (hī-pō-thər-mē-ah)–Decreased body temperature.

hypothesis (hī-pohth-eh-sihs)–A statement of research supposition.

hypothyroidism (hī-pō-thī-royd-ihzm)–An abnormal condition of deficient thyroid secretion resulting in decreased metabolic rate, lethargy, and increased sensitivity to the cold.

hypotropia (hī-pō-trō-pē-ah)–Deviation of one eye downward.

hypovolemia (hī-pō-vō-lē-mē-ah)–Decreased circulating blood volume.

hypoxia (hī-pohck-sē-ah)–Below normal oxygen levels.

hypsodont (hihps-ō-dohnt)–A type of tooth characterized by long crowns and progressive continuous eruption.

hysterectomy (hihs-tər-ehck-tō-mē)–Surgical removal of the uterus.

hystricomorph (hihs-trihck-ō-mōrf)–Type of rodent that includes guinea pigs, chinchillas, and porcupine.

I

iatrogenic (ī-aht-rō-jehn-ihck)–Produced by treatment.

iatrogenic disease (ī-aht-rō-jehn-ihck dih-zēz)–A disorder caused by treatment.

icterus (ihck-tər-uhs)–Jaundice; yellow discoloration of the skin and mucous membranes due to greater than normal levels of bilirubin.

ictus (ihck-tuhs)–An attack; the period of an actual seizure.

identical twins (ī-dehn-tih-kahl twihnz)–Twins that develop from a single fertilized egg that separates into two parts shortly after fertilization.

idiopathic (ihd-ē-ō-pahth-ihck)–Of unknown cause; a disease peculiar to an individual; not likely to be seen in others.

ileectomy (ihl-ē-ehck-tō-mē)–Surgical removal of the ileum.

ileitis (ihl-ē-ī-tihs)–Inflammation of the ileum.

ileostomy (ihl-ē-ohs-tō-mē)–Surgical production of an artificial opening between the ileum and abdominal wall.

ileum (ihl-ē-uhm)–The distal or aboral portion of the small intestine; located between the jejunum and large intestine.

ileus (ihl-ē-uhs)–Cessation of intestinal peristalsis.

ilium (ihl-ē-uhm)–The cranial bones of the pelvis.

imbricate (ihm-brih-kāt)–To tighten with sutures.

imbrication (ihm-brih-kā-shuhn)–The overlapping of apposing surfaces.

immobilization (ihm-mō-bihl-ih-zā-shuhn)–The act of holding, suturing, or fastening a bone in a fixed position, usually with a bandage or cast; to prevent movement.

immunity (ihm-yoo-nih-tē)–The state of being resistant to a specific disease.

immunize (ihm-yoo-nīz)–To render an animal resistant to disease by vaccination or inoculation.

immunizing agent (ihm-yoo-nī-zihng ā-jehnt)–A substance that, when introduced into the body of an animal, will build up antibodies in the blood that will resist or overcome an infection to which most of the same genus or species are susceptible.

immunofluorescence (ihm-yoo-nō-floo-rehs-ehns)–A method of tagging antibodies with a luminating dye to detect antigen-antibody complexes.

immunoglobulin (ihm-yoo-nō-glohb-yoo-lihn)–An antibody made by plasma cells; there are five distinct immunoglobulins.

immunology (ihm-yoo-nohl-ō-jē)–The study of the immune system.

immunosuppressant (ihm-yoo-nō-suhp-prehs-ahnt)–A substance that prevents or decreases the body's reaction to invasion by disease or foreign material.

immunosuppression (ihm-yoo-nō-suhp-prehsh-uhn)–The reduction or decrease in the state of resistance to disease.

impaction (ihm-pahck-shuhn)–An obstruction of an area usually with feed that is too dry.

implant (ihm-plahnt)–A material inserted or grafted into the body.

implantation (ihm-plahn-tā-shuhn)–Attachment and embedding of the zygote within the uterus.

impotence (ihm-pō-tehns)–Temporary or permanent loss of reproductive power or virility.

impregnate (ihm-prehg-nāt)–To fertilize a female animal.

imprinting (ihm-prihnt-ihng)–A kind of behavior common to some newly hatched birds or newborn animals that causes them to adopt the first animal, person, or object they see as their parent.

impulse (ihm-puhlz)–A wave of excitation transmitted through nervous tissue.

in foal (ihn fōl)–Designating a pregnant mare.

in hand (ihn hahnd)–Designating an animal under control or immediately available.

in heat (ihn hēt)–Designating a female animal at a period when she will accept coitus with a male. Also called estrus. See estrus.

in season (ihn sē-szohn)–See estrus, in heat.

in situ (ihn sih-too)–At the normal site.

in utero (ihn yoo-tər-ō)–Within the uterus.

in vitro (ihn vē-trō)–Outside living organisms; in test tubes or other laboratory glassware.

in vitro fertilization (ihn vē-trō fər-tihl-ih-zā-shuhn)–Fertilizing an egg with sperm in a test tube or petri dish, then implanting the fertilized egg into the uterus. See artificial insemination, embryo transfer.

in vivo (ihn vē-vō)–Inside living organisms.

in wear (ihn wehr)–A condition where a tooth has risen to the masticatory level.

in-and-in breeding (ihn ahnd ihn brē-dihng)–The breeding together of closely related animals for a number of successive generations to improve or eliminate certain characteristics. Also called breed in and in, line breeding, inbreeding.

inappetence (ihn-ahp-eh-tehns)–Lacking the desire to eat; not the same definition as anorexia.

inappropriate urination (ihn-ah-prō-prē-aht yoo-rih-nā-shuhn)–Eliminating urine either at the wrong time or in the wrong place.

inbred (ihn-brehd)–(1) An individual with parents who show 50% or more of common ancestry in their pedigree. (2) Of, or pertaining to, an animal produced by breeding between close relatives.

inbreeding (ihn-brēd-ihng)–The mating of very closely related animals such as mother and son, father and daughter, brother and sister.

incidence (ihn-sih-dehns)–The number of new cases of a disease occurring in a given population during a specific period, divided by the total number of individuals at risk of developing the disease during that same period.

incise (ihn-sīz)–To surgically cut into.

incision (ihn-sih-zhuhn)–A cut.

incisional biopsy (ihn-sih-shuhn-ahl bī-ohp-sē)–Cutting into and removing part of a mass, tissue, or organ for examination.

incisors (ihn-sīz-ōrz)–The front teeth of the dental arcade used for cutting; abbreviated I in dental formula.

inclusion (ihn-klū-zhuhn)–A nonliving substance or particle in a cell.

incompatibility (ihn-kohm-paht-ih-bihl-ih-tē)–A condition in animals in which the viable male gamete will not fertilize the viable female gamete.

incomplete dominance (ihn-kohm-plēt dohm-ih-nahns)–A kind of inheritance where a gene does not completely cover up or modify the expression of its allele; also may be known as codominance or blending inheritance.

incontinence (ihn-kohn-tihn-ehns)–The inability to control excretory function.

incubation (ihn-kyoo-bā-shuhn)–The process of a fertilized poultry egg developing into a newly hatched bird.

incubator (ihn-kū-bā-tehr)–An apparatus or chamber that provides favorable environmental conditions for the development of embryos, the hatching of eggs, or the growth of cultures.

incus (ihng-kuhs)–The auditory ossicle known as the anvil.

independent culling levels (ihn-dē-pehnd-ehnt kuhl-ihng lehv-ehlz)–Selection of culling based on cattle meeting specific levels of performance for each trait included in the breeder's selection program.

index (ihn-dehcks)–A system for comparing animals within a herd, or area, based on the average of the group.

indigestion (ihn-dī-jehs-chuhn)–A condition of the digestive system of animals in which the normal digestive process is halted or disturbed. There are many causes, such as infection, spoiled food, overeating, etc.

indiscriminate breeder (ihn-dihs-krihm-ih-nahnt brē-dər)–An animal that will breed with any animal of the same type and the opposite sex.

induced ovulator (ihn-dooəd ohv-yoo-lā-tər)–A species that ovulates only as a result of sexual activity (cats, rabbits, ferrets, llamas, camels, mink).

inert ingredient (ihn-ərt ihn-grē-dē-ehnt)–A substance in a feed, pesticide, etc., that does not act as a feed, pesticide, etc. The substance may serve a purpose but is usually used as a filler, vehicle, etc.

infarct (ihn-fahrckt)–A localized area of necrosis caused by an interrupted blood supply.

infect (ihn-fehckt)–To cause disease by the introduction of germs, parasites, or fungi. See infection.

infection (ihn-fehck-shuhn)–Invasion of the tissues of the body of a host by disease-producing organisms in such a way that

injury results; the presence of multiplying parasites, bacteria, viruses, etc., within the body of a host. See infestation.

infectious (ihn-fehck-shuhs)–Designating a communicable disease.

infectious disease (ihn-fehck-shuhs dih-zēz)–Disorder caused by pathogenic organisms. A disease caused by bacteria, protozoa, viruses, or fungi entering the body. It is not necessarily contagious or spread by contact.

infective (ihn-fehck-tihv)–Capable of entering and establishing itself in a host; able to infect a susceptible plant or animal.

inferior (ihn-fēr-ē-ər)–Lowermost, below, or toward the tail.

infertile (ihn-fehr-tihl)–Designating that which is incapable of reproduction, e.g., a barren female animal, a male animal with nonviable spermatozoa, an unfertilized egg.

infest (ihn-fehst)–To assail, attack, overrun, annoy, disturb; as ticks infest a cow.

infestation (ihn-fehs-tā-shuhn)–The act of infesting, or state of being attacked, molested, vexed, or annoyed by large numbers of insects, etc., as an animal may be subject to an infestation of parasites, such as fleas, ticks, mites, etc.

inflammation (ihn-flah-mā-shuhn)–A localized protective response to destroy, dilute, or wall off injury; classic signs are heat, redness, swelling, pain, and loss of function.

infundibulum (ihn-fuhn-dihb-yoo-luhm)–A funnel-shaped opening or passageway.

infuse (ihn-fūz)–To inject or introduce a liquid or medicinal agent, as to infuse the udder with a drug.

ingest (ihn-jehst)–To eat, or take in food for digestion by way of the mouth.

ingesta (ihn-jehst-ah)–Material taken orally.

ingluvies (ihn-gloo-vēz)–The crop or craw of a chicken or other bird; also the rumen or first stomach of a ruminating animal, as a cow.

inguinal canal (ihng-gwih-nahl kah-nahl)–The opening in the abdominal wall through which the testes pass from the body cavity into the scrotum.

inguinal hernia (ihng-gwih-nahl hər-nē-ah)–A protrusion of the bowel through a weakened area in the groin.

inhalants (ihn-hah-lahntz)–Medicinal preparations that are inhaled or drawn into the lungs.

inhalation (ihn-hah-lā-shuhn)–The act of breathing in; also called inspiration.

inheritance (ihn-hehr-ih-tahns)–The transmission of genetic factors from parent to offspring.

inherited characteristic (trait) (ihn-hehr-ih-tehd kahr-ahk-tər-ihs-tihck)–Character, the expression of which is determined by a particular gene or genes.

inhibit (ihn-hihb-iht)–To slow or stop.

inject (ihn-jehckt)–To introduce a substance into the body of an animal or plant by mechanical means.

innervate (ihn-nər-vāt)–To provide nerve stimulation.

innervation (ihn-nər-vā-shuhn)–To supply or stimulate a body part through the action of nerves.

inoculation (ih-nohck-ū-lā-shuhn)–Introduction into healthy animal tissue of microorganisms to produce a mild form of the disease, followed by immunity. (2) Process of taking a small amount of bacteria to start a new culture.

inotrope (ihn-ō-trōp)–A substance affecting muscle contraction.

insect (ihn-sehckt)–An air-breathing animal (phylum Arthropoda) that has a distinct head, thorax, and abdomen.

insect growth regulator (ihn-sehckt grōth rehg-yoo-lā-tər)–A chemical that interferes with the normal growth pattern of insects causing normal development and thus death.

insecticide (ihn-sehck-tih-sīd)–A substance that kills insects by chemical action, as a stomach poison, contact poison, or fumigant.

inseminate (ihn-sehm-ih-nāt)–To place semen in the vagina of an animal during coitus, or, in the practice of artificial insemination, to introduce semen into the vagina by a method other than coitus.

inseminating tube (ihn-sehm-ih-nāt-ihng toob)–A rubber, glass, or metal tube, usually with a syringe attachment, used in artificial insemination to introduce the semen into the vagina of a female animal.

inseminator (ihn-sehm-ih-nāt-ōr)–The technician, in the employ of an artificial breeder's unit, who brings the prepared semen to the farmer's herd and performs the technical service of inseminating the cows.

insertion (ihn-sihr-shuhn)–The muscle ending that is the movable end or portion away from midline.

insidious disease (ihn-sihd-ē-uhs dih-zēz)–A disease that develops slowly in a stealthy, subtle manner over a long period of time.

inspissation (ihn-spihs-sā-shuhn)–The process of rendering dry or thick by evaporation.

instinct (ihn-stihnckt)–The ability of an animal based upon its genetic makeup to respond to an environmental stimulus; it does not involve a mental decision.

insufficiency (ihn-sah-fihsh-ehn-sē)–The inability to perform at proper level of function.

insufflation (ihn-suh-flā-shuhn)–The blowing of a powder or vapor into a cavity usually for administering medication.

insulin (ihn-suh-lihn)–Pancreatic origin or synthetic hormone that transports blood glucose into body cells.

insulinoma (ihn-suh-lihn-ō-mah)–A neoplasm of the islets of Langerhans.

intact (ihn-tahckt)–(1) Capable of reproduction. (2) No cuts, scrapes, openings, or alterations.

integumentary (ihn-tehg-yoo-mehn-tah-rē)–The skin, hair, nails, sweat glands, and sebaceous glands.

interatrial (ihn-tər-ā-trē-ahl)–Between the atria.

interbreeding (ihn-tər-brē-dihng)–In livestock breeding, breeding closely within a family or strain for the purpose of fixing type and desired characters.

intercellular (ihn-tər-sehl-yoo-lahr)–Between cells.

intercostal (ihn-tər-kohs-tahl)–Between the ribs.

intercrossing (ihn-tər-krohs-ihng)–Crossbreeding.

interdigital (ihn-tər-dihg-ih-tahl)–Between the toes.

interfering (ihn-tehr-fēr-ihng)–The striking of the fetlock or cannon by the opposite foot that is in motion.

intermediate host (ihn-tər-mē-dē-aht hōst)–An animal other than the primary host that a parasite uses to support part of its life cycle.

intermingling color (ihn-tehr-mihng-glihng kuhl-ər)–A coat color pattern of animals in which the separate colors merge where they meet. It is usually an objectionable color pattern.

interphalangeal (ihn-tər-fah-lahn-jē-ahl)–Between the phalanges.

intersex (ihn-tehr-sehcks)–Designating an organism that displays primary and secondary sexual characteristics intermediate between male and female.

interspecific (ihn-tehr-speh-sihf-ihck)–Referring to events or relationships that occur between individuals of different species.

interspecific hybrid (ihn-tər-speh-sihf-ihck hī-brihd)–A cross between individuals of different species. Taxonomically identified by listing both species separated by an x.

interstitial (ihn-tər-stih-shahl)–The space within a tissue or organ.

interstitial cystitis (ihn-tər-stihsh-ahl sihs-tī-tihs)–Inflammation of the wall of the urinary bladder.

interstitial fluid (ihn-tər-stihsh-ahl flū-ihd)–A clear, colorless tissue fluid that leaves the capillaries and flows in the spaces between the cells of a tissue or organ.

interventricular (ihn-tər-vehn-trihck-yoo-lahr)–Between the ventricles.

intervertebral disc (ihn-tər-vər-tēh-brahl dihsk)–The cartilage pad that separates and cushions the vertebrae from each other; disc may be spelled disk.

intestinal crypt (ihn-tehs-tih-nahl krihpt)–The valley of intestinal mucous membrane lining the small intestine.

intestinal flora (ihn-tehs-tih-nahl flō-rah)–The normal microorganisms residing in the gastrointestinal tract.

intestinal villus (ihn-tehs-tih-nahl vihl-uhs)–A small protrusion from the mucous membranes lining the small intestine to increase surface area and absorption.

intestine (ihn-tehs-tihn)–The part of the digestive tract between the stomach and anus.

intra op (ihn-trah ohp)–A common term for during or within surgery; intraoperatively.

intracapsular extraction (ihn-trah-kahp-soo-lahr-ehcks-trahck-shuhn)–Cataract removal that includes the surrounding capsule.

intracellular (ihn-trah-sehl-ū-lahr)–Within, inside of, a cell.

intracervical method (ihn-trah-sehr-veh-kahl mehth-uhd)–A method of artificial insemination whereby semen is placed directly in the cervix or uterus and not in the vagina.

intracranial (ihn-trah-krā-nē-ahl)–Within the portion cranium.

intradermal (ihn-trah-dər-mahl)–Within the true skin.

intramedullary pins (ihn-trah-mehd-yoo-lahr-ē pihnz)–Metal rods that are inserted into the medullary cavity of long bones to repair stable fractures.

intramuscular (ihn-trah-muhs-kyū-lahr)–Within the muscle.

intranasal (ihn-trah-nā-zahl)–Within the nose or nostrils.

intranasal instillation (ihn-trah-nā-zahl ihn-stih-lā-shuhn)–The placing of fluid or medicine inside the nose.

intraocular (ihn-trah-ohck-yoo-lahr)–Inside the eyeball.

intraocular vaccination (ihn-trah-ohck-yoo-lahr vahck-sih-nā-shuhn)–The placement of a vaccine directly into the eye.

intraperitoneal (ihn-trah-pehr-ih-tō-nē-ahl)–Within the peritoneal cavity.

intraspecific (ihn-trah-speh-sihf-ihck)–Referring to events or relationships that occur between individuals of the same species.

intraspecific hybrid (ihn-trah-speh-sihf-ihck hī-brihd)–A cross between individuals within the same species, but of different genotypes.

intratracheal (ihn-trah-trā-kē-ahl)–Within the trachea, or windpipe.

intrauterine (ihn-trah-yoo-tər-ihn)–Within the uterus.

intravenous (ihn-trah-vēn-uhs)–Within the vein.

intravenous pyelogram (ihn-trah-vē-nuhs pī-eh-lō-grahm)–A radiographic study of the kidney and ureters in which a dye is injected into a vein to define structures more clearly; abbreviated IVP.

introduced (ihn-trō-doosd)–Designating a plant, animal, disease, etc., that is not indigenous to an area, but is brought in purposely or accidentally.

introgastric (ihn-trō-gahs-trihck)–In the stomach.

introgression hybridization (ihn-trō-grehs-shuhn hī-brihd-ih-zā-shuhn)–Long-continued interspecific hybridization leading to an infiltration of genes from one species into another.

intromission (ihn-trō-mihsh-uhn)–Insertion of one part or instrument into another.

intussusception (ihn-tuhs-suhs-sehp-shuhn)–Telescoping of one part of the intestine into an adjacent part.

inversion (ihn-vər-shuhn)–Turning inward.

invertebrate (ihn-vehr-teh-brāt)–Any animal with no spinal column.

inverted nipple (ihn-vehrt-ehd nihp-uhl)–A nonfunctional teat (usually refers to a sow's teat) that appears to be inverted toward the udder.

involucrum (ihn-voh-loo-kruhm)–The covering or sheath that contains a sequestrum of bone.

involution (ihn-vō-lū-shuhn)–The return of an organ to normal size after a time of enlargement, as in the case of the uterus after parturition.

ion (ī-ohn)–An atom or a group of atoms carrying an electrical charge, which may be positive or negative.

ionized (ī-ohn-īzd)–Electrically charged.

ionophore (ī-ohn-ō-fōr)–A feed substance or additive that makes the digestive processes in ruminants more efficient.

ipsilateral (ihp-sē-lah-tər-ahl)–On the same side.

iridectomy (ihr-ih-dehck-tō-mē)–Surgical removal of a portion of the iris.

iridocorneal (ihr-ihd-ō-kōr-nē-ahl)–The iris and cornea.

iris (ī-rihs)–The pigmented muscular layer that surrounds the pupil; part of the choroid.

iritis (ī-rī-tihs)–Inflammation of the iris.

iron (ī-ərn)–Fe; a metallic element that is an essential constituent of blood hemoglobin where it functions to transport oxygen.

iron gray (ī-ərn grā)–A coat color term for a horse that (a) in the United States, denotes a white coat with a high percentage of black hairs; (b) in England, a gray coat with reddish hairs throughout.

ischemia (ihs-kē-mē-ah)–A deficiency in blood supply.

ischium (ihs-kē-uhm)–The caudal bones of the pelvis.

islets of Langerhans (ī-lehts ohf lahng-ər-hahnz)–Specialized pancreatic cells that secrete insulin, glucagon, and pancreatic polypeptide.

isoechoic (ī-sō-eh-kō-ihck)–A tissue that has the same ultrasonic appearance as that of the surrounding tissue.

isogamy (ī-sohg-ah-mē)–Morphological similarity of fusing gametes.

isolate (ī-sō-lāt)–To place an animal in confinement away from other animals to prevent breeding or spread of disease.

isthmus (ihs-muhs)–(1) In biology, a narrow piece of tissue that connects two larger parts; the part of the fallopian tube between the ampulla and the uterus. (2) In poultry, that part of the oviduct where the two shell membranes are added to the egg.

J

jack (jahck)–A male donkey.

jaundice (jawn-dihs)–A yellow discoloration of the skin and mucous membranes due to greater than normal levels of bilirubin; also called icterus (ihck-tər-uhs).

jejunum (jeh-joo-nuhm)–The middle portion of the small intestine; located between the duodenum and ileum.

jenny (jehn-ē)–A female donkey. Also called jennet.

jet black (jeht blahck)–A horse color that is a shiny black.

jib(ber) (jihb[ər])–A balky horse.

jig (jihg)–An uneven gait of a horse, closely associated with prancing and weaving.

jill (jihl)–An intact female ferret.

jog (johg)–A slow trot.

joint (joynt)–The place of bone union; the articulation between the two bones.

jowl (jowl)–(1) The cheek of a pig. (2) Meat taken from the cheeks of a hog.

jug (juhg)–(1) A small claiming pen where ewes with newborn lambs are kept until the ewe has nursed and claimed her young. (2) A small pen, part of a corral, used to funnel animals down a narrow alley (chute) to facilitate sorting, loading, or catching; used primarily with cattle.

jughead (juhg-hehd)–A horse with poor characteristics, little sense, and not amenable to training.

jumper bull (juhmp-ər buhl)–A male bovine around maturity that is intended for breeding.

jumpers (juhm-pərz)–Horses that compete at shows by jumping and are judged on height, time, and faults.

juvenile (joo-veh-nīl)–Pertaining to the young of any animal.

K

karyotype (kahr-ē-ō-tīp)–A picture or diagram of the chromosomes of a particular cell as they appear in the metaphase of mitosis arranged in pairs by size and location in the centromeres.

keel (kēl)–(1) The part of a fowl's body that extends backward from the breast. Also called breastbone. (2) In ducks, the pendant fold of skin along the entire underside of the body. (3) In geese, the pendant fold of flesh from the legs forward on the underpart of the body.

keeled (kē-uhld)–Of, or pertaining to, a ridge (like the bottom of a boat) on an animal or plant part.

keet (kēt)–The young guinea fowl.

keratectomy (kehr-ah-tehck-tō-mē)–Surgical removal of part of the cornea.

keratin (kehr-ah-tihn)–A fibrous protein that composes hair, epidermis, and nails.

keratitis (kehr-ah-tī-tihs)–Inflammation of the cornea.

keratocentesis (kehr-ah-tō-sehn-tē-sihs)–A puncture of the cornea to allow for aspiration of aqueous humor.

keratoconjunctivitis (kehr-ah-tō-kohn-juhnck-tih-vī-tihs)–Inflammation of the cornea and conjunctiva.

keratoplasty (kehr-ah-tō-plahs-tē)–Surgical repair of the cornea (may include corneal transplant).

keratosis (kehr-ah-tō-sihs)–An abnormal condition of excessive keratin formation causing thickening; plural is keratoses (kehr-ah-tō-sēz).

keratotomy (kehr-ah-toh-tō-mē)–An incision into the cornea.

ketone (kē-tōn)–A by-product of fat metabolism. Also called ketone bodies.

ketosis (kē-tō-sihs)–A metabolic disease of excessive ketone buildup in the blood and tissues.

ketonuria (kē-tō-nū-rē-ah)–The presence of abnormal substances produced during increased fat metabolism in the urine.

kicking hobble (kihck-ihng hohb-uhl)–A rope or strap fastened to the rear legs or feet of a horse, ass, or cow to prevent kicking.

kicking strap (kihck-ihng strahp)–A strap fastened to the crupper strap of the harness in breaking colts. It extends down each side of the shaft of the carriage and is used to prevent kicking.

kid (kihd)–A young goat.

kid crop (kihd krohp)–The number of kids produced by a given number of does, usually expressed as a percentage of kids weaned of does bred.

kidding (kihd-ihng)–To give birth to a kid.

kidney (kihd-nē)–A paired urinary system organ that filters blood and removes waste products from the blood.

kilogram (kihl-ō-grahm)–A unit of weight that is one thousand grams; one kilogram is approximately 2.2 pounds.

kindle (kihn-dl)–Pregnancy in rabbits.

kindling (kihnd-lihng)–Giving birth to rabbits or ferrets.

kine (kīn)–Cows or cattle.

kip (kihp)–A hide taken from a very young calf.

kit (kiht)–A young rabbit (blind and deaf) or ferret.

kitten (kiht-tehn)–A young cat.

knee (nē)–In ungulates, the carpus; in dogs and cats may be used by lay people to describe the stifle joint.

knee banger (nē bahn-gər)–A horse that strikes its knees with the opposite front foot while walking or running.

knee spavin (nē spahv-ihn)–A chronic inflammation of the small bones of the carpal joint of horses, characterized by lameness, bony enlargement, a dragging of the toe of the affected leg, and a fusion of the carpal bones.

knob (nohb)–(1) The horny protuberance at the juncture of the head and upper bill in African and Chinese geese. (2) A deformity growth on the breastbone, usually at the front, sometimes found in chickens and turkeys; a defect. (3) The rounded protuberant part of the skull in crested fowl.

knock-kneed (nohck nēd)–An undesirable conformation or weakness of the front legs of horses in which the legs are not straight due to the knees coming too close together.

knuckle (nuhck-ehl)–The joint of two bones often used in making soup stock.

L

labial surface (lā-bē-ahl sər-fihs)–The tooth surface facing the lips.

labile (lā-bīl)–Unstable.

labium (lā-bē-uhm)–A lip or fleshy border; plural is labia (lā-bē-ah).

labyrinth (lahb-ih-rihnth)–The inner ear.

laceration (lahs-ər-ā-shuhn)–An accidental cut into skin.

lacrimal canaliculi (lahck-rih-mahl kahn-ah-lihck-yoo-lī)–The ducts that collect tears and drain them into the lacrimal sac located at the inner corner of each eye.

lacrimal gland (lahck-rih-mahl glahnd)–The gland that secretes tears.

lacrimal sac (lahck-rih-mahl sahck)–An enlargement at the upper portion of the tear duct; also called dacryocyst (dahck-rē-ō-sihst).

lacrimation (lahck-rih-mā-shuhn)–A condition of normal tear secretion.

lacromotomy (lahck-rō-moh-tō-mē)–An incision into the lacrimal gland or duct.

lactation (lahck-tā-shuhn)–The process of forming and secreting milk.

lacteals (lahck-tē-ahls)–Specialized lymph vessels that transport fats and fat-soluble vitamins located in the small intestine.

lactiferous duct (lahck-tihf-ər-uhs duhckt)–The tube that carries milk to the nipple.

lactiferous glands (lahck-tihf-ər-uhs glahndz)–The glands located in the mammary tissue that produce milk.

lactogenic (lahck-tō-jehn-ihck)–Capable of stimulating milk production.

lactogenic hormone (lahck-tō-jehn-ihck hōr-mōn)–Any hormone that stimulates lactation, especially prolactin.

lactometer (lahck-tohm-eh-tehr)–An instrument that determines the density of milk. It acts as a specific gravity indicator, from which information on the percentage of milk solids may be calculated.

lactose (lahck-tōs)–$C_{12}H_{22}O_{11}$; a white crystalline disaccharide made from whey and used in pharmaceuticals, bakery products, and confections; also called milk sugar.

lagomorph (lāg-ō-mōrf)–Animals such as rabbits or hares that have a second pair of incisors in the maxilla.

lamb (lahm)–(1) The young of sheep, specifically, (a) a young ovine that has not yet acquired the front pair of permanent incisor teeth; technically, (b) in stock judging, a sheep born in the same calendar year as shown. (2) To give birth to a lamb.

lamb crop (lahm krohp)–The number of lambs produced by a given number of ewes, usually expressed as a percentage of lambs weaned of ewes bred.

lambing (lahm-ihng)–Giving birth to sheep.

lambing loop (lahm-ihng loop)–A length of smooth or plastic-coated wire used as an aid in difficult lambing.

lambing pen (lahm-ihng pehn)–A specially equipped, isolated pen in the sheep barn in which a ewe is placed just before she gives birth to her young.

lambing percentage (lahm-ihng pehr-sehnt-ahj)–The numbere of lambs produced per 100 ewes.

lambing time (lahm-ihng tīm)–The season of the year, usually late winter or early spring, when sheep produce their young.

lamella (lah-mehl-ah)–One of the layers of a cell wall; a thin layer in a shell, like a leaf in a book.

lameness (lām-nehs)–Any soreness, tenderness, or unsoundness that is indicated by the unnatural action of feet or legs of animals.

lamina (lahm-ih-nah)–(1) One part of the dorsal portion of a vertebra. (2) Tissue that attaches the hoof to the underlying foot structures.

laminectomy (lahm-ihn-ehck-tō-mē)–Surgical removal of the lamina of the vertebral body to relieve pressure on the spinal cord.

laminitis (lahm-ihn-ī-tihs)–Inflammation or congestion of the sensitive tissues, the laminae, which lie immediately below the outer horny wall of a horse. Also called founder.

lane (lān)–A passageway from the barnyard to the pasture or field areas; a narrow road confined between fences.

lanolin (lahn-ō-lihn)–A fatlike substance secreted by the sebaceous glands of sheep.

laparotomy (lahp-ah-roht-ō-mē)–Surgical incision through the abdominal wall into the abdominal cavity.

lapin (lahp-ihn)–A castrated male rabbit.

lariat (lahr-ē-eht)–Rawhide, horsehair, or hemp rope, generally 35 to 45 feet long and arranged in a coil on the saddle. Used to lasso livestock for branding or restraining purposes. See lasso.

larva (lahr-vah)–The immature insect hatching from the egg and up to the pupal stage in orders with complex metamorphosis; the six-legged first instar of mites and ticks.

larvicide (lahr-vah-sīd)–A chemical used to kill the larval or preadult stages of parasites.

laryngectomy (lār-ihn-jehck-tō-mē)–Surgical removal of the voice box.

laryngitis (lahr-ihn-jī-tihs)–Inflammation of the voice box.

laryngopharynx (lah-rihng-gō-fār-ihnkz)–The portion of the throat which opens into the voice box and esophagus below the epiglottis.

laryngoplasty (lah-rihng-ō-plahs-tē)–Surgical repair of the voice box.

laryngoplegia (lahr-ihng-gō-plē-jē-ah)–Paralysis of the voice box.

laryngoscope (lahr-ihng-ō-skōp or lahr-ihn-gō-skōp)–An instrument used to visually examine the voice box.

laryngoscopy (lahr-ihng-gohs-kō-pē)–Endoscopic examination of the voice box.

laryngospasm (lah-rihng-ō-spahzm)–A sudden fluttering and/or closure of the voice box.

larynx (lār-ihnkz)–The voice box; part of the respiratory tract between the pharynx and trachea.

laser (lā-zər)–A device that transfers light into an intense beam for various purposes; acronym for Light Amplification by Stimulated Emission of Radiation.

lasso (lahs-ō or lah-soo)–(1) To throw a lariat in such a manner as to catch an animal by the horns, neck, or legs in the loop at the end, generally for restraining purposes. (2) A rope or line with a running loop used for catching animals.

latent (lā-tehnt)—Designating an infection that is present but which is not manifest in the host under consideration.

latent infection (lā-tehnt ihn-fehck-shuhn)–A condition that may not be clinically noticed, but under stress or poor health will develop into a recognizable disease state.

lateral (laht-ər-ahl)–Away from or farther from the midline; the side.

lateral canter (lah-tər-ahl kahn-tər)–A horse's gait characterized by a simultaneous beat of corresponding front and rear feet, as opposed to the diagonal front and rear feet.

lateral canthus (lah-tər-ahl kahn-thuhs)– The corner of the eye farthest away from the nose; also called the outer canthus.

lateral lacunae (lah-tər-ahl lah-kū-nā)–The two grooves that border the horny frog of a horse's hoof.

lateral projection (lah-tər-ahl prō-jehck-shuhn)–An x-ray beam that passes from side to side with the patient at right angles to the film.

lateral recumbency (laht-ər-ahl rē-kuhm-behn-sē)–Lying on the side (usually the term is preceded by right or left, depending on the side the animal is lying on).

lateral ringbone (lah-tər-ahl rihng-bōn)–A bony growth on the side of the pastern bone in a horse's leg. See ringbone.

lateral ventricle (lah-tər-ahl vehn-trih-kuhl)–A cavity in each cerebral hemisphere that connects with the third ventricle.

lather (lah-thər)–The accumulation of sweat on a horse's body.

lavage (lah-vahj)–The irrigation of tissue with copious amount of fluid.

laxative (lahck-sah-tihv)–A mild medicine used to relieve constipation.

laxity (lahcks-ih-tē)–The degree of looseness; not tight.

lay (lā)–To ovulate and deposit an egg or eggs produced within the female generative organs, as a hen lays eggs.

layer (lā-ər)–A mature female fowl that is kept for egg-laying purposes, especially one in current egg production.

lead cattle (lēd kah-tuhl)–The cattle at the head of a moving herd.

lead line (lehd līn)–A bluish line at the margin of the gums of an animal that indicates chronic lead poisoning, usually as a result of ingestion of paint or spray materials.

lead rope (lēd rōp)–A piece of rope, leather, or nylon that is attached via a clasp to a halter.

leader (lē-dehr)–The front animals of a tandem hitch.

leads (lēdz)–The conductors on an electrocardiograph.

lean (lēn)–Designating an animal lacking in condition of flesh or finish.

lean to fat ratio (lēn too faht rā-shō)–The amount of lean meat in a carcass compared to the amount of fat.

leather (lehth-ər)–(1) The cured, tanned skins of animals, especially of the bovines. (2) A pad of leather placed between the shoe and the foot of a horse with a sensitive sole.

left shift (lehft shihft)–A common term for an alteration in the distribution of leukocytes in which there are increases in band forms usually in response to infection.

leg (lehg)–(1) Any of the limbs of animals that support and move the body, such as foreleg, front leg, hind leg. (2) To haul or drag a sheep from the pen to the shearing board by a hind leg.

leg cues (lehg kyooz)–Signals given to the horse through movement of the rider's legs.

leggy (lehg-ē)–Designating an animal, usually very young, as a colt, whose legs are disproportionally long in relation to its body size.

legume (lehg-yoom)–Roughage plants that have nitrogen-fixing nodules on their roots; examples include alfalfa and clover.

leiomyositis (lī-ō-mī-ō-sī-tihs)–Inflammation of smooth muscle.

lens (lehnz)–A clear, flexible, curved capsule located behind the iris and pupil.

lensectomy (lehn-sehck-tō-mē)–Surgical removal of a lens (usually performed on cataracts).

leptomeninges (lehp-tō-meh-nihn-gēz)–A collective term for the pia mater and arachnoid membranes of the brain.

lesion (lē-zhuhn)–A pathologic change in tissue.

lethal (lē-thahl)–Causing death.

lethal dose (lē-thahl dōs)–The lethal dose (written as LD_{10}, LD_{50}, LD_{100}, or any percentage) median is a measurement of the milligrams of toxicant per kilogram of body weight that kills 10, 50, or 100 percent of the target species.

lethal gene (lē-thahl jēn)–A gene that can cause the death of an individual when it is allowed to express itself.

lethargy (lehth-ahr-jē)–A condition of drowsiness, listlessness, or indifference.

letting down (leht-ihng down)–The adjusting period, with some loss of flesh, which takes place when beef cattle are changed from heavy, drylot to feeding to open pastures or range feeding.

leucocyte (loo-kō-sīt)–The white cells of the blood that destroy disease germs and help remove foreign matter from the bloodstream or tissue. Found predominantly in pus.

leukemia (loo-kē-mē-ah)–An abnormal increase in the number of malignant white blood cells.

leukocytopoiesis (loo-kō-sīt-ō-poy-ē-sihs)–The production of white blood cells; also called leukopoiesis (loo-koo-poy-ē-sihs).

leukocytosis (loo-kō-sī-tō-sihs)–An abnormal increase in the number of white blood cells.

leukopenia (loo-kō-pē-nē-ah)–An abnormal decrease in the number of white blood cells.

levator palpebrae muscles (lē-vā-tər pahl-pē-brā muhs-uhlz)–The muscles that raise the upper eyelid.

Leydig's cells (lih-dihgz sehlz)–Specialized interstitial cells that secrete testosterone; also called interstitial cells.

libido (lih-bih-dō or lih-bē-dō)–Sexual drive.

lice (līs)–Small, nonflying, biting or sucking insects that are true parasites of humans, animals, and birds. Biting lice belong to the order Mallophaga; the sucking or true lice belong to the order Anoplura.

life cycle (līf sī-kuhl)–A life history; the changes in the form of life that an organism goes through.

ligament (lihg-ah-mehnt)–A band of fibrous connective tissue that connects bone to bone or organ to organ.

ligation (lī-gā-shuhn)–The act of tying.

ligature (lihg-ah-chər)–A substance used to tie a vessel or strangulate a part.

light bay (līt bā)–A horse color; light tan with a black mane and tail.

light chestnut (līt chehst-nuht)–A horse color; light, yellow-gold.

light dun (līt duhn)–A horse color; a dun imposed on a sandy bay or sorrel.

light feeder (līt fē-dər)–(1) An animal that is being fed for maintenance and normal growth but not for quick finish or fattening. (2) An animal that does not eat as much as most animals of its class.

light-handed (līt-hahnd-ehd)–Denoting a rider who handles a horse well within a minimum of tugging or jerking on the bridle.

limb (lihm)–The leg or wing of an animal.

limbic (lihm-bihck)–A border.

limbic system (lihm-bihck sihs-stehm)–The portion of brain located in the telencephalon, diencephalon, and mesencephalon that affects emotional activity and behavior.

limbus (lihm-buhs)–The corneoscleral junction.

limited feeding (lihm-ih-tehd fē-dihng)–(1) The feeding of livestock to maintain weight and growth but not to fatten or increase production. (2) Restricting an animal to less than maximum weight increase.

line (līn)–(1) The reins of a harness. (2) In marketing, the whole of a herd of sheep. (3) A group of animals that retain their uniform appearance in succeeding generations.

line breeding (līn brē-dihng)–Mating of selected members of successive generations among themselves in an effort to maintain or fix desirable characteristics.

line test (līn tehst)–(1) A test in which a series of samples are taken from the milk supply of the whole herd during milking or milk processing; used in the bacteriological control examination of the milk supply. (2) A test used to determine the vitamin D content of milk. (3) A test to determine the degree of calcification of a growing bone; a measure for rickets.

linea alba (lihn-ē-ah ahl-bah)–A fibrous band of connective tissue on the ventral abdominal wall that is the center attachment of the abdominal muscles; "white line."

lingual surface (lihng-gwahl sər-fihs)–The tooth surface that faces the tongue.

liniment (lihn-ih-mehnt)–A preparation used for bathing or rubbing sprains, bruises, etc., usually containing a counterirritant.

linkage (lihngk-ihj)–The association of characters from one generation to the next due to the fact that the genes controlling the characters are located on the same chromosome linkage group.

lip (lihp)–(1) In a dam, a small wall on the downstream end of the apron to break the flow from the apron. (2) One of the edges of a wound. (3) Either of the two external fleshy folds of the mouth opening. (4) Either one of the inner or outer fleshy folds of the vulva.

lip strap (lihp strahp)–The small strap running through the curb chain from one side of the bit shank to the other. Its primary function is to keep the horse from taking the shank or the bit in its teeth.

lipase (lī-pās or lihp-ās)–A fat-digesting enzyme.

lipemia (lī-pē-mē-ah)–Excessive amount of fats in the blood.

lipemic serum (lī-pē-mihck sēr-uhm)–Fats from blood that have settled in the fluid portion (serum); clinically the serum will appear cloudy and white.

lipoid (lihp-oyd)–Concerning fat or fatty tissue.

lipolysis (lih-pohl-ih-sihs)–The hydrolysis of fats by enzymes, acids, alkalis, or other means, to yield glycerol and fatty acids.

lipoma (lī-pō-mah)–A benign growth of fat cells.

lipophilic (lihp-ō-fihl-ihck)–Fat loving; nonionized form.

liquid manure (lih-kwihd mahn-oo-ər)–The liquid excrement from animals, mainly urine, collected from the gutters of barns into large tanks and hauled to the fields.

listless (lihst-lehs)–Lethargic, lifeless, lacking energy. A listless animal would lie around and appear weak and not be aroused by the stimuli that usually arouse it.

liter (lē-tər)–A metric unit of volume, equal to 0.2642 gallons.

lithotripsy (lihth-ō-trihp-sē)–The destruction of a stone using ultrasonic waves traveling through water.

litter (liht-tər)–The multiple offspring born during the same labor; also a substance used by animals that is appropriate for urination and defecation.

litter floor (lih-tər flōr)–The floor of a poultry house that is composed of straw and shavings or ground corncobs, droppings, and other waste materials, and builds up during one laying year.

livability (lihv-eh-bihl-ih-tē)–The inherited stamina, strength, and ability to live and grow; an important character of all young animals, such as baby chicks, lambs, pigs, etc.

live delivery (līv deh-lihv-ər-ē)–A common guarantee by hatcheries of delivering live chicks by the addition of two or more chicks per order of one hundred.

live foal guarantee (lihv fōl gahr-ehn-tē)–A common guarantee by stallion owners that a live foal will result from the breeding service. In case of failure, the breeeding fee is not charged or, if already charged, is refunded.

live virus (līv vī-ruhs)–A virus whose ability to infect has not been altered.

live weight (līv wāt)–The gross weight of a live animal as compared with the dressed weight after slaughter.

liver (lih-vehr)–A glandular organ in people and other animals that secretes bile and performs certain metabolic functions. The liver of various animals and fowls is edible, and its extracts are used medicinally, especially for anemia.

liver chestnut (lih-vehr chehst-nuht)–A horse color; a dark shade of chestnut.

livestock (līv-stohck)–Farm animals raised to produce milk, meat, work, and wool; includes beef and dairy cattle, swine, sheep, horses, and goats; may also include poultry.

livetin (lī-veh-tihn)–A water-soluble protein found in egg yolk.

llama (lah-mah)–(1) *Auchenia llama;* a ruminant quadruped of South America, allied to the camel, but humpless, smaller (about 3 feet high at the shoulder) and with a long, wooly coat; used in the Andes as a beast of burden. (2) The wool of llamas or material made from it.

loading chute (lō-dihng shoot)–An inclined chute that is used for animals to walk from the ground into a truck or trailer.

loafing barn (lō-fihng bahrn)–A light type of building usually attached to the main dairy barn or milking parlor where cows can be turned loose after milking for exercise and comfort.

lobe (lōb)–A well defined portion of an organ.

lobectomy (lō-behck-tō-mē)–Surgical removal of a lobe (the lobe of a lung, liver, or any well defined area).

lobes of cerebrum (lōbz ohf sər-ē-bruhm)–Well defined portions of the cerebrum; the lobes of the cerebrum are occipital, frontal, temporal, and parietal.

local anesthesia (lō-kahl ahn-ehs-thē-sē-ah)–The absence of sensation after chemical injection to an adjacent area.

local effect (lō-kahl eh-fehckt)–An effect that a toxic substance causes at its original contact point with the body, e.g., eye damage.

localized (lō-kahl-īzd)–Within a well-defined area.

lock (lohck)–(1) A small tuft of cotton, wool, flax, fibers, etc. (2) Small bits of wool that are packed separately for market (Australia). (3) The single cavity in an ovary.

locus (lō-kuhs)–The position on or region of a chromosome where a gene is located.

lofty (lohf-tē)–Designating wool or a woolen fabric that has springiness when compressed, is bulky in comparison with its weight, is light in condition, and has an even, distinct crimp.

loggering (lohg-ehr-ihng)–A riding posture in which the rider holds the horn of the saddle rather than sitting free and erect in the saddle.

loin (loyn)–The lumbar region of the back, between the thorax and pelvis.

long feed (lohng fēd)–(1) Any coarse, unchopped feed for livestock, such as fodder, hay, straw. (2) Feeding for a long period of time as contrasted with a short feed period.

long pastern bone (lohng pahs-tərn bōn)–In ungulates, phalanx I.

long yearling (lohng yēr-lihng)–(1) An animal more than a year old. (2) A senior yearling of livestock-show class.

long-coupled (lohng cuhp-uhld)–Too much space between the last rib and the point of the hip of an animal.

longe (luhng)–The act of exercising a horse on the end of a long rope, usually in a circle.

longevity (lohn-jehv-ih-tē)–The length of life; a long duration of life.

loop of Henle (loop of Hehn-lē)–A U-shaped turn in the convoluted tubule of the kidney located between the proximal and distal convoluted tubules.

loose hay (loos hā)–Hay stored in the hay mow or stack without chopping, baling, or compressing.

loose housing (loos hows-ihng)–A management system for cattle wherein the adult animals are given unrestricted access to a feeding area, water, a resting area, and an adjoining open lot.

loose rein (loos rān)–A condition in riding or driving in which the reins on the harness of a horse are generally relaxed.

loose side (loos sīd)–The left side of a beef carcass. See tight side.

lope (lōp)–In a horse, a slow gallop.

lophodont (lō-fō-ō-dohnt)–Animals with teeth that have ridged occlusal surfaces, i.e., equine.

lordosis (lōr-dō-sihs)–The concave dorsal curvature of the spine that may be associated with estrus.

lot (loht)–Any particular grouping of animals, plants, seeds, fertilizers, etc., without particular regard to number, as a seed lot.

louse (lows)–A type of parasitic insect; plural is lice (līs).

lower GI (lō-ər gē-ī)–A contrast radiograph to visualize the structures of the lower gastrointestinal tract in which an enema is used to introduce contrast material into the colon; also called barium enema (bār-ē-uhm ehn-ah-mah).

lower respiratory tract (lō-ər rehs-pih-tōr-ē trahck)–The part of the respiratory system that consists of the bronchial tree and lungs.

lugger (luhg-ehr)–A horse that pulls at the bit.

lumbar (luhm-bahr)–The region of the back between the thorax and the pelvis; refers to the loins.

lumbar vertebrae (luhm-bahr vər-teh-brā)–The third set of vertebrae that are found in the loin area; abbreviated L in vertebral formula.

lumbosacral (luhm-bō-sā-krahl)–The connection between the lumbar and sacral vertebrae.

lumbosacral plexus (luhm-bō-sā-krahl plehck-suhs)–A network of intersecting nerves that arise from the last few lumbar and first one, two, or three sacral nerves (species dependent) that innervate the hind limb.

lumen (loo-mehn)–An opening within a vessel or organ.

luminous (loo-mih-nuhs)–Giving off a soft glowing light.

lumpectomy (luhmp-ehck-tō-mē)–A general term for the surgical removal of a mass.

lung (luhng)–The main organ of respiration.

lunker (luhn-kər)–(1) A very big, awkward, heavy-boned horse. (2) Any animal that is considerably larger than the average.

luster (luhs-tər)–Shine.

luteinizing hormone (loo-tē-ihn-ī-zihng hōr-mōn)–In animals, a hormone of the anterior pituitary gland that stimulates ovulation and development of the corpus luteum in females and secretion of testosterone by the interstitial cells in males.

luxation (luhck-sā-shuhn)–Dislocation or displacement of a bone from its joints; also called dislocation (dihs-lō-kā-shuhn).

lymph (lihmf)–A nearly colorless fluid made up of the liquid portion of the blood and the white corpuscles but free from the red corpuscles.

lymph nodes (lihmf nōdz)–Small bean-shaped structures that filter lymph and store B and T lymphocytes.

lymphadenectomy (lihm-fahd-ehn-ehck-tō-mē)–Surgical removal of lymph node.

lymphadenitis (lihm-fahd-eh-nī-tihs)–Inflammation of the lymph nodes; also called swollen glands.

lymphadenopathy (lihm-fahd-eh-nohp-ah-thē)–Disease of the lymph nodes; an example of incorrect use of "aden/o" because nodes are not glands.

lymphangiography (limh-fahn-jē-ohg-rah-fē)–A radiographic examination of the lymphatic vessels following injection of contrast material.

lymphangioma (lihm-fahn-jē-ō-mah)–An abnormal collection of lymphatic vessels forming a mass (usually benign).

lymphatic (lihm-fah-tihck)–Referring to the system of vessels returning lymph from the tissues to the bloodstream.

lymphatic vessels (lihm-fah-tihck vehs-ulz)–The valved tubes that carry lymph from tissues toward the thoracic cavity.

lymphocyte (lihm-fō-sīt)–A class of agranulocytic leukocyte that has phagocytic and antibody formation functions.

lymphocytosis (lihm-fō-sī-tō-sihs)–Increased numbers of lymphocytic leukocytes in the blood.

lymphoid (lihm-foyd)–Lymph or tissue of the lymphatic system.

lymphoma (lihm-fō-mah)–A general term for neoplasm composed of lymphoid tissue (usually malignant); also called lymphosarcoma; abbreviated LSA.

lymphopenia (lihm-fō-pē-nē-ah)–Decreased numbers of lymphocytic leukocytes in the blood.

lysin (lī-sihn)–An antibody that dissolves or disintegrates cells: bacteriolysin dissolves bacteria, hemolysin dissolves red blood cells.

lysis (lī-sihs)–The destruction or breakdown.

lysozyme (lī-sō-zīm)–An enzyme present in certain body secretions that can destroy certain kinds of bacteria.

M

macrocephaly (mahck-rō-sehf-ah-lē)–An abnormally large skull.

macrocytic (mahck-rō-siht-ihck)–A large cell.

macrophage (mahck-rō-fahj or mahck-rō-fāj)–A large cell that destroys by eating.

macroscopic (mahck-rō-skohp-ihck)–Visible to the human eye without the aid of a microscope.

macula (mahck-yoo-lah)–A centrally depressed, clearly defined yellow area in the center of the retina; also called the fovea (fō-vē-ah) or macula lutea (mahck-yoo-lah loo-tē-ah).

macular degeneration (mahck-yoo-lahr dē-jehn-ər-ā-shuhn)–A condition of central vision loss.

macule (mahck-yool)–A flat, discolored lesion less than 1 cm in diameter.

maggot (mā-goht)–Fly larva found especially in dead or decaying tissue.

magnet (māg-neht)–A charged metal device that is used to prevent hardware disease (traumatic reticuloperitonitis); it is given orally and placed in the reticulum.

magnetic resonance imaging (māg-neh-tihck reh-sohn-ahns ih-mah-gihng)–A procedure in which radio waves and a strong magnetic field pass through the patient and show the internal body structures in three-dimensional views.

magnum (mahg-nuhm)–(1) Great. (2) The part of the oviduct of a bird located between the infundibulum and the isthmus; the source of the albumin of an egg.

maiden (mā-dehn)–An unbred female.

maiden mare (mā-dehn mār)–An intact female horse that has never bred.

maintenance ration (mā-tehn-ahns rā-shuhn)–The amount of feed needed to support an animal when it is doing no work, yielding no product, and gaining no weight.

malabsorption (mahl-ahb-sōrp-shuhn)–The impaired uptake of nutrients from the intestines.

malaise (mahl-āz)–General bodily discomfort.

male (māl)–Designating an animal capable of producing spermatozoa, or male sex cells.

male process (māl proh-schs)–Genital eminence.

malformation (mahl-fōr-mā-shuhn)–Any unusual, abnormal growth, organ, or part of an animal.

malignant (mah-lihg-nahnt)–Tending to spread, becoming progressively worse, life-threatening.

malleolus (mah-lē-ō-luhs)–A bony protuberance on each side of the distal fibula or distal tibia.

malleus (mahl-ē-uhs)–The auditory ossicle known as the hammer.

malnutrition (mahl-nū-trihsh-uhn)–An unhealthy condition resulting from either poor feed or lack of feed.

malocclusion (mahl-ō-kloo-shuhn)–Abnormal contact between teeth.

malpresentation (mahl-prē-sehn-tā-shuhn)–An abnormal position of a fetus just prior to parturition.

mammary glands (mahm-mah-rē glahndz)–The milk-producing gland in females; also called udder, breast, or mammae.

mammary system (mahm-mah-rē sihs-stehm)–The udder, blood vessels, and teats of an animal, especially of the dairy cow.

mammary vein (mahm-mah-rē vān)–Any of the numerous, long, tortuous, prominent, large veins or blood vessels found just under the skin along the belly to the udder of a dairy cow.

mammitis (mahm-ī-tihs)–See mastitis.

mammoplasty (mahm-ō-plahs-tē)–Surgical repair of the mammary gland or breast.

mandible (mahn-dih-buhl)–The lower jaw bone, the only movable bone of the skull.

mane (mān)–The region of long course hair at the dorsal border of the neck and terminating at the poll.

mange (mānj)–A common term for a skin disease caused by mites.

manger (mān-jər)–A trough or other receptacle of metal, wood, concrete, or stone, in which fodder is placed for cattle to eat.

manipulation (mahn-ihp-yoo-lā-shuhn)–The attempted realignment of a fractured or dislocated bone; also known as reduction (rē-duhck-shuhn).

manubrium (mah-nū-brē-uhm)–The cranial end of the sternum.

manure (mahn-oo-ər)–The excreta of animals, dung and urine (usually with some bedding), used to fertilize land.

manure pit (mahn-oo-ər piht)–A storage unit in which accumulations of manure are collected before subsequent handling or treatment, or both, and ultimate disposal.

mare (mār)–An intact female horse four years old and over.

mare mule (mār mūl)–A female mule.

marking (mahrk-ihng)–Castrating, docking, branding, and ear marking.

marrow (mahr-ō)–A soft, yellow or red fatty tissue that fills the cavities of most bones.

martingale (mahr-tihn-gāl)–A leather strap attached to the girth of a horse's saddle or harness on one end, with the other divided and passing between the front legs and tied to the noseband of the bridle. It stops the horse from rearing.

mash (mahsh)–A ground feed of cereals and malt, etc., fed in a wet or dry form to livestock and poultry.

masked (mahskd)–Designating disease symptoms that are hidden.

mass selection (mahs seh-lehck-shuhn)–In animal breeding, the selection for breeding purposes of animals on the basis of their individual performances, type, or conformation.

massive (mahs-ihv)–Indicating a large-sized and deeply muscled animal; designates an ideal quality in a draft horse.

mast cell (mahst sehl)–The functional unit of tissue that responds to insult by producing and releasing histamine and heparin.

mastectomy (mahs-tehck-tō-mē)–Surgical removal of the mammary gland or breast.

masticate (mahs-tih-kāt)–To chew; to prepare food for swallowing and digestion.

mastication (mahs-tih-kā-shuhn)–Breaking down food into smaller parts and mixing it with saliva; also called chewing.

mastitis (mahs-tī-tihs)–An infectious or noninfectious inflammation of the udder. Also called garget, felon quarter, weed, bad quarter, mammitis.

mate (māt)–(1) To pair off two animals of opposite sexes for reproduction. Mating may be for a single season or for life. (2) In plants, to be cross pollinated.

maternal breeding value (mah-tər-nahl brē-dihng vahl-ū)–A prediction of how the daughters of a bull will milk based on weaning weight information. The accuracy figure is the amount of reliability that can be placed on the breeding value.

maternity pen (mah-tər-nih-tē pehn)–A special pen in a barn where animals about to bring forth their young may be isolated from the rest of the herd.

mating season (mā-tihng sē-szohn)–The season of the year when animals naturally breed and when conception is normally high. It varies with different species.

matroclinous (maht-rō-klī-nuhs)–Resembling the female parent. The opposite is patroclinous.

matron (mā-trohn)–A mare that has produced a foal.

matter (maht-ər)–(1) Purulent discharge; pus. (2) To discharge matter; to generate pus.

maturation (mahch-ū-rā-shuhn)–(1) Becoming mature or ripe. (2) Changes in cell division, especially in reproductive elements, in which the number of chromosomes in the nucleus of the new cells is half that of the original.

maturity (mah-chuhr-rih-tē)–(1) A state of full or complete growth development. (2) Animals old enough to reproduce. (3) A designation of the age of a beef animal at the time of slaughter.

maverick (mahv-ehr-ihck)–(1) Any unbranded animal, particularly a calf. Named after Samuel A. Maverick, a lawyer and a nonconformist Texan (1803–1870) who refused to brand his cattle. (2) A motherless calf (western United States).

maxilla (mahck-sih-lah)–The upper jaw bone of the skull.

mealy gray (mē-lē grā)–The rusty blue color of a horse's coat.

meat (mēt)–(1) The edible flesh of an animal. (2) The entire egg except the shell.

meatus (mē-ā-tuhs)–An opening.

meaty (mē-tē)–(1) Designating a well fleshed animal. (2) Any cut of the meat that shows a high percentage of good muscle structure.

mecate (meh-kāt)–A special kind of riding horse reins often made of horse hair (western United States).

mechanical buffer (meh-kahn-ih-kahl buhf-ər)–A machine used in the final operation of plucking feathers from poultry slaughtered for market. Also called buffer.

mechanical manipulation (meh-kahn-ih-kahl mahn-ihp-yoo-lā-shuhn)–Collecting semen for artificial breeding purposes by massaging the male's genital organs or by using the hand to produce ejaculation.

mechanical poultry picking (meh-kahn-ih-kahl pōl-trē pihck-ihng)–The removal of feathers in dressing poultry for the market by the use of machines in contrast to removal by hand labor. See mechanical buffer.

meconium (meh-kō-nē-uhm)–The first stools of a newborn that consist of material collected in the intestine of the fetus.

media (mē-dē-ah)–The middle coat of an artery. (2) The plural of medium. (3) Material for artificial propagation of various microorganisms.

medial (mē-dē-ahl)–Toward or nearer the midline.

medial canthus (mē-dē-ahl kahn-thuhs)–The corner of the eye nearest the nose; also called inner canthus.

median lethal dose (mē-dē-ahn lē-thahl dōs) (LD_{50})–The amount of concentration of a toxic substance that will result in the death of 50 percent of a group of test (target) organisms upon exposure (by ingestion, application, injection, or in their surrounding environment) for a specified period of time. (Complementary to median tolerance limit, TL_{50}).

median plane (mē-dē-ahn plān)–An imaginary line dividing the body into equal right and left halves; also called midsagittal plane (mihd-sahdj-ih-tahl).

median tolerance limit (mē-dē-ahn tohl-ər-ahns lihm-iht) (TL_{50})–The concentration of some toxic substance at which just 50 percent of the test (target) animals are able to survive for a specified period of exposures. (Complementary to median lethal dose, LD_{50}).

mediastinum (mē-dē-ahs-tī-nuhm)–The space in the thorax between the lungs; this space contains the heart, aorta, esophagus, trachea, bronchial tubes, and thymus.

medication (mehd-ih-kā-shuhn)–The application of medicines, salves, etc., to an injured or sick animal.

medium (mē-dē-uhm)–Any of a number of natural or artificial substances, pastelike or liquid, in or on which microorganisms, such as bacteria and fungi, can be cultured.

medium-spectrum antibiotic (mē-dē-uhm spehck-truhm ahn-tih-bī-oh-tihck)–An antibiotic that attacks a limited number of gram-positive and gram-negative bacteria.

medulla (meh-doo-lah)–The inner layer or region.

medulla oblongata (meh-dool-ah ohb-lohng-goh-tah)–The part of the myelencephalon or brainstem that controls basic life functions.

medullary cavity (mehd-yoo-lahr-rē kahv-ih-tē)–The innermost portion of long bones located in the shaft.

megacolon (mehg-ah-kō-lihn)–An abnormally large colon.

megaesophagus (mehg-ah-ē-sohf-ah-guhs)–An abnormally large esophagus.

megakaryocyte (mehg-ah-kahr-ē-ō-sīt)–A large nucleated cell found in the bone marrow from which platelets are formed.

megaspore (mehg-ah-spōr)–A spore that has the property of giving rise to a gametophyte (embryo sac) bearing a female gamete. One of the four cells produced by two meiotic divisions of the megaspore-mother-cell (megasporocyte).

megaspore-mother-cell (mehg-ah-spōr moh-thehr sehl)–The cell that undergoes two meiotic divisions to produce four megaspores.

meibomian glands (mī-bō-mē-ahn glahndz)–The sebaceous organ on the margins of each eyelid; also called tarsal (tahr-sahl) glands.

meiosis (mī-ō-sihs)–The cell division of sex cells in which the cell receives half the chromosomes from each parent.

melanin (mehl-ah-nihn)–The black or dark brown pigment found in the skin and hair cells.

melanocyte (mehl-ah-nō-sīt)–A cell that contains black pigment.

melanoma (mehl-ah-nō-mah)–A neoplasm composed of melanin-pigmented cells.

melena (mehl-lē-nah)–Black feces containing digested blood.

membrane (mehm-brān)–A thin layer of tissue that covers a surface, lines a cavity, or divides a space or organ.

menace response (mehn-ahs rē-spohnz)–A diagnostic test to detect vision in which movement is made toward the animal to test if it will try to close its eyelids.

menadione (mehn-ah-dī-ōn)–Vitamin K_3; an antihemorrhagic vitamin that is essential for all animals and people to control bleeding.

meninges (meh-nihn-jēz)–The three layers of connective tissue membranes that surround the brain and spinal cord.

meningioma (meh-nihn-jē-ō-mah)–A benign tumor of the meninges (usually originates from the arachnoid).

meningitis (mehn-ihn-jī-tihs)–Inflammation of the meninges.

meningocele (meh-nihng-gō-sēl)–A protrusion of the meninges through a defect in the skull or vertebrae.

meningoencephalitis (meh-nihn-gō-ehn-sehf-ah-lī-tihs)–Inflammation of the meninges and brain.

meningoencephalomyelitis (meh-nihn-gō-ehn-cehf-ah-lō-mī-eh-lī-tihs)–Inflammation of the meninges and spinal cord.

meniscus (meh-nihs-kuhs)–The curved fibrous cartilage found in some synovial joints.

mesencephalon (mēz-ehn-sehf-ah-lohn)–The portion of the brain that is the midbrain.

mesentery (mehs-ehn-tehr-ē) or (mehz-ehn-tehr-ē)–The layer of the peritoneum that suspends parts of the intestine within the abdominal cavity.

mesial contact surface (mē-zē-ahl cohn-tahckt sər-fihs)–The contact surface of a tooth that is closest to midline.

mesoderm (meh-sō-dərm)–The middle layer of an embryo.

mesophiles (mehs-ō-fīlz)–Parasites and often pathogenic bacteria that grow best at a body temperature of 98.6°F (37°C).

mesothelium (mēs-ō-thē-lē-uhm)–A cellular covering that forms the lining of serous membranes like the pleura and peritoneum.

metabolic (meht-ah-bohl-ihck)–Designating the chemical changes that take place in living plant and animal cells whereby one compound is converted to one or more other compounds.

metabolism (meh-tahb-ō-lihzm)–The processes involved in the body's use of nutrients.

metabolizable energy (meh-tahb-ō-līz-ah-bl ehn-ər-jē)–The total amount of energy in feed less the losses in feces, combustible gases, and urine. Also called available energy.

metacarpals (meht-ah-kahr-pahlz)–The bones distal to the carpus and proximal to the phalanges (number of bones varies with species).

metamorphosis (meht-ah-mohr-fō-sihs)–A process by which an organism changes in form and structure in the course of its development, as many insects do.

metanephric (meht-ah-nehf-rihck)–An embryonic-like kidney.

metaphase (meht-ah-fāz)–That stage of cell division in which the chromosomes are arranged in an equatorial plate or plane. It precedes the anaphase stage.

metaphysis (meht-tahf-ih-sihs)–The wider portion of the long bone shaft that is found between the epiphysis and diaphysis; "to grow beyond."

metaplasia (meht-ah-plā-zē-ah)–The abnormal transformation of an adult

(mature), full differentiated tissue of one kind into differentiated tissue of another kind.

metastasis (meh-tahs-tah-sihs)–Pathogenic growth distant from the primary disease site; "beyond control"; plural is matastases (meh-tahs-tah-sēz).

metastasize (meh-tahs-tah-sīz)–An invasion by the pathogenic growth to a point distant from the primary disease.

metatarsals (meht-ah-tahr-sahlz)–The bones distal to the tarsus and proximal to the phalanges (number of bones varies with species).

metencephalon (meht-ehn-sehf-ah-lohn)–The portion of the brain that is the cerebellum and pons.

meter (mē-tər)–A metric unit of length, equal to 1.09 yards.

metestrus (meht-ehs-truhs)–The phase of the estrous cycle of nonprimates following estrus and characterized by the development of the corpus luteum and the preparation of the uterus for pregnancy.

methane (mehth-ān)–CH_4; an ordorless, colorless, and asphyxiating gas that can explode under certain circumstances; can be produced by manures or solid waste undergoing anaerobic decomposition, as in the rumen and in silos.

methemoglobin (meht-hē-mō-glō-bihn)–A rust-colored substance, formed when oxygen unites with an oxidized product of hemoglobin. Methemoglobin results from blood decomposition or from poisoning of blood by nitrates, nitrites, and other substances.

methemoglobinemia (meht-hē-mō-glō-bihn-ē-mē-ah)–(1) A lack of oxygen in the blood due to oxidation of iron from the ferrous to the ferric state. (2) The presence of methemoglobin in the blood resulting in cyanosis. This can be induced by excessive nitrates, nitrites, and certain drugs, or to a defect on the enzyme NADH.

methionine (meh-thī-ō-nihn)–A sulfur-containing essential amino acid; indispensable for animals.

methyl red (mehth-ihl rehd)–An acid/base indicator; a dye stain used in determining the classification or taxonomy of certain bacteria such as *Coliform bacilli*.

methylene blue reduction test (mehth-eh-lēn bloo rē-duhck-shuhn tehst)–A test for the bacteriological quality of milk that consists of placing methylene blue thiocyanate in the milk under controlled conditions. The more rapidly the methylene blue changes color, the greater the number of bacteria in the milk.

metritis (meh-trī-tihs)–Inflammation of the uterus.

microbes (mī-krōbz)–Minute plant or animal life. Some microbes cause disease, but others are beneficial.

microbiologist (mī-krō-bī-ohl-ō-jihst)–A scientist concerned with the study of plant and animal microorganisms.

microcephaly (mī-krō-sehf-ah-lē)–An abnormally small skull.

microcytic (mī-krō-siht-ihck)–A small cell.

microglia (mī-krō-glē-ah)–Phagocytic cells of the nervous system.

microorganism (mī-krō-ōr-gahn-ihzm)–A living organism of microscopic dimensions.

microphthalmia (mī-krohf-thahl-mē-ah)–Abnormally small eyes.

microphthalmos (mī-kroph-thahl-mōs)–Abnormally small eye(s).

micropyle (mī-krō-pīl)–The minute necklike opening in the integuments of an ovule, where the sperms enter.

microspore (mī-krō-spōr)–One of the four cells produced by the two meiotic divisions (mitoses) of the microspore mother-cell (microsporocyte). It gives rise to a gametophyte bearing only male gametes. Also called pollen grain.

micturition (mihck-too-rihsh-uhn)–Excreting urine (implies neurologic control of urination).

midbrain (mihd-brān)–The part of the mesencephalon or brainstem that extends from the cerebrum to the pons.

middle milk (mihd-uhl mihlk)–The milk obtained from a cow during the middle of the milking period, as opposed to that obtained in the beginning and at the end. It has a higher fat content and a lower bacteriological count than the other two.

mild (mīld)–Designating a gentle animal.

milk (mihlck)–(1) The natural, whitish or cream-colored liquid discharged by the mammary glands of mammals. (2) To draw milk from the udder of a cow.

milk cistern (mihlck sihs-tərn)–Part of a cow's udder; the cavity holding about one-half a pint that is drained by the teat ducts.

milk duct (mihlck duhckt)–The cavity in the teat that leads into the milk cistern. Also called teat cistern.

milk-fat basis (mihlck faht bā-sihs)–A method of paying for milk at receiving stations on the basis of price per pound of butterfat. See butterfat.

milk-fed (mihlck fehd)–Designating animals fed largely on dairy products.

milk fever (mihlck fē-vər)–See parturient paresis.

milk flow (mihlck flō)–The amount of milk produced by a cow or by the entire herd in a day, week, season, etc.

milk replacer (mihlck rē-plās-ər)–A powder that, when mixed with water, is fed to young animals as the milk portion of their diet.

milk sample (mihlck sahmp-uhl)–A small amount of milk, normally collected at a farm each time the milk is collected; used for making certain tests at the dairy plant to determine butterfat content, bacterial count, etc.

milk solids (mihlck sohl-ihdz)–The portion of milk that is left after water is removed; includes protein and fat.

milk stone (mihlck stōn)–In dairy manufacture, a grayish-white, thin, chalky deposit that sometimes accumulates on heating surfaces, coolers, utensils, and freezers when improper washing methods have been used.

milk substitute (mihlck suhb-stih-toot)–In animal or dairy husbandry, any of a number of gruel feeds or mixtures fed to calves or pigs as a substitute for milk.

milk teeth (mihlck tēth)–The temporary deciduous teeth of a young animal that are much whiter and smaller than the permanent teeth.

milk vein (mihlck vān)–The subcutaneous, abdominal veins that are a continuation of the mammary veins of the udder of a cow. The largeness and tortuousness of the veins is used in judging dairy cattle.

milk well (mihlck wehl)–The depression in the cow's ventral underline where the milk veins enter the body.

milk yield (mihlck yē-uhld)–The amount of milk produced in a given period.

milking (mihlck-ihng)–(1) The quantity or quality of milk removed from an animal's udder at one time. (2) The act of removing milk from an animal's mammary gland. (3) Pertaining to milking, as in milking machine.

milking machine (mihlck-ihng mah-shēn)–A mechanical device replacing hand labor in the milking of cows. The essential parts are teat cups, a vacuum pump, and a milker pail or milk line. Milk is drawn from the udder by application of alternate vacuum and atmospheric pressure.

milking parlor (mihlck-ihng pahr-lōr)–In a dairy, a specially arranged and equipped room where cows are separately fed concentrates and milked by mechanical milking equipment.

milt (mihlt)–A ductless gland near the stomach and intestine of fowls. It contributes to the formation of new blood cells. See spleen.

mineral (mihn-ehr-ahl)–(1) A chemical compound or elements of inorganic origin. (2) Designating the inorganic nature of a substance.

mineral feeder (mihn-ehr-ahl fē-dər)–Any of a number of box or bin devices that automatically supplies a mineral or mineral ration in the feeding of livestock.

mineral mixture (mihn-ər-ahl mihcks-chuhr)–Any feed containing salt, limestone, phosphates, minor elements, etc.

mineralized salt (mihn-ər-eh-līzd sahlt)–A salt compound for animal feeding that is composed of a common salt base with the trace mineral elements, such as manganese, copper, iron, iodine, and zinc, added to it.

miniature horse (mihn-ah-chuhr hōrs)–A breed of horse that is small (usually less than 8.2 hands).

miosis (mī-ō-sihs)–Pupillary constriction.

miotic agent (mī-oh-tihck ā-jehnt)–A substance used to constrict pupils.

mites (mītz)–Arachnids except ticks; very small arthropods belonging to the order Acarina.

miticide (mīt-tih-sīd)–Any poisonous substance used to kill mites.

mitosis (mī-tō-sihs)–Cell division involving the formation of chromosomes, spindle fibers, and the division of chromosomes by a process of longitudinal splitting. Each of the resulting daughter cells thus has a full set of chromosomes as distinguished from reduction division or meiosis, in which the daughter cells have half the somatic number. See meiosis.

mitral valve (mī-trahl vahlv)–The membranous fold located between the left atrium and left ventricle; also termed the bicuspid valve because it is shaped with two points; also called the left atrioventricular valve (ā-trē-ō-vehn-trihck-yoo-lahr).

mitral valve prolapse (mī-trahl vahlv prō-lahps)–An abnormal protrusion of the left atrioventricular valve that results in incomplete closure of the valve.

mix (mihcks)–(1) A formula; a combination of two or more ingredients, blended together for a specific purpose, such as feedstuff for livestock. (2) To cross or interbreed animals, plant varieties, etc., by chance.

mixed breeds (mihcksd brēdz)–Farm animals that result from crossing different breeds of livestock.

mixed calves (mihcksd kahlvz)–A market class for a group of slaughter calves that are not uniform in grade.

mixing pens (mihcks-ihng pehnz)–In range management, a large pen in which a number of animals are confined. In handling goats, the does and kids are placed together until such time as the kids are able to follow the mothers in grazing.

modified live virus (mohd-ih-fīd līv vī-ruhs)–A virus that has been changed by passage through an unnatural host, such as passing hog cholera virus through rabbits, so that it no longer possesses the disease-producing characteristics but so that it will stimulate antibody production and immunity when injected into susceptible animals.

modifier (mohd-ih-fī-ər)–(1) Any element that is added to, or taken from a substance that alters its normal appearance or function. (2) A substance that can alter the course of carcinogenesis.

modifying gene (mohd-ih-fī-ihng jēn)–A gene that changes the expression of the chief gene, or genes, controlling a character.

molar (mō-lahr)–The most caudally located cheek teeth in the dental arcade; abbreviated M in dental formula.

mold (mōld)–Fungi distinguished by the formation of a mycelium (a network of filaments or threads), or by spore masses; usually saprophytes.

molecule (mohl-eh-kyool)–The smallest part of a substance that can exist separately and still retain its chemical properties and characteristic composition; the smallest combination of atoms that will form a given chemical compound.

molt (mōlt)–The casting off of old feathers, skin, horns, etc., before a new growth, as with the normal annual renewal of plume in adult chickens, turkeys, etc. Also called the castoff covering (British); moult.

molting (mōlt-ihng)–An interruption of egg laying, either natural or forced.

monestrous (mohn-ehs-truhs)–Having one estrous or "heat" cycle per season.

mongrel (mohn-grehl)–A mixed breed of any animal.

monkey mouth (mohn-kē mowth)–A condition in animals in which the lower teeth protrude over and beyond the upper teeth.

monochromatism (mohn-ō-krō-mah-tihzm)–The lack of ability to distinguish colors; also called color blindness.

monocyte (mohn-ō-sīt)–A class of agran-ulocytic leukocyte that has phagocytic function.

monocytopenia (moh-nō-sīt-ō-pē-nē-ah)–A decrease in the number of monocytic leukocytes in the blood.

monocytosis (moh-nō-sī-tō-sihs)–Increased numbers of monocytic leukocytes in the blood.

monofilament (mohn-ō-fihl-ah-mehnt)–A single strand of material.

monogamous (moh-noh-goh-muhs)–Pairing with one mate.

monogastric (mohn-ō-gahs-trihck)–An animal with one stomach compartment.

monohybrid (mohn-ō-hī-brihd)–A hybrid whose parents differ in a single character.

monohybrid cross (mohn-ō-hī-brihd krohs)–A cross between two individuals that are heterozygous for one pair of genes; an example is Aa ∞ Aa.

monophagia (mohn-ō-fā-jē-ah)–(1) Eating only one kind of food. (2) Eating only one meal a day.

monoplegia (mohn-ō-plē-jē-ah)–Paralysis of one limb.

monorchid (mohn-ōr-kihd)–An animal with only one testicle descended into the scrotum; also called unilaterally cryptorchid (krihp-tōr-kihd).

monovalent (mohn-ō-vā-lehnt)–A vaccine, antiserum, or antitoxin developed specifically for a single antigen or organism.

morbid (mōr-bihd)–Afflicted with disease.

morbidity (mōr-bihd-ih-tē)–The ratio of diseased animals to well animals in a population.

moribund (mōr-ih-buhnd)–Near death.

morphogenesis (mohr-fō-jehn-eh-sihs)–The developmental history of organisms or of their parts.

morphology (mōr-foh-lō-jē)–A branch of biologic science that deals with the forms, rather than the functions, of plants and animals.

Morrison standard (mohr-ih-sohn stahnd-ahrd)–A feeling standard formulated by F. B. Morrison of the University of Wisconsin, author of *Feeds and Feeding,* which is adapted to practical feeding conditions and is used extensively.

mortality (mōr-tahl-ih-tē)–The ratio of diseased animals to diseased animals that die.

mossy (mohs-ē)–Designating irregular, dark markings that spoil an otherwise desirable color contrast on the feathers of domestic birds.

motile (mō-tihl)–Capable of active motion.

motility (mō-tihl-ih-tē)–Active movement in artificial insemination of the sperm in a male's semen.

motor neurons (mō-tər nū-rohnz)–Nervous system cells that send impulses to muscles or glands.

mottle (moht-l)–A spot or blotch of a color different from the mass color of a surface.

mottled (moht-ld)–(1) Designating bird feathers marked with white tips at the ends or spotted with colors or shades at variance with the ideal. (2) Irregularly marked with spots of different colors.

mount (mownt)–(1) A horse for riding. (2) A riding seat. (3) To copulate, as a male animal. (4) To get into a saddle on the back of a riding horse. (5) The preparatory step to animal mating that involves one animal climbing on top of another animal or object (used as an indicator of heat).

mouse dun (mows duhn)–A coat color of some animals, especially horses. It is a dun color imposed on black, seal brown, dark mahogany bay, or dark liver chestnut giving a smoky effect. Sometimes called smoky.

mouth speculum (mowth spehck-yoo-luhm)–An instrument used by veterinarians for holding an animal's mouth open to facilitate inspection of the mouth, dental work, or the administering of medicines. Also called jaw spreader.

mouthing (mow-thihng)–Determining the approximate age of a horse by examining the teeth.

mouton (moo-tohn)–A modern fur made by the chemical treatment and processing of sheep pelts carrying designated lengths and qualities of wool on the skin.

mucoid (myoo-koyd)–Resembling mucus.

mucolytic (myoo-kō-liht-ihck)–Breaking down mucus.

mucopurulent (myoo-kō-pər-ū-lehnt)–Containing mucus and pus.

mucor (mū-kōr)–A widely distributed genus of molds of the family Mucoraceae, mainly saphrophytic species, abundant in soil, decaying vegetable matter, dung, etc.

mucosa (myoo-kōs-ah)–See mucous membrane.

mucous membrane (myoo-kuhs mehm-brān)–A specialized form of epithelial tissue that secretes mucus.

mucus (myoo-kuhs)–A slimelike substance that is composed of glandular secretion, salts, cells, and leukocytes.

muffs (muhfz)–The whiskerlike crop of feathers on either side of the face below the eyes of certain fowls.

mugging (muhg-ihng)–Bulldogging or throwing a calf.

mule (mūl)–The hybrid offspring of a jackass and a mare; used as a draft animal. Mules are sterile.

mule jack (mūl jahck)–A jack bred to a mare in contrast to one used for jack stock perpetuation. See Jack.

muley (mūl-lē)–A cow without horns; generally, any polled or hornless beef or dairy cattle; also spelled mulley.

multifilament (muhl-tī-fihl-ah-mehnt)–Several strands that are twisted together.

multiparous (muhl-tihp-ah-ruhs)–Given birth two or more times.

multiple alleles (muhl-tih-puhl ah-lēlz)–More than two different genes that can occupy the same locus on homologous chromosomes.

multiple farrowing (muhl-tih-puhl fār-ō-ihng)–Arranging the breeding program so that groups of sows farrow at regular intervals throughout the year.

multiplication (muh-tih-plih-kā-shuhn)–Reproducing.

mummification (muhm-ih-fih-kā-shuhn)–In animal reproduction, the drying up and shriveling of the unborn young.

murine (moo-rēn)–Mice and rats.

murmur (mər-mər)–An abnormal slushing auscultatory sound that represents abnormal blood flow.

muscle (muhs-uhl)–Connective tissue composed of fibers that contract to allow for movement.

mushy (muhsh-ē)–Wool that is dry and wasty in manufacturing.

musk glands (muhsk glahndz)–Four secretory organs that open lateroventrally near the carapace edge of turtles; in the male goat they are mainly situated just in front of the horns.

mustang (muhs-tahng)–(1) In the south-western United States, a wild descendant of the Spanish horse, generally smaller and inferior to the domestic Spanish horses. (2) A small domestic pony that is largely of Spanish breeding. (3) A range horse that is somewhat wilder than the average. Also called mestang. See feral.

mutable (myoo-tah-bl)–Changeable.

mutagenic (mū-tah-jehn-ihck)–Producing change or creating genetic abnormalities.

mutagen (myoo-tah-jehn)–A chemical, physical, and/or radioactive agent that interacts with DNA to cause a permanent, transmissable change in the genetic material of a cell. See teratogen.

mutant, mutation (myoo-tahnt, myoo-tā-shuhn)–A variant, differing genetically and often visibly from its parent or parents and arising rather suddenly or abruptly.

mutton (muh-tihn)–Adult sheep meat.

mutualism (myoo-tū-ahl-ihzm)–The dependency of two organisms upon each other. Among insects, an example is furnished by the cornfield ant and the corn-root aphid.

mutualistic (myoo-tū-ahl-ihs-tihck)–Designating a mutually beneficial relationship between organisms; symbiosis, e.g., *Rhizobium* bacteria and compatible legumes.

muzzle (muh-zuhl)–The two nostrils (including the skin and fascia) and the muscles of the upper and lower lip; also used to describe an appliance placed over the mouth of an animal to prevent biting.

myasthenia (mī-ahs-thē-nē-ah)–Muscle weakness.

mycology (mī-kohl-ō-jē)–The science dealing with fungi.

mycoplasma (mī-kō-plahz-mah)–An organism that is between a virus and a bacterium in size. It may possess characteristics of a virus and is not visible under a light microscope.

mycosis (mī-kō-sihs)–A disease caused by the growth of fungi in plants or animals.

mycotic disease (mī-kō-tihck dih-zēz)–Disease caused by a fungus.

mycotoxins (mī-kō-tohck-sihnz)–Chemical substances produced by fungi that may result in illness and death of animals and humans when food or feed containing them is eaten.

mydriasis (mih-drī-ah-sihs)–Pupillary dilation.

mydriatic agent (mihd-rē-ah-tihck ā-jehnt)–A substance used to dilate the pupils.

myectomy (mī-ehck-tō-mē)–Surgical removal of muscle or part of a muscle.

myelencephalon (mī-lehn-sehf-ah-lohn)–The portion of the brain that is the medulla oblongata.

myelin (mī-lihn)–A protective covering over some nerve cells; present in parts of the spinal cord, white matter of the brain, and most peripheral nerves; also called myelin sheath.

myelitis (mī-eh-lī-tihs)–Inflammation of the spinal cord or bone marrow.

myelogram (mī-eh-lō-grahm)–A record of the spinal cord after injection of contrast material into the subarachnoid space; also used to represent a differential white blood cell count.

myelography (mī-eh-lohg-rah-fē)–A diagnostic study of the spinal cord after injection of contrast material into the subarachnoid space.

myeloma (mī-eh-lō-mah)–A tumor composed of cells derived from hematopoietic tissues of the bone marrow.

myelopathy (mī-eh-loh-pahth-ē)–Disease of the spinal cord or bone marrow.

myiasis (mi-yah-sihs) Fly larvae parasitism of living tissue.

myocarditis (mī-ō-kahr-dī-tihs)–Inflammation of the myocardium.

myocardium (mī-ō-kahr-dē-uhm)–The middle and thickest layer of heart muscle.

myoclonus (mī-ō-klō-nuhs)–Repetitive, rhythmic contraction of skeletal muscles (usually the result of distemper viral infection or encephalitis in dogs); also called chorea (kōr-ē-ah).

myometrium (mī-ō-mē-trē-uhm)–The muscular layer of the uterus.

myoparesis (mī-ō-pahr-ē-sihs)–Muscle weakness.

myopathy (mī-ohp-ah-thē)–An abnormal condition or disease of muscle.

myoplasty (mī-ō-plahs-tē)–Surgical removal of muscle.

myosarcoma (mī-ō-sahr-kō-mah)–A malignant neoplasm composed of muscle.

myositis (mī-ō-sī-tihs)–Inflammation of muscles.

myotomy (mī-oht-ō-mē)–A surgical incision into or dividing a muscle.

myotonia (mī-ō-tō-nē-ah)–Delayed relaxation of a muscle after contraction.

myringectomy (mihr-ihn-jehck-tō-mē)–Surgical removal of all or part of the eardrum; also called tympanectomy (tihm-pahn-ehck-tō-mē).

myringitis (mihr-ihn-jī-tihs)–Inflammation of the eardrum.

myxoma (mihcks-ō-mah)–A tumor of connective tissue.

N

nag (nahg)–A horse or pony of nondescript breed.

nanny (nahn-ē)–An incorrect term for a female goat.

nape (nāp)–The back of the neck of an animal.

narcolepsy (nahr-kō-lehp-sē)–A syndrome of recurrent uncontrollable sleep episodes.

naris (nār-ihz)–An opening of the nasal cavity; nostril; plural is nares (nār-ēz).

narrow-spectrum antibiotic (nār-ō spehck-truhm ahn-tih-bī-oh-tihck)–An antibiotic whose activity is restricted to either gram-negative or gram-positive bacteria.

nasal septum (nā-zahl sehp-tuhm)–A wall of cartilage that divides the nasal cavity longitudinally.

nasogastric (nā-zō-gahs-trihck)–The nose and stomach.

nasogastric intubation (nā-zō-gahs-trihck ihn-too-bā-shuhn)–The placement of a tube through the nose into the stomach.

nasolacrimal duct (nā-zō-lahck-rih-mahl duhckt)–A passageway that drains tears into the nose.

nasopharynx (nā-zō-fār-ihnckz)–The portion of throat posterior to the nasal cavity and above the soft palate.

native (nā-tihv)–(1) Designating animals, such as cattle, hogs, and horses, that, though originally introduced into a region, have lost some of their original characteristics or have gone wild; a scrub or mongrel. (2) Designating an unbranded beef hide. See feral.

native disease (nā-tihv dih-zēz)–A disease caused by an indigenous organism.

natural cross (naht-uhr-ahl krohs)–Interbreeding or hybridizing that takes place in nature without assistance from people.

natural enemy (naht-uhr-ahl ehn-eh-mē)–In nature, any organism that preys or feeds upon another.

natural immunity (naht-uhr-ahl ihm-yoo-nih-tē)–Immunity to disease, infestation, etc., that results from qualities inherent in plants, animals, or people.

natural selection (naht-uhr-ahl seh-lehck-shuhn)–A natural process by which less-vigorous plants and animals tend to be eliminated from a population without leaving enough descendants to perpetuate their traits.

natural selection theory (naht-uhr-ahl seh-lehck-shuhn thēr-ē)–The theory of evolution, propounded by Charles Darwin in the nineteenth century, that postulates that the distinctive characteristics of fitness can be inherited. Also called the survival of the fittest theory.

nausea (naw-sē-ah)–Stomach upset or a sensation of the urge to vomit; difficult to use descriptively in animals.

navel (nā-vehl)–In *Mammalia* the point of connection between the umbilical cord and the fetus.

navicular bone (nah-vihck-yoo-lahr bōn)–A small oblong bone in the lower part of a horse's leg between the second and third phalanx; the sesamoid of the third phalanx.

near side (nēr sīd)–The left side of horse; the side from which to mount in riding. See off side.

nebulization (nehb-yoo-lih-zā-shuhn)–The process of making a fine mist; a method of drug administration.

neck (nehck)–(1) In animals, the connecting link between the head and the body. (2) To tie cattle neck-to-neck (western United States).

neck rein (nehck rān)–To turn a saddle horse in a desired direction by laying the reins on the neck rather than pulling on the bit.

neck rope (nehck rōp)–(1) A rope tied around the neck of a horse being trained for use in lassoing cattle. (2) A rope used in picketing horses.

necropsy (nehck-rohp-sē)–An examination of the internal organs of a dead body to determine the cause of death. Also called autopsy, post-mortem.

necrosis (neh-krō-sihs)–An abnormal condition of dead tissue.

necrotic (nch-kroh-tihck)–Dead tissue.

needle biopsy (nē-dahl bī-ohp-sē)–The insertion of a sharp instrument (needle) into a tissue for extraction of tissue to be examined.

needle teeth (nē-duhl tēth)–The eight sharp teeth present in newborn piglets. Especially in large litters, these teeth can cause injury to other piglets or the sow's udder and should be clipped.

neonate (nē-ō-nāt)–A newborn usually under four weeks old (species dependent).

neonatology (nē-ō-nāt-ohl-ō-jē)–The study of the newborn.

neoplasm (nē-ō-plahz-uhm)–Any abnormal new growth of tissue in which the multiplication of cells is uncontrolled, more rapid than normal, and progressive.

nephrectomy (neh-frehck-tō-mē)–Surgical removal of a kidney.

nephritis (neh-frī-tihs)–Inflammation of the kidney.

nephrolith (nehf-rō-lihth)–A kidney stone or renal calculus.

nephrolithiasis (nehf-rō-lih-thī-ah-sihs)–A disorder characterized by the presence of kidney stones.

nephromalacia (nehf-rō-mah-lā-shē-ah)–Abnormal softening of the kidney.

nephron (nehf-rohn)–The functional unit of the kidney.

nephropathy (neh-frohp-ah-thē)–Disease of the kidney(s).

nephropexy (nehf-rō-pehcks-sē)–Surgical fixation of a kidney to the abdominal wall.

nephrosclerosis (nehf-rō-sckleh-rō-sihs)–An abnormal hardening of the kidney.

nephrosis (neh-frō-sihs)–An abnormal condition of the kidney(s).

nephrotoxin (nehf-rō-tohcks-ihn)–A poison having destructive effects of the kidney(s).

nerve (nərv)–One or more bundles of impulse-carrying fibers that connect the central nervous system to the other parts of the body.

nerve trunk (nərv truhnck)–A bundle of nerve fibers that run together.

net-energy value (neht ehn-ehr-jē vahl-ū)–The amount of energy that remains after deducting from a feed's total energy value the amount of energy lost in feces, urine, combustible gases, and heat increment. Sometimes called work of digestion.

neuralgia (nū-rahl-jē-ah)–Nerve pain.

neurectomy (nū-rehck-tō-mē)–Surgical removal of a nerve.

neuritis (nū-rī-tihs)–Inflammation of the nerve(s).

neuroanastomosis (nū-rō-ahn-ahs-tō-mō-sihs)–Connecting nerves together.

neuroblastoma (nū-rō-blahs-tō-mah)–A malignant neoplasm composed of nervous tissue origin.

neuroglia (nū-rohg-lē-ah)–The supportive cells of the nervous system; also called glial cells; consist of astrocytes, microglia, oligodendroglia, and Schwann cells.

neurohypophysis (nū-rō-hī-pohf-ih-sihs)–The posterior pituitary gland; the portion of the pituitary gland with neurologic function.

neurolemma (nū-rō-lehm-ah)–A tubelike membrane that covers nerve fibers, which may be myelinated or unmyelinated; also spelled neurilemma (nū-rih-lehm-ah).

neurologist (nū-rohl-ō-jihst)–A specialist who studies the nervous system.

neurology (nū-rohl-ō-jē)–The study of nerves.

neuromuscular (nū-rō-muhs-kū-lahr)–The area between the nerve endings and muscles or the area of the nerves and muscles.

neuron (nū-rohn)–The basic unit of the nervous system.

neuroplasty (nū-rō-plahs-tē)–Surgical repair of a nerve.

neurorrhaphy (nū-rōr-ah-fē)–Suturing the ends of a severed nerve.

neurotomy (nū-roht-ō-mē)–Surgical incision or dissection of a nerve.

neurotoxicity (nū-rō-tohck-sihs-ih-tē)–The state or condition of being poisonous to the brain and nerves of the body.

neurotransmitter (nū-rō-trahnz-miht-ər)–A chemical substance that allows an impulse to move across the synapse from one neuron to another.

neuter (nū-tər)–To sexually alter an animal; usually implies male sexual alteration.

neutropenia (nū-trō-pē-nē-ah)–A decreased number of neutrophilic leukocytes in the blood.

neutrophil (nū-trō-fihl)–A class of granulocytic leukocyte that has phagocytic function; also called polymorphonuclear leukocyte (poh-lē-mōr-fō-nū-klē-ahr) or PMN.

niacin (nī-ah-sihn)–A vitamin of the B-complex group. Also called nicotinic acid, antipellagra vitamin.

niche (nihch)–A term used to describe the status of an animal in its community, that is, its biotic, trophic, and abiotic relationships; all the components of the environment with which an organism or population interacts.

nick (nicking) (nihck)–A term used by livestock producers when the offspring is better than its parents.

nicked (nihckd)–(1) Descriptive of a horse's tail in which some of the muscles have been severed so that the tail is carried upward. (2) Designating a mating of animals that results in an offspring superior to either of the parents.

nicker (nihck-ər)–To neigh softly (as a horse).

nictitating membrane (nihck-tih-tāt-ihng mehm-brān)–The conjunctival fold attached at the medial canthus that moves across the cornea when the eyelids close; also called the third eyelid or nictitans (nihck-tih-tahnz) or haws (hawz).

night milk (nīt mihlck)–The milk obtained at the evening or night milking.

night pasture (nīt pahs-chər)–A pasture in which domestic animals graze during the night, usually close by the barn or dwelling.

night stool (nīt stool)–A rabbit's nocturnal feces that is looser than normal and contains vitamins and nutrients that the rabbit consumes.

nippers (nihp-pərz)–(1) The two central incisor teeth of a horse. (2) Small pincers used for holding, breaking, or cutting.

nipple (nihp-l)–The protuberance of the udder of a mammal that contains the mammary gland. See teat.

nipple pail (nihp-uhl pāl)–A pail with a tube and rubber nipple fastened at the bottom, used in calf feeding, weaning, etc.

nipple waterer (nihp-uhl wah-tər-ər)–An automatic watering system in which the animal pushes a nipple in the end of a pipe to get water.

nit (niht)–The egg of a louse.

nitrogen (nī-trō-jehn)–N; a gas that occurs naturally in the air and soil, where it is converted into usable forms for plant use by bacteria and other natural processes.

nitrogen-free extract (nī-trō-jehn frē ehcks-trahckt)–The portion of a feed made up primarily of starches and sugars; nitrogen-free extract is determined by subtracting the ether extract, crude fiber, crude protein, ash, and water from the total weight of the feed sample.

niveus (nihv-eh-uhs)–Snowy white.

nociceptive (nō-sē-sehp-tihv)–Causing pain; also spelled nocioceptive (nō-sē-ō-sehp-tihv).

nociceptor (nō-sē-sehp-tər)–A receptor stimulated by injury or pain.

nocturia (nohck-too-rē-ah)–Excessive urination at night.

nodder (nohd-ər)–A horse that character-istically nods its head when walking, such as the Tennessee walking horse.

nodes of Ranvier (nōdz of rohn-vē-ā)–Areas along a myelinated nerve where ionic exchange takes place.

nodule (nohd-ūl)–A small knot, lump, or roundish mass of abnormal tissue.

noil (noyl)–The short fibers that are removed from the staple wool in combing. Noil is satisfactory for the manufacture of felts and woolens.

nonadditive genes (nohn-ahd-ih-tihv jēnz)–Genes that express themselves in a dominant or epistatic fashion.

noncontagious (nohn-kohn-tā-juhs)–A disease that cannot be spread to another animal by contact or contact with an infected object.

nonessential amino acid (nohn-eh-sehn-shahl ah-mē-nō ah-sihd)–Amino acids that can be synthesized by the animal's body.

noninfectious disease (nohn-ihn-fehck-shuhs dih-zēz)–A disorder not caused by organisms (examples include genetic, traumatic, or iatrogenic).

nonionized (nohn-ī-ohn-īzd)–Not charged electrically.

nonmotile (nohn-mō-tihl)–Not capable of locomotion.

nonnutritive additive (nohn-nū-trih-tihv ahd-ih-tihv)–An additive that has no nutritive or food value, e.g., certain drugs or preservatives.

nonparenteral (nohn-pah-rehn-tər-ahl)–Administration via the gastrointestinal tract.

nonpathogenic (nohn-pahth-ō-jehn-ihck)–Not capable of producing disease.

nonprotein nitrogen (NPN)–(nohn-prō-tēn nī-trō-jehn)–The nitrates, amides, and amino acids that are the forerunners of protein; toxic to livestock in some of these forms.

nonspecific immunity (nohn-speh-sih-fihck ihm-yoo-nih-tē)–An increase of antibodies or the production of immunity resulting from the injection of some nonspecific antigen.

nontoxic (nohn-tohcks-ihck)–Not poisonous to plant or animal. See toxic, toxicity.

nose (nōz)–The part of the face of people and animals that covers the nostrils; snout; muzzle.

nose clamp (nōz klahmp)–A device that may be fitted tightly on the nose of an animal and used to control it during shoeing, surgical operations, or various types of training. See twitch, nose twitch.

nose lead (nōz lēd)–A removable metal ring that is snapped into a bull's nose. A rope is attached to the ring for leading and controlling the animal. This device is required in most livestock shows to show bulls.

nose ring (nōz rihng)–(1) A metal ring fastened through the cartilage of the nose of a bull for safe control of the animal. A staff, or metal rod, about 6 feet long may be snapped into the ring for handling the animal. See nose lead. (2) A metal ring fastened in the nose of a hog to prevent it from rooting.

nose twitch (nōz twihch)–A looped, light rope or heavy cord attached to a short stick. The loop is placed over the end of the nose of the horse and when twisted acts as a control in breaking, training, or leading. See nose clamp.

nosebag (nōz-bahg)–A feed bag hung with the open end over a horse's nose and attached by straps over the head and back of the ears.

nosocomial infection (nōs-ō-kō-mē-ahl ihn-fehck-shuhn)–A disorder caused by pathogenic organisms contracted in a facility or clinic.

nostril (nohs-trihl)–One of the two outer openings of the nose that serves as a passage for air in breathing.

notching (nohch-ihng)–Cutting dents on the ears of animals for identification.

notifiable disease (nō-tih-fī-ah-bl dih-zēz)–Any disease that must be reported to the government health authorities. Also called reportable disease.

nuclear sclerosis (nū-klē-ahr sklehr-ō-sihs)–The drying out of the lens with age.

nucleoplasm (nū-klē-ō-plahzm)–The protoplasm within the nucleus.

nucleus (nū-klē-uhs)–The cellular structure that contains the nucleoplasm, chromosomes, and surrounding membrane; also a mass of gray matter in the brain or spinal cord.

nulligravida (nuhl-ih-grahv-ih-dah)–A female that has never been pregnant.

nullipara (nuhl-ihp-ahr-ah)–A female that has never borne a viable offspring.

numdah (noom-dah)–A thick, belt blanket placed under a saddle to absorb sweat. Also called namda, nammad.

nursling (nuhrs-lihng)–A calf that is still suckling the cow.

nutrient (nū-trē-ehnt)–(1) A substance that favorably affects the nutritive processes of the body; a food. (2) In stock feeding, any feed constituent or group of feed constituents of the same general composition that aids in the support of life, as proteins, carbohydrates, fats, minerals, and vitamins.

nutrient level (nū-trē-ehnt leh-vehl)–Specifically, the concentration of any particular nutrient in the ration of animals.

nutriment (noo-trih-mehnt)–Nourishment; nutritious substances; food.

nutrition (noo-trihsh-uhn)–The sum of the processes by which an organism utilizes the chemical components of food through metabolism to maintain the structural and biochemical integrity of its cells, thereby ensuring its viability and reproductive potential.

nutritive additive (noo-trih-tihv ahd-ih-tihv)–An additive that has some food value, such as a vitamin or mineral.

nutritive ratio (noo-trih-tihv rā-shō)–In animal feeds, the ratio or proportion between the digestible protein and the digestible nonnitrogenous nutrients found by adding the digestible carbohydrates plus the digestible fat multiplied by 2.25, and dividing the sum by the digestible protein. The ratio is an expression of the energy value of a ration against its body-building power.

nutritive value (noo-trih-tihv vahl-ū)–The relative capacity of a given feed to furnish nutrition for livestock. (Usually prefixed by high, low, etc.)

nyctalopia (nihck-tah-lō-pē-ah)–The inability to seeing at night or difficulty seeing at night; also called night blindness.

nymphomania (nihm-fō-mā-nē-ah)–An abnormal sexual desire in a female.

nystagmus (nī-stahg-muhs)–The involuntary, constant, rhythmic movements of the eye.

O

obesity (ō-bē-siht-ē)–Excessive fat accumulation in the body.

obligate aerobe (ohb-lih-gāt ahr-ōb)–An organism that lives only in the presence of free oxygen.

obligate anaerobe (ohb-lih-gāt ahn-ahr-ōb)–An organism that lives only in the absence of free oxygen.

obligate parasite (ohb-lih-gāt pahr-ah-sīt)–An organism that develops and lives only as a parasite, and is confined to a specific host.

obligate saprophytes (ohb-lih-gāt sahp-rō-fītz)–Microorganisms not related to living cells that secure their nutrients from dead organic tissue or inorganic materials.

obligate symbiont (ohb-lih-gāt sihm-bī-ohnt or sihm-beh-ohnt)–An organism that is dependent on mutual relations with another for its existence. Also spelled symbiotic.

oblique projection (ō-blēk prō-jehck-shuhn)–An x-ray beam that passes through the body on an angle.

obstetrical chain (ohb-steht-reh-kahl chān)–A metal chain, 30 to 60 inches long, used by veterinarians to assist animals having difficulty in the delivering of newborn (usually calves or foals). It is used to apply a small amount of traction to the unborn.

obstetrics (ohb-steht-rihckz)–The branch of medicine that deals with the birth of the young and management practices during pregnancy.

obstruction (ohbs-truhck-shuhn)–A complete stoppage or impairment to passage.

obtunded (ohb-tuhn-dehd)–Depressed.

occlusal surface (ō-klū-zahl sər-fihs)–The chewing surface of teeth.

occlusion (ō-kloo-shuhn)–Any contact between the chewing surfaces of the teeth; also blockage in a vessel or passageway in the body.

occult (ō-kuhlt)–Hidden or difficult to see.

occults spavin (oh-kuhltz spahv-ihn)–A type of spavin that occurs between the bones at the hock joint of a horse. See spavin.

ocular (ohck-yoo-lahr)–Pertaining to the eye.

odontomata (ō-dohn-tō-mah-tah)–A tumor arising in a tissue that normally produces teeth; sometimes found in horses and cattle. Also known as odontoma.

off feed (ohf fēd)–(1) Designating an animal that has digestive disturbances due to excessive or improper feeding. (2) Designating an animal that fails, from any cause, to eat the normal feed.

off side (ohf sīd)–The right side of horse.

offal (aw-fuhl)–The inedible visceral organs and unusable tissue removed from the carcass of a slaughtered animal.

offspring (ohf-sprihng)–The young produced by animals.

oil gland (oyl glahnd)–The gland at the base of the tail of a chicken or other bird that secretes an oil used by the bird in preening its feathers. The gland is removed in dressing chickens for the market. Also called preen gland, uropygial gland.

ointment (oynt-mehnt)–A salve; an unguent; a medicinal preparation that has a base of some kind of fat or soft unctuous substance, such as the petroleum product, petrolatum.

old-crop lambs (ohld krohp lahmz)–Sheep over one year of age that have lamb teeth.

olecranon (ō-lehck-rah-nohn)–The proximal projection of the ulna.

olfactory (ohl-fahck-tōr-ē)–Pertaining to the sense of smell.

oligodendrocyte (ohl-ih-gō-dehn-drō-sīt)–A type of neuroglial cell with few dendritic branches.

oligospermia (ohl-ih-gō-spər-mē-ah)–A deficient amount of sperm in semen.

oligotrophia (ohl-ih-gō-trō-fē-ah)–A condition of inadequate nutrition.

oliguria (ohl-ih-goo-rē-ah)–Scanty or little urination.

omasum (ō-mās-uhm)–The third compartment of the ruminant stomach. Contains a mass of suspended, parallel, rough-surfaced leaves that grind ingesta to a fine consistency.

omentum (ō-mehn-tuhm)–A fold of the peritoneum connecting the stomach to the adjacent organs. The fat deposited in the folds of the great omentum of cattle, sheep, and swine is in part the source of the suet of commerce.

omnivore (ohm-nē-vōr)–An animal that sustains life by eating plant and animal products.

on feed (ohn fēd)–(1) Designating livestock kept on farms and ranches that are being fattened for the market. (2) Designating any animal being fed grain for milk or meat production.

on full feed (ohn fuhl fēd)–In feeding poultry and livestock, feeding the animal all the food necessary for optimum gain.

on pasture (ohn pahs-chər)–Designating livestock that are grazing on pasture in contrast to those feeding in the barn or feedlot.

oncology (ohng-kohl-ō-jē)–The study of tumors.

ontogeny (ohn-tohj-eh-nē)–(1) The development of an individual tissue, organ, or organism. (2) The complete developmental history of an organism from the egg, spore, bud, etc., to the adult stage.

onychectomy (ohn-ih-kehck-tō-mē)–Surgical removal of the claw; also known as declaw.

onychomycosis (ohn-ih-kō-mī-kō-sihs)–A superficial fungal infection of the claw, nail, or hoof.

oocyst (ō-ō-sihst or oo-sihst)–The encysted or encapsulated stage of parasites, known as coccidia, passed with droppings of infected animals.

oocyte (ō-ō-sīt)–Ovicyte; one of the intermediate cells in the process of ovigenesis.

oocyte-egg-mother-cell (ō-ō-sīt ehg moh-thehr sehl)–The cell that undergoes two meiotic divisions, oogenesis, to form the egg cell, as primary oocyte, the stage before completion of the first meiotic division, and secondary oocyte, after completion of the first meiotic division.

oogenesis (ō-ō-jehn-eh-sihs)–The process by which germ cells are produced by the female.

oogonium (ō-ō-gō-nē-uhm)–Ovigonium; the first or primary germ cell from which the female gamete is produced.

oon egg (ō-ōn ehg)–An egg, expelled from the vent of a hen without a shell.

oophorectomy (ō-ohf-ō-rehck-tō-mē)–Surgical removal of an ovary (or ovaries).

opaque (ō-pāk)–The condition of not being able to pass light through.

open (ō-pehn)–Designating a female animal that is not pregnant.

open breed (ō-pehn brēd)–A breed in which entry is not restricted to progeny of animals registered in that breed (i.e., opposite of a closed breed).

open-faced (ō-pehn fāsd)–Sheep that naturally have little or no wool covering about the face and eyes.

open formula (ō-pehn fōrm-yoo-lah)–On feed containers, the statement on the tag about the number of pounds of each ingredient in a ton of mixed feed.

open heifer (ō-pehn hehf-ər)–A non-pregnant heifer. See bred heifer.

open herd (ō-pehn hərd)–A group of animals in which animals from other groups are allowed to join the existing group.

open herding (ō-pehn hərd-ihng)–Allowing a band of sheep or goats to spread freely while grazing.

open range (ō-pehn rānj)–An extensive range area where grazing is unrestricted. Also, ranges that have not been fenced into management units.

ophthalmia (ohf-thahl-mē-ah)–Inflammation of the eyeball or conjunctiva.

ophthalmic (ohp-thahl-mihck)–Pertaining to the eye, for example, ophthalmic ointment, one used in the eye.

ophthalmic tuberculin test (ohf-thahl-mihck too-behr-ku-lihn tehst)–A test for tuberculosis in which a small quantity of a concentrated tuberculin is placed within the lower eyelid of the animal to be tested.

ophthalmologist (ohp-thahl-moh-lō-jihst)– One who studies the eye.

ophthalmology (ohp-thahl-moh-lō-jē)–The study of the eye.

ophthalmoplegia (ohp-thahl-mō-plē-gē-ah)–Paralysis of the eye muscles.

ophthalmoscope (ohp-thahl-mō-skōp)–An instrument used to examine the interior eye structures; may be direct or indirect.

opisthotonos (ohp-ihs-thoht-ō-nohs)–A tetanic spasm in which the head and tail are bent dorsally and the abdomen is bent ventrally.

opportunistic (ohp-ər-too-nihs-tihck)–The ability to cause disease due to debilitation when normally disease would not be produced.

optic chiasm (ohp-tihck kī-ahzm)–The crossing of the vision nerves on the ventral surface of the brain.

optic disk (ohp-tihck dihsck)–A region in the eye where nerve endings of the retina gather to form the optic nerve; also called blind spot.

optimum condition (ohp-tih-muhm kohn-dih-shun)–The ideal environment, with regard to nourishment, light, temperature, etc., for an organism's growth and reproduction.

oral (ōr-ahl)–Pertaining to the mouth.

orb (ōrb)–The eyeball.

orbit (ōr-biht)–The bony cavity of the skull that contains the eyeball.

orchidectomy (ōr-kih-dehck-tō-mē)–Surgical removal of the testis (testes); also called orchectomy (ōr-kehck-tō-mē) or orchiectomy (ōr-kē-ehck-tō-mē).

orchitis (ōr-kī-tihs)–Inflammation of the testes; also called testitis (tehs-tī-tis).

organ (ōr-gahn)–A part of the body that performs a special function or functions.

organ of Corti (ōr-gahn ohf kōr-tē)–The spiral organ of hearing located within the cochlea that receives and relays vibrations.

organic (ōr-gahn-ihck)–(1) Produced by animals; of animal origin. (2) More inclusively, designating chemical compounds that contain carbon.

organism (ōr-gah-nihzm)–Any living individual whether plant or animal.

organogen (ōr-gahn-ō-jehn)–Any of certain chemical elements without which organisms cannot exist: oxygen, carbon, nitrogen, phosphorus, etc.

organoleptic (ohr-gah-nō-lehp-tihck)–Concerning the sensory impressions, such as temperature, taste, smell, feel, sweet, sour, and salt, associated with eating and drinking.

orifice (ōr-ih-fihs)–An entrance or outlet from a body cavity.

origin (ōr-ih-jihn)–A muscle beginning that is usually the more fixed attachment or the portion toward the midline.

orogastric intubation (ōr-ō-gahs-trihck ihn-too-bā-shuhn)–Clinically, the passage of a tube from the mouth to the stomach; also called stomach tube.

orogastric (ōr-ō-gahs-trihck)–The mouth and stomach.

oropharynx (ō-rō-fār-ihnkz)–The portion of the throat between the soft palate and epiglottis.

orthopedist (ōr-thō-pē-dihst)–A specialist in the preservation and restoration of the skeletal system.

os penis (ohs pē-nihs)–The bone found within the canine penis.

osmophyllic (ohz-mō-fihl-ihck)–Organisms that grow in solutions with high osmotic pressure.

osmosis (ohz-mō-sihs)–The movement of water across a cell membrane.

osmotic pressure (ohz-moh-tihck prehs-shuhr)–The hydrostatic pressure required to stop osmosis or prevent the diffusion of molecules of a dilute solution from passing through the walls of a semipermeable membrane into a more concentrated solution.

ossification (ohs-ih-fih-kā-shuhn)–The production or formation of bone or bony material.

ossolets (ohs-ih-lehtz)–Inflammation of the periosteum that results in a soft, warm, sensitive swelling over the front and sometimes the sides of the fetlock joint in the horse.

ostealgia (ohs-tē-ahl-jē-ah)–Bone pain.

ostectomy (ohs-tehck-tō-mē)–Surgical removal of bone.

osteitis (ohs-tē-ī-tihs)–Inflammation of bone.

osteoarthritis (ohs-tē-ō-ahr-thrī-tihs)–A noninflammatory degenerative joint disease involving cartilage, bone, and the synovial membrane.

osteoblast (ohs-tē-ō-blahst)–A bone forming cell.

osteochondrosis (ohs-tē-ō-kohn-drō-sihs)–Degeneration or necrosis of bone and cartilage followed by regeneration or recalcification.

osteoclast (ohs-tē-ō-klahst)–A bone remodeling cell; cell associated with the absorption and removal of bone.

osteocyte (ohs-tē-ō-sīt)–A bone cell, particularly one encased in hard bone.

osteofibrosis (ohs-tē-ō-fī-brō-sihs)–A loss of calcium salts from the bones that causes them to become fragile. A condition chiefly affecting horses, it may also appear in pigs, goats, and dogs.

osteology (ohs-tē-ohl-ō-jē)–The study of bone.

osteomalacia (ohs-tē-ō-mah-lā-shē-ah)–Abnormal softening of bone. Also known as creeping sickness, loin disease of cattle, down in the back, adult rickets.

osteomyelitis (ohs-tē-ō-mī-eh-lī-tihs)–Inflammation of bone and bone marrow.

osteonecrosis (ohs-tē-ō-neh-krō-sihs)–The death of bone tissue.

osteophyte (ohs-tē-ō-fīt)–A knoblike projection found at cartilage degeneration sites; free floating osteophytes are commonly called joint mice.

osteoporosis (ohs-tē-ō-pō-rō-sihs)–An abnormal condition of marked loss of bone density and an increase in bone porosity.

osteosarcoma (ohs-tē-ō-sahr-kō-mah)–A malignant neoplasm composed of bone.

osteotomy (ohs-tē-oht-ō-mē)–The surgical incision or sectioning of bone.

otalgia (ō-tahl-gē-ah)–Ear pain.

otic (ō-tihck)–Pertaining to the ear.

otitis (ō-tī-tihs)–Inflammation of the ear; usually accompanied by a second term that describes location.

otitis externa (ō-tī-tihs ehcks-tər-nah)–Inflammation of the outer ear.

otitis interna (ō-tī-tihs ihn-tər-nah)–Inflammation of the inner ear.

otitis media (ō-tī-tihs mē-dē-ah)–Inflammation of the middle ear.

otolith (ō-tō-lihth)–A stone in the utricle and saccule of the inner ear.

otomycosis (ō-tō-mī-kō-sihs)–A fungal infection of the ear.

otopathy (ō-tohp-ah-thē)–Disease of the ear.

otoplasty (ō-tō-plahs-tē)–Surgical repair of the ear.

otopyorrhea (ō-tō-pī-ō-rē-ah)–Pus discharge from the ear.

otorrhea (ō-tō-rē-ah)–Ear discharge.

otoscope (ō-tō-skōp)–An instrument used to examine the ear.

ototoxic (ō-tō-tohcks-ihck)–Poisonous to the ear.

outbred (owt-brehd)–Offspring from unrelated parents; also called random bred.

outbreeding (owt brē-dihng)–The mating of animals distinctly unrelated, usually with diverse type or production traits.

outcrossing (owt-krohs-ihng)–The mating of individuals that are less closely related than the average of the breed.

outlaw (owt-lahw)–(1) A horse that refuses or fails to become tractable during the breaking process. (2) A wild, unbroken horse.

ova (ō-vah)–The female eggs or gametes. Singular is ovum.

oval window (ō-vahl wihn-dō)–The membrane that separates the middle and inner ear.

ovarian cyst (ō-vahr-ē-ahn sihst)–A collection of fluid or solid material within the female gonad.

ovarian follicle (ō-vahr-ē-ahn fohl-lih-kl)–The small cystlike structures in ovaries that, when fully developed, contain a mature ovum. Also called Graafian follicle.

ovariohysterectomy (ō-vahr-ē-ō-hihs-tər-ehck-tō-mē)–Surgical removal of the ovaries, oviducts, and uterus; also called spay; abbreviated OHE or OVH.

ovary (ō-vah-rē)–The female gonad that produces eggs and estrogen.

overo (ō-vehr-ō)–Denotes a horse that is basically white in color and the spotting is usually roan and extends upward from the belly. The darker areas are usually small or rather ragged patches; the mane and tail are usually a mixture of color giving a roan effect.

overreach (ō-vehr-rēch)–In a horse, to strike the heel of the forefoot with the front of the hind foot.

overshot jaw (ō-vehr-shoht jaw)–A condition where the maxilla is longer than the mandible; also called parrot mouth.

overstocked (ō-vehr-stohckd)–Designating a pasture, range, or grazing game area that has more animals on it than the vegetation of the area will support.

oversummer (ō-vehr-suhm-ehr)–To live through the summer.

over-the-counter drug (ō-vər the kownt-ər druhg)–Medication that can be purchased without a prescription.

ovicide (ō-vih-sīd)–Any substance that kills parasites or other organisms in the egg stage.

ovicyte (ō-vih-sīt)–An intermediate cell in the process of ovigenesis. See oocyte.

oviducts (ō-vih-duhckts)–The tubes that extend from the uterus to the ovary; also called fallopian tubes or uterine tubes.

ovigenesis (ō-vih-jehn-eh-sihs)–Oogenesis; the process of producing the female gamete.

ovigonium (ō-vih-gō-nē-uhm)–The primary germ cell from which the female gamete is produced.

ovine (ō-vīn)–Species name for sheep.

oviparous (ō-vihp-ahr-uhs)–Producing eggs from which young are hatched.

oviposition (ō-vih-pō-zihsh-uhn)–The process of laying an egg.

ovoid (ō-voyd)–Referring to a solid body with the shape of a hen's egg; the point of attachment, if any, is at the broader end.

ovotestis (ō-vō-tehs-tihs)–A gonad, part of which is composed of ovarian tissue and testicular tissue.

ovoviviparous (ō-vō-vī-vihp-ah-ruhs)–Refers to animals who produce eggs that are incubated inside the body of the dam and hatch inside the body or shortly after laying. See oviparous, viviparous.

ovulation (oh-vū-lā-shuhn)–The process of egg maturation and release.

ovule (ō-vūl)–The body that, after fertilization, becomes the seed; the egg-containing unit of the ovary.

ovum (ō-vuhm)–A female gamete; egg; plural is ova (ō-vah).

ox (ohcks)–Any species of the bovine family of ruminants. Specifically, the domesticated and castrated male bovines used for work purposes, as distinguished from steers used for meat, or the uncastrated bulls used for breeding; plural is oxen (ohcks-ehn).

oxen (ohcks-ehn)–See ox.

oxidase (ohck-sih-dās)–Oxidizing enzymes. See enzyme.

oxidation (ohck-sih-dā-shuhn)–Any chemical change that involves the addition of oxygen or its chemical equivalent.

oxyhemoglobin (ohck-sē-hē-mō-glō-bihn)–A substance formed by the union of oxygen with the hemoglobin of the blood, the union being of such a nature that the oxygen is readily given up at times and in places where it is needed.

oxytocin (ohck-sē-tō-sihn)–The hormone released from the posterior pituitary of the female that causes contractions of the uterus at the time of breeding, aids in parturition, and causes milk let down.

ozena (ō-zē-nah)–A fetid nasal discharge accompanied by chronic inflammation of the mucous membranes, associated with a disease of bones in the nose.

P

pace (pās)–A fast, two-beat gait in which the front and hind feet on the same side start and stop at the same time.

pacer (pā-sər)–A horse, one of whose gaits is a pace. Also called side-wheeler.

pachymeninx (pahck-ē-meh-nihcks)–The dura mater; plural is pachymeninges (pahck-ē-meh-nihn-jēz).

pack (pahck)–(1) A group of canines. (2) The load that is carried by a pack animal. (3) The sheep given to the shepherd as his share for tending the flock.

pack animal (pahck ahn-ih-mahl)–A burro, mule, or horse used for carrying packs and equipment over rough areas and to places inaccessible by other means.

paddling (pahd-lihng)–(1) A term used to describe a horse that throws the front feet outward as they are picked up. This condition is predisposed in horses with too-narrow or pigeon-toed standing positions. (2) A term describing the leg movements of a seizuring animal.

paddock (pah-dohck)–A small fenced-in area; also called corral.

paint (pānt)–A color pattern in horses involving white patches on a dark background.

paint roan (pānt rōn)–A color coat of horses in which the roan color seems to be imposed on other colored areas.

pair (pār)–(1) A male and female of a species. (2) A team of horses or mules frequently spoken of as a well matched pair regardless of sex.

palatability (pahl-ah-teh-bihl-ih-tē)–The degree to which a feed is liked or acceptable to an animal.

palatable (pahl-ah-teh-buhl)–Describing feed that an animal prefers and selects over other feeds.

palate (pahl-aht)–The roof of the mouth separating the oral cavity from the nasal cavity.

palatitis (pahl-ah-tī-tihs)–A painfully swollen condition of the palate of a horse's mouth that may be caused by eating hard food. Also called lampas, lampers.

palatoplasty (pahl-ah-tō-plahs-tē)–Surgical repair of a cleft palate.

palliative (pah-lē-ah-tihv)–Relief of a condition, but not a cure.

palmar (pahl-mahr)–The bottom of a front foot or hoof.

palmitic acid (pahl-mih-tihck ah-sihd)–One of the fatty acids of butterfat.

palpate (pahl-pāt)–To test or examine by feel. Used to determine pregnancy in cattle.

palpation (pahl-pā-shuhn)–To examine by feeling.

palpebra (pahl-pē-brah)–The eyelid; plural is palpebrae (pal-pē-brā).

palpebral (pahl-pē-brahl)–Pertaining to the eyelid.

palpitation (pahl-pih-tā-shuhn)–Pounding with or without irregularity in rhythm.

pancreas (pahn-krē-ahs)–The gland located near the proximal duodenum that has both exocrine (digestive enzymes) and endocrine (insulin) functions.

pancreatectomy (pahn-krē-ah-tehck-tō-mē)–Surgical removal of the pancreas.

pancreatic juice (pahn-krē-ah-tehck jūs)–A secretion by the pancreas containing ferments that contribute to digestion of foods.

pancreatitis (pahn-krē-ah-tī-tihs)–Inflammation of the pancreas.

pancreatotomy (pahn-krē-ah-toht-ō-mē)–A surgical incision into the pancreas.

pancytopenia (pahn-sīt-ō-pē-nē-ah)–A deficiency of all types of blood cells.

pandemic (pahn-dehm-ihck)–Occurring over a large geographic area.

panophthalmitis (pahn-ohp-thahl-mih-tī-tihs)–Inflammation of all eye structures.

panotitis (pahn-ō-tī-tihs)–Inflammation of all ear parts.

pantothenic acid (pahn-tō-thehn-ihck ah-sihd)–A vitamin of the B-complex group; required by poultry and swine.

panzootic (pahn-zō-oht-ihck)–Referring to a widespread epidemic among animals.

pap (pahp)–The teat of a cow.

papilla (pah-pihl-ah)–A small protrusion or elevation; plural is papillae (pah-pihl-ā).

papilledema (pahp-ehl-eh-dē-mah)–Swelling of the optic disc.

papilloma (pahp-ih-lō-mah)–A benign epithelial growth that is lobed, raised, fleshy, or warty; also known as a wart of fibropapilloma.

papule (pahp-yool)–A small, raised skin lesion less than 0.5 cm in diameter.

paracostal incision (pahr-ah-koh-stahl ihn-sihz-shuhn)–A surgical cut oriented parallel to the last rib.

paralysis (pah-rahl-ih-sihs)–Immobility or loss of motor function.

paramedian incision (pahr-ah-mē-dē-ahn ihn-sihz-shuhn)–A surgical cut lateral and parallel to the ventral midline, but not on the midline.

paraphimosis (pahr-ah-fih-mō-sihs)–Retraction of the skin of the prepuce causing a painful swelling of the glans penis that prevents the penis from being retracted.

paraplegia (pahr-ah-plē-jē-ah)–Paralysis of the caudal body or hind limbs in quadrupeds; paralysis of the legs in bipeds.

parasite (pahr-ah-sīt)–An organism that lives on or within another living organism.

parasiticide (pahr-ah-siht-ih-sīd)–An agent that kills parasites.

parasympathetic nervous system (pahr-ah-sihm-pah-theh-tihck nər-vuhs sihs-tehm)–The portion of the autonomic nervous system that returns the body to a normal state after stressful or emergency situations.

parathormone (pahr-ah-thōr-mōn)–The hormone of the parathyroid glands. It helps to maintain the calcium level of an animal's blood by removing calcium from the bones; also called parathyroid hormone.

parathyroid gland (pahr-ah-thī-royd glahnd)–Four glands located on the surface of the thyroid gland that help regulate blood calcium levels.

parathyroidectomy (pahr-ah-thī-royd-ehck-tō-mē)–Surgical removal of one or more parathyroid glands.

parenchyma (pahr-ehnck-ih-mah)–The functional elements of a tissue or organ.

parental generation (pahr-rehn-tahl jehn-ər-ā-shuhn)–The P_1 generation; the first generation in a series of crosses; usually involves homozygotes for different alleles.

parenteral (pah-rehn-tər-ahl)–Administration through routes other than the gastrointestinal tract.

paresis (pah-rē-sihs)–Slight or incomplete paralysis that affects the ability of an animal to move.

paresthesia (pahr-ehs-thē-zē-ah)–An abnormal sensation.

paries (pā-rē-ehs)–A wall of a cavity or hollow body organ in plants and animals.

parietal (pahr-ī-eh-tahl)–Borne on or belonging to a wall or walls of a cavity.

parotid (pah-roht-ihd)–Located near the ear.

parotid gland (pah-roht-ihd glahnd)–A gland that secretes saliva. This is a paired gland located behind the jaws and somewhat below the ear.

parous (pah-ruhs)–Having previously given birth.

parovarian (pahr-ō-vahr-ē-ahn)–Near the ovary.

paroxysm (pahr-ohck-sihzm)–A sudden convulsion or spasm.

parrot mouth (pār-oht mowth)–A condition in animals in which the upper teeth protrude over and beyond the lower teeth.

parthenogenesis (pahr-theh-nō-jehn-eh-sihs)–The development of an individual from an unfertilized egg cell.

partial dominance (pahr-shahl dohm-ih-nahns)–A kind of interaction between alleles where one gene is not completely dominant to its allele but where the appearance of the heterozygote is more similar to one of the homozygotes than to the other.

parts per million (pahrtz pehr mihl-yuhn)–The number of weight or volume units of a solution or a mixture; a measure of concentration, especially of chemicals in solution; one milligram per liter. Abbreviated ppm.

parturient paresis (pahr-too-rē-ahnt pahr-ē-sihs)–A hypocalcemic metabolic disorder of ruminants seen in late pregnancy or early lactation due to mobilization or large amounts of calcium to produce milk to feed the newborn. This causes blood calcium levels to drop below the point necessary for impulse transmission along the nerve tracts. Commonly called milk fever.

parturition (pahr-tyoo-rihsh-uhn)–The act of giving birth; also called labor.

passeriformes (pahs-ər-ih-fōrmz)–A group of perching birds that includes finches, sparrows, mynahs, and canaries.

passive immunity (pahs-ihv ih-mū-nih-tē)–A kind of immunity acquired by animals when they are injected with antibodies against some disease.

pastern bone (pahs-tərn bōn)–Phalanx I (long pastern) or phalanx II (short pastern) in livestock.

pastern joint (pahs-tərn joynt)–A common name for the connection between phalanx I and II in livestock.

pasteurization (pahs-tər-ih-zā-shuhn)–The process of destroying all or most of the vegetative bacteria in a substance, such as milk or fresh fruit juices, by application of heat of from 140°F to 185°F and then cooling.

pasture (pahs-chər)–An area for grazing animals; also means grass or other forage that grazing animals eat.

pasture-bred (pahs-chər-brehd)–A cow serviced by a bull in the pasture.

pasture mating (pahs-chər mā-tihng)–Natural breeding; also called natural cover.

pasture off (pahs-chər ohf)–To remove a crop by grazing; a common practice on grass seedings and grain fields in late summer or fall.

pasture rotation (pahs-chər rō-tā-shuhn)–The practice of moving the herd from one field to another after a few days of intensive pasturing.

pastured (pahs-chərd)–Designating an area or crop that has been grazed off by livestock.

patch (pahch)–A localized skin color change greater than 1 cm in diameter.

patella (pah-tehl-ah)–The kneecap; sesamoid bone over the stifle joint.

patent (pā-tehnt)–Open, unobstructed, or not closed.

paternal (pah-tər-nahl)–Pertaining to the male parent. See maternal.

pathobiology (pahth-ō-bī-ohl-ō-jē)–The study of disease processes; the biology of disease.

pathogen (pahth-ō-jehn)–Producing disease.

pathogenic (pahth-ō-jehn-ihck)–Producing disease.

pathology (pahth-ohl-ō-jē)–The study of the nature, causes, and development of abnormal conditions (disease).

pathophysiology (pahth-ō-fihz-ē-ohl-ō-jē)–The study of changes in function caused by disease.

patroclinous (paht-rō-klī-nuhs)–Resembling the male parent.

paunch (pawnch)–Another name for the rumen.

paunching (pawnch-ihng)–Removing the entrails from a carcass.

pecking order (pehck-ihng ōr-dər)–The order in which same poultry and wild birds within a flock may peck others without being pecked in return.

pectoral (pehck-tōr-ahl)–The breast or chest area.

pedal (pē-dahl)–Pertaining to the foot.

pedigree (pehd-ih-grē)–A list of an individual animal's ancestors, usually only those of the five closest generations.

pedunculated (peh-duhnck-yoo-lā-tehd)–Having a peduncle or stalk.

peewees (pē-wēz)–Small or stunted lambs.

peg pony (pehg pō-nē)–A saddle horse trained to change direction rapidly.

pellet (pehl-eht)–(1) Mixtures of ground ingredients pressed to form aggregates of a convenient size to feed livestock. (2) A mass of indigestible hair and bones regurgitated by carnivorous birds or mammals.

pelleted feed (pehl-eht-ehd fēd)–A pill-like or cubical type of animal feed made by forcing the loose, bulky, or dusty feeds into small, uniform pellets by the use of grinding, molding, and compressing machinery.

pelleted hay (pehl-eht-ehd hā)–Hay that has been highly compressed by passing through a pelleting machine. It is easy to handle and is free from dust.

pelt (pehlt)–(1) The natural, whole skin covering, including the hair, wool, or fur of the smaller animals, such as sheep, foxes, etc. A large pelt is more often called a hide. (2) To remove the whole skin or pelt from the body of an animal.

pelvic capacity (pehl-vihck kah-pah-sih-tē)–The dimensions of a female's pelvic area that is an indication of its ability to give birth easily.

pelvic cavity (pehl-vihck kahv-ih-tē)–The hollow space that contains the reproductive and some excretory system organs; formed by the pelvic bones.

pelvis (pehl-vihs)–The hip.

pen (pehn)–(1) In poultry shows, a male and four female birds of the same variety. (2) A small space enclosed by any kind of fence, used for confining pigs, cows, and other animals.

pendulous udder (pehn-joo-luhs uh-dər)–A low-hanging, poorly attached udder.

penetrant (pehn-eh-trahnt)–An adjuvant that aids a liquid's ability to enter the pores of a surface. See adjuvant.

penicillin (pehn-ih-sihl-ihn)–An antibiotic extracted from cultures of certain molds of the genera of *Penicillium* and *Aspergillus* that have been grown on special media. Penicillin is also produced synthetically.

penis (pē-nihs)–The male sex organ that carries the reproductive and urinary products out of the body.

pepsin (pehp-sihn)–A digestive enzyme secreted by glands in the stomach. It is commercially obtained from the lining of the pyloric end of a pig stomach to be used in medicine as an aid to protein digestion.

peptide (pehp-tīd)–A compound made up of a series of amino acids; an intermediate in the synthesis or digestion of a protein.

peptones (pehp-tōnz)–Products obtained by the digestive action of enzymes on albuminous matter. Peptones of different kinds are produced from lean muscle tissue or by the action of yogurt bacteria on lactalbumin in milk, and serve as easily assimilated forms of proteins or as cultural media.

per os (pehr ohs)–Orally; abbreviated po.

peracute (pehr-ah-kūt)–Excessively acute, e.g., when symptoms of a disease occur much earlier than usual and are well marked.

percent (pər-sehnt)–The number of parts per one hundred parts.

perception (pər-sehp-shuhn)–The ability to recognize sensory stimulus.

percussion (pər-kuh-shuhn)–A diagnostic procedure used to determine density in which sound is produced by tapping various body surfaces with the finger or an instrument (the sound produced over the chest where air is present varies from an area where fluid is present).

percutaneous (pehr-kyoo-tā-nē-uhs)–Through the skin.

perforating ulcer (pər-fōr-āt-ihng uhl-sihr)–Erosion through the entire thickness of a surface.

perfusion (pər-fū-shuhn)–Blood flow through tissue.

perianal (pehr-ih-ā-nahl)–Around or surrounding the anus.

pericardial (pehr-ih-kahr-dē-ahl)–The membrane surrounding the heart.

pericardial fluid (pehr-ih-kahr-dē-ahl flū-ihd)–The liquid found between the layers of the pericardium.

pericarditis (pehr-ih-kahr-dī-tihs)–Inflammation of the pericardium.

pericardium (pehr-ih-kahr-dē-uhm)–A double-walled membranous sac enclosing the heart.

perilymph (pehr-eh-lihmf)–The fluid that fills two of the semicircular canals (vestibular and tympanic).

perimetrium (pehr-ih-mē-trē-uhm)–The membranous outer layer of the uterus.

perineal urethrostomy (pər-ih-nē-ahl yoo-rē-throhs-tō-mē)–The surgical creation of a permanent opening between urethra and perineum.

perineum (pehr-ih-nē-uhm)–The region between the scrotum or vulva and the anus.

periocular (pehr-ē-ohck-yoo-lahr)–Around the eye.

period of lactation (pehr-ē-ohd ohf lahck-tā-shuhn)–The period of milk production in an animal; the time between the beginning and end of the milk flow.

periodontitis (pehr-ē-ō-dohn-tī-tihs)–Inflammation of the tissues surrounding and supporting the teeth; also called periodontal disease.

periople (pehr-ē-ō-puhl)–A varnish-like coating that holds moisture in the hoof and protects the hoof wall.

periorbita (pehr-ih-ōr-bih-tah)–The eye socket.

periosteum (pehr-ē-ohs-tē-uhm)–A tough, fibrous tissue that forms the outermost covering of bone.

periostitis (pehr-ē-ohs-tī-tihs)–Inflammation of the periosteum.

peripheral nervous system (pehr-ihf-ər-ahl nər-vuhs sihs-tehm)–That portion of the nervous system that consists of the cranial and spinal nerves; abbreviated PNS.

peristalsis (pehr-ih-stahl-sihs)–A series of wavelike contraction of the smooth muscles.

peritoneum (pehr-ih-tō-nē-uhm)–The membrane covering the walls of the abdominal and pelvic cavities and some organs in this area.

peritonitis (pehr-ih-tō-nī-tihs)–Inflammation of the peritoneum.

perivascular (pehr-ih-vahs-kyoo-lahr)–Around the vessels.

permanent pasture (pehr-mah-nehnt pahs-chər)–Grazing land in farms occupied by perennial grasses and legumes. It is not part of a regular rotation of fields and usually remains unplowed for long periods.

pernicious (pehr-nihsh-uhs)–Harmful or fatal.

peroxidase (peh-rohck-sih-dās)–A heart-resisting enzyme present in milk, especially in abnormal milk.

pessary (pehs-ah-rē)–A vaginal suppository for administration of drugs.

pest (pehst)–Anything, such as an insect, animal, or plant that causes injury, loss or irritation to a crop, stored goods, an animal, or people.

pesticide (pehs-tih-sīd)–A substance used to control insect, plant, or animal pests. Pesticides include insecticides, herbicides, fungicides, nematocides,and rodenticides.

pet (peht)–Any animal, such as a cat, dog, lamb, bird, etc., that is kept for affection and companionship.

petechia (peh-tē-kē-ā)–A small, pinpoint hemorrhage; plural is petechiae (peh-tē-kē-ā).

petechiae (pē-tē-kē-ā)–Small, purple, or red spots in the skin of animals caused by a small hemorrhage in which blood is released from its normal channel in the blood vessels into very minute areas of the surrounding body tissues.

pH–A numerical measure of the acidity or hydrogen ion activity of a substance. The neutral point is a pH of 7.0. All pH values below 7.0 are acid and all above 7.0 are alkaline.

phagocyte (fā-gō-sīt)–A cell that eats; a leukocyte that ingests foreign material.

phagocytosis (fāg-ō-sī-tō-sihs)–An abnormal condition of engulfing or eating cells.

phagomania (fāg-ō-mā-nē-ah)–An insatiable craving for food.

phagophobia (fāg-ō-fō-bē-ah)–A morbid fear of eating.

phalanx (fā-lahncks)–A bone of the digit; plural is phalanges (fā-lahn-jēz).

pharmaceutical (fahr-mah-sū-tih-kahl)–Any substance used to enhance the health of humans or animals.

pharmacokinetics (fahrm-ah-kō-kihn-eht-ihcks)–The movement of drugs or chemicals; consists of absorption, distribution, biotransformation, and elimination.

pharmacology (fahrm-ah-kohl-lō-jē)–The study of the nature, uses, and effects of drugs.

pharyngitis (fār-ihn-jī-tihs)–Inflammation of the throat.

pharyngoplasty (fār-rihng-ō-plahs-tē)–Surgical repair of the throat.

pharyngostomy (fār-ihng-ohs-tō-me)–Surgical creation of an artificial opening into the throat.

pharyngotomy (fār-ihng-oht-ō-mē)–A surgical incision into the throat.

pharynx (fār-ihncks)–The cavity in the caudal oral cavity that joins the respiratory and gastrointestinal systems; also called the throat.

phase (fās)–(1) The view that an object presents to the eye. (2) Any one of the varying aspects or stages through which a disease or process may pass.

phase feeding (fās fē-dihng)–Changes in an animal's diet to adjust for age, stage of production, season of the year, temperature change, differences in body weight, and nutrient requirements of different strains of animals, or to adjust one or more nutrients as other nutrients are changed for economic or availability reasons.

phenotype (fē-nō-tīp)–The observed character of an individual without reference to its genetic nature. Individuals of the same phenotype look alike but may not breed alike. See genotype.

pheochromocytoma (fē-ō-krō-mō-sī-tō-mah)–A tumor of the adrenal medulla resulting in increased secretion of epinephrine and norepinephrine.

pheromone (fehr-ō-mōn)–A substance secreted to the outside of the body by an individual organism that causes a specific reaction by another organism of the same species.

phimosis (fih-mō-sihs)–A narrowing of the skin of the prepuce so it cannot be retracted to expose the glans penis.

phlebitis (fleh-bī-tihs)–Inflammation of a blood vessel that returns blood to the heart.

phlegm (flehm)–A thick mucus secreted by the respiratory lining.

phobia (fō-bē-ah)–An extreme fear.

phonation (fō-nā-shuhn)–The act of producing sound.

photophobia (fō-tō-fō-bē-ah)–A fear of intolerance of light.

phrenic nerve (frehn-ihck nərv)–The nerve that innervates the diaphragm.

phylogeny (fī-lohj-eh-nē)–The developmental history of a species.

phylum (fī-luhm)–The highest grouping in the taxonomy of the plant and animal kingdoms, based on assumed common ancestry.

physiological (fihz-ē-ō-lohj-ih-kahl)–(1) Referring to or concerning the science of physiology or the branch of biology that deals with life processes and functions. (2) Referring to the functions of the organs of plants and animals.

physiological age (fihz ē ō lohj-ih-kahl āj)–The age of an animal that is determined by an examination of the carcass.

physiology (fihz-ē-ohl-ō-jē)–The study of body functions.

physis (fī-sihs)–A bone segment that involves growth in length of the bone in young animals; also means growth.

phytobezoar (fī-tō-bē-zōr)–A concentration of plant material.

pia matter (pē-ah mah-tər)–The third and deepest layer of the meninges.

piaffe step (pī-ahf stehp)–A high action step to which horses are trained for exhibition.

pica (pī-kah)–Eating and licking abnormal substances; depraved appetite.

pick (pihck)–(1) To pluck the feathers from a fowl in dressing it for the market. (2) To nibble at food.

picket (pihck-eht)–To tether or control the grazing range of an animal by a rope.

picking machine (pihck-ihng mah-shēn)–In poultry dressing, a machine consisting of a revolving drum equipped with flexible rubber fingers that pick the feathers from carcasses.

piebald (pī-bahld)–Designating a horse with a black-and-white or dark coat; a pinto. Also called pied.

piebaldism (pī-bahld-ihzm)–A skin condition in which the skin is partly white (albinism) and partly brown (vitiligo). A common example is a piebald horse.

pig (pihg)–(1) A young swine weighing less than 120 pounds. (A few markets, such as Chicago, IL, set 130 pounds as the maximum weight for an animal of this class.) (2) Any young, unweaned swine.

pig-eating sow (pihg ē-tihng sow)–A sow that devours her young at farrowing time presumably as a result of a ration deficiency, an unsuitable environmental situation, or exciting disturbances at farrowing time.

pig eye (pihg ī)–In a horse, a small, retracted eye that may cause imperfect vision.

pig teeth nipper (pihg tēth nih-pər)–A clipperlike instrument used for clipping the canine teeth (temporary or milk teeth) of suckling pigs.

pigeon (pihdg-ohn)–Any bird of the genus *Columba.*

pigeon loft (pihdg-ohn lohft)–A dovecote elevated house, or pen atop buildings for raising pigeons.

pigeon-toed (pihdg-ohn tōd)–Designating an animal or person whose feet turn inward.

piggin-string (pihg-ihn strihng)–A short rope used for tying down animals, as in calf roping.

pig-guard rail (pihg gahrd rāl)–A rail about 8 to 10 inches above the floor and 8 to 10 inches from the wall of a farrowing pen, beneath which recently farrowed pigs can move and thus be protected from crushing by the sow.

piglet (pihg-leht)–A young pig of either sex. Also called pig, shoat, or shote.

pigment (pihg-mehnt)–Any of the natural coloring materials in the cells and tissues of plants and animals. In fruit and vegetables, the green pigment is chlorophyll; orange to red pigments are carotenoids; red to blue colors are anthocyanins; light-yellow pigments are flavoners and flavonols. In meat, the chief pigment producing the pink or red color is myoglobin.

pigskin (pihg-skihn)–The tanned hides of hogs.

piles (pī-uhlz)–A common term for a prolapsed rectum in swine.

pilgrim (pihl-grihm)–A worn-out, decrepit horse.

piloerection (pī-lō-ē-rehck-shuhn)–Hair standing upright.

pin (pihn)–In poultry slaughter, to remove the pin feathers that are embedded in the skin.

pin bone (pihn bōn)–The region on each side of the tail head on the hindquarters of a bovine. These should be wide apart and well-defined for good conformation.

pin feather (pihn fehth-ər)–The young feather embedded in, or just emerging from, the skin of a fowl.

pin nipples (pihn nihp-uhlz)–Small, underdeveloped, usually nonfunctional nipples on the teats of a pig.

pincers (pihnch-ərz)–The incisor teeth of a horse. Also called nippers.

pinch (pihnch)–A common term for a bloodless castration using an emasculatome.

pincher (pihn-shər)–A tool used to remove horseshoes.

pineal gland (pī-nē-ahl glahnd)–The gland that secretes hormones that affect circadian rhythm and other unknown functions located in the central portion of the brain.

pinealectomy (pīn-ē-ahl-ehck-tō-mē)–Surgical removal of the pineal gland.

pinealopathy (pīn-ē-ah-loph-ah-thē)–A disorder of the pineal gland.

pinion (pihn-yehn)–(1) The outermost part of a bird's wing including the carpus, metacarpus, and phalanges; the part of a wing, corresponding to the forearm, on which the primary flight feathers are borne. (2) To cut off the pinion on one (or both) wings to permanently prevent a bird from flying. (3) A gear that has the teeth formed on the inside of the hub.

pinkies (pihnk-ēz)–Newborn mice.

pinna (pihn-ah)–The external portion of the ear; also called auricle (aw-rih-kuhl).

pinning (pihn-ihng)–(1) The insertion of a metal rod into the medullary cavity of a long bone. (2) The sticking of a young lamb's tail to its anus.

pinny (pihn-ē)–In poultry marketing, designating a bird carcass that has an excessive amount of pin feathers, and is therefore of low grade and marketable only at a reduced price.

pinocytosis (pihn-ō-sī-tō-sihs)–The engulfing or absorption of fluids by cells.

pins (pihnz)–A protrusion of the ischial bones just lateral to the tail head in ruminants. Also called pin bones.

pinto (pihn-tō)–Designating a horse that has a spotted or piebald coat color.

pip (pihp)–(1) A horny, dried condition of the tip of the tongue that develops in poultry in cases of infectious coryza when mouth breathing is long continued. (2) The raised crown or individual rootstock of a plant.

pipped egg (pihpd ehg)–An egg through which the chick has forced its beak in the first step of breaking out of the shell during incubation.

pipping (pihp-ihng)–The process of breaking the eggshell by a chick before hatching.

pirouette (pihr-oo-eht)–A high-school exercise for horses in which the forelegs are held more or less in place while the horse moves his hindquarters around them.

pit silo (piht sī-lō)–A shallow pit of variable size for storing silage, which is dug in well-drained soil and is frequently walled with lumber or concrete if of a permanent nature.

pitch (pihch)–The jumping action of a horse in its attempt to unseat its rider.

pithing (pihth-ihng)–The destruction of the brain and spinal cord by thrusting a blunt needle into the cranium or vertebral column.

pituitarism (pih-too-ih-tār-ihzm)–Any disorder of the pituitary gland.

pituitary gland (pih-too-ih-tār-ē glahnd)–The gland located at the base of the brain that responds to hypothalamic stimuli; its function is to maintain appropriate blood levels of hormones.

pivot (pihv-eht)–A leg action in which a horse pivots around his hindquarters while holding one leg in place and side stepping with the other hindfoot.

pizzle (pihz-l)–The penis, especially of a bull.

placebo (plah-sē-bō)–An inactive substance that is given for its suggestive effects or a substance used as a control in experimental settings.

placenta (plah-sehn-tah)–The female organ of mammals that develops during pregnancy and joins mother and offspring for exchange of nutrients, oxygen, and waste products; also called afterbirth.

placental retention (plah-sehn-tahl rē-tehn-shuhn)–The undue retention of the placenta and other fetal membranes after the young is born. Also called retained placenta or cleaning a cow.

placentome (plahs-ehn-tōm)–The part of placenta that includes the caruncle and cotyledon.

plantar (plahn-tahr)–The bottom of a rear foot or hoof.

plantar cushion (plahn-tahr coosh-uhn)–A pad of fibro-fatty tissue in the foot to the rear of and under the navicular and coffin bones of animals. In horses, it is pronounced and overlies the frog. Also called digital cushion.

plantation gait (plahn-tā-shuhn gāt)–A natural horse gait, somewhat faster than a walk, which is easy on the horse and rider.

plantigrade (plahnt-ih-grād)–Walking on the entire foot surface (sole).

plaque (plahck)–A small, differentiated, raised area on a body surface.

plasma (plahz-mah)–A straw-colored fluid portion of blood that transports nutrients, hormones, and waste products.

plasma cell (plahz-mah sehl)–An immune cell that produces and secretes a specific antibody for a specific antigen; also called plasmocyte (plahz-mō-sīt).

plastron (plahs-trohn)–The ventral region of a turtle shell.

platelet (plāt-leht)–An anucleate clotting cell.

pleasure horse (pleh-zuhr hōrs)–A classification for horses that includes those used for riding, driving, or racing.

pleiotropy (plī-oht-rō-pē)–A situation where one gene affects more than one trait.

pleura (ploor-ah)–The multilayered membrane that covers the lung; plural is pleurae (ploor-ā).

pleural effusion (ploor-ahl eh-fū-zhuhn)–An abnormal accumulation of fluid in the airtight space between the folds of the pleura; note that a small amount of lubricating fluid in this space is normal.

pleural space (ploor-ahl spās)–An airtight area containing lubricating fluid located between the folds of the pleura; also called pleural cavity.

pleurectomy (ploor-ehck-tō-mē)–Surgical removal of all or part of the pleura.

pleurisy (ploor-ih-sē)–Inflammation of the pleura; also called pleuritis (ploor-ī-tihs).

pleurodont (pluhr-ō-dohnt)–Teeth attached by one side on the inner jaw surface.

pleuropneumonia (ploor-ō-nū-mō-nē-ah)–An abnormal condition of the pleura and the lung (usually involves inflammation and congestion).

plexus (plehck-suhs)–A network of intersecting nerves or vessels; also called rete (reh-tē).

pluck (pluhck)–(1) In hog and cattle slaughtering, the organs that lie in the thoracic cavity consisting of the heart, lungs, gullet, and windpipe. (2) In poultry slaughter, to remove the feathers.

plumage (ploo-mihj)–The feathers of a fowl.

plumb line (pluhm līn)–The line formed when a weight is placed on the end of a string to measure the perpendicularity of something.

plume (ploom)–Down feathers.

plump (pluhmp)–In poultry slaughter, to create an effect of plumpness by dipping the picked carcass in scalding water. It shrinks the skin and draws up the legs, wings, and neck closer to the body, thus imparting a false plumpness.

pneumatic tonometer (nū-mah-tihck tō-noh-meh-tər)–An instrument that measures intraocular pressure by blowing a puff of air against the cornea to flatten it slightly.

pneumocentesis (nū-mō-sehn-tē-sihs)–Puncture of the lung to remove fluid or air.

pneumocystogram (nū-mō-sihs-tō-grahm)–A radiographic film of the urinary bladder after air has been placed into the urinary bladder via a urinary catheter.

pneumonectomy (nū-mō-nehck-tō-mē)–Surgical removal of lung tissue.

pneumonia (nū-mō-nē-ah)–An abnormal condition of the lung (usually involves inflammation and congestion of the lung).

pneumothorax (nū-mō-thōr-rahckz)–An accumulation of air or gas in the pleural space.

pod (pohd)–A flock of animals, birds, etc.

poikilocytosis (poy-kē-lō-sī-tō-sihs)–A condition of irregular cells; clinically means varied shapes of erythrocytes.

pointing (poyn-tihng)–A stride in which extension is more pronounced than flexion.

points (poyntz)–(1) A black coloration from the knees and hocks down in bays and browns (may include the ear tips). (2) The nose (mask), ears, tail, and feet of an animal.

poison (poy-zuhn)–Any substance ingested, inhaled, or developed within the body that causes or may cause damage or disturbance of function of plants, animals, or people. See toxin.

poke (pōk)–A yoke with an attached sharp spur pointing forward that is placed on the neck of an animal to prevent it from crawling through or jumping a fence.

polioencephalomalacia (pō-lē-ō-ehn-sehf-ah-lō-mah-lā-shah)–Abnormal softening of the gray matter of the brain.

polioencephalomyelitis (pō-lē-ō-ehn-sehf-ah-lō-mī-eh-lī-tihs)–Inflammation of the gray matter of the brain and spinal cord.

poliomyelitis (pō-lē-ō-mī-eh-lī-tihs)–Inflammation of the gray matter of the spinal cord.

poll (pōl)–The top of the head; the occiput.

poll evil (pōl ē-vihl)–A fistula of the poll between the ears that may follow a severe injury to that part, a common affliction of horses and mules.

pollakiuria (pōl-lahck-ē-yoo-rē-ah)–Frequent urination.

pollard (pohl-ahrd)–A hornless, ox, sheep, or goat.

polled (pōld)–Naturally hornless.

polychromasia (poh-lē-krō-mah-zē-ah)–A condition of many colors; clinically means erythrocytes with varied or multicolored staining qualities.

polycythemia (poh-lē-sī-thē-mē-ah)–A condition of many cells; clinically means excessive erythrocytes.

polydactyly (pohl-ē-dahck-tih-lē)–More than normal number of digits.

polydipsia (pahl-ē-dihp-sē-ah)–Increased thirst or drinking; abbreviated PD.

polyestrous (pohl-ē-ehs-truhs)–Having many estrous or "heat" cycles per season.

polygamous (poh-lihg-ah-muhs)–Having multiple mates.

polygastric (pohl-ē-gahs-trihck)–Having many stomach compartments as in the ruminant animal (such as cattle).

polymorphic (pohl-ē-mōr-fihck)–Having two or more forms.

polymorphonuclear (pohl-ē-mōr-fō-nū-klē-ahr)–A multishaped nucleus.

polymorphonuclear leucocytes (pohl-ē-mōr-fō-nū-klēr-ahr lū-kō-sītz)–White blood cells in which the nuclei are constricted into irregular shapes.

polyneuritis (pohl-ē-nū-rī-tihs)–Inflammation of many nerves.

polyp (pohl-uhp)–A small growth from a body surface (mucous membrane or cutaneous surface).

polyphagia (pohl-ē-fā-jē-ah)–Increased eating or swallowing.

polyploid (pohl-ē-ployd)–An organism with more than two sets of the basic or haploid number of chromosomes, e.g., triploid, tetraploid, pentaploid, hexaploid, heptaploid, octaploid.

polypnea (pohl-ehp-nē-ah)–Rapid or panting respiration. More commonly called tachypnea.

polyspermy (pohl-ē-spehr-mē)–The entrance of many sperm cells into the ovum at fertilization.

polytocous (poh-liht-ō-kuhs)–Giving birth to multiple offspring.

polyuria (pohl-ē-yoo-rē-ah)–Increased urination; abbreviated PU.

polyvalent (pohl-ē-vā-lehnt)–A vaccine, antiserum, or antitoxin, that is active against multiple antigens or organisms; "mixed vaccine."

pommel (pohm-ehl)–(1) The knob, ball, or protuberant part serving as a means of grip on anything, such as the high, forward part of a saddle. (2) A block of hard wood grooved like a crimping board and employed by curriers to render leather supple and impart a grain to it.

pons (pohnz)–The bridge at the base of the brain that is part of the metencephalon or brainstem.

pony (pōn-ē)–An equine that is between 8.2 and 14.2 hands when mature.

pop-eyed (pohp īd)–Refers to a horse whose eyes are generally prominent or bulge out a little more than normal; also to a horse that is "spooky" or attempts to see everything that goes on and is often frightened.

pop-hole (pohp hōl)–A hole small enough to allow piglets to creep through to reach their mother.

popliteal (pohp-liht-ē-ahl)–The posterior surface of the stifle.

popped knee (pohpd nē)–In livestock, a general term describing inflammatory conditions affecting the knees, so named because of the sudden swelling that accompanies it.

porcine (pōr-sīn)–Species name for swine or pigs or hogs.

pore (pōr)–In animal membranous tissues, minute openings for the absorption and transpiration of matter.

pose (pōz)–In the show ring, a special stance or position that a horse may be trained to assume. A posed horse stands with his front feet extended and his rear feet back.

positioning (pō-sih-shuhn-ihng)–A specified body position and the part of the body closest to the film.

post (pōst)–The proper position and balance a rider assumes in the saddle in riding a horse while in the trot gait, rising with each second beat of the one-two rhythm of the trot.

post cenam (pohst sē-nahm)–After feeding (in scientific writing often abbreviated pc).

post op (pōst ohp)–A common term for after surgery; postoperatively.

posterior (pohs-tēr-ē-ōr)–The rear of the body (in quadrupeds and arthropods the tail is on the posterior end of the body, in bipeds the dorsal surface is the rear of the body).

posterior chamber (pohs-stēr-ē-ōr chām-bər)–The aqueous-containing space located in the eye between the iris and the lens.

postictal (pōst-ihck-tahl)–The period after a seizure.

post-legged (pōst-lehg-ehd)–An animal having extremely straight hind legs.

postmortem (pōst-mōr-tuhm)–An examination of an animal carcass or human body after death.

postnatal (pōst-nāt-ahl)–Subsequent to birth; relating to an infant immediately after birth.

postpartum (pōst-pahr-tuhm)–The period immediately following parturition (giving birth).

postprandial (pōst-prahn-dē-ahl)–After eating.

postrenal (pōst-rē-nahl)–After the kidney.

pot-bellied (poht behl-ēd)–Designating any animal that has developed an abnormally large abdomen, usually because of improper feeding, nutrition, or genetics.

potency (pō-tehn-sē)–(1) The power of a medicine to produce the desired effects. (2) The ability of an embryo to develop into a viable destiny. (3) The ability of the male of any animal species to fertilize the female germ cells.

poult (pōlt)–A young turkey or chicken.

poultice (pōl-tihs)–A hot, wet dressing applied to an injury or swelling for its softening and soothing properties.

poultry band (pōl-trē bahnd)–A strip of plastic or aluminum marked with numbers, etc., attached to the leg or wing of a bird for identification.

poultry (pōl-trē)–Any domesticated fowl raised for meat, eggs, or feathers.

pound (pownd)–(1) A unit of weight. The standard British unit of weight equals 453.59 grams. (2) An enclosure in which stray animals are legally confined. (3) An enclosure in which groups of animals, such as flocks of sheep, may be gathered for shelter, etc. (4) An enclosure used to trap wild animals.

pound of gain (pownd of gān)–In animal feeding, the net gain of weight in pounds derived from a particular number of pounds of feed fed.

pounding (pownd-ihng)–In the movement of a horse, the heavy contact of the foot with the ground that usually accompanies a high stride.

prance (prahns)–To walk with a swagger or nervous action, such as a horse.

pre op (prē ohp)–The common term for before surgery; preoperatively.

precipitin (prē-sihp-ih-tihn)–An antibody developed in the blood serum of an animal that has been injected previously with a foreign protein.

preconditioning (prē-kohn-dihsh-ohn-ihng)–Cattle that have been weaned, castrated, dehorned, vaccinated for several diseases, wormed, treated for grubs, and taught to eat from bunk-type feeders before being shipped to the feedlot.

precooling (prē-kool-ihng)–(1) The preliminary cooling of milk immediately after a milking to prevent spoilage. (2) The cooling of meats after slaughter and before cutting.

predicted difference (PD) (prē-dihck-tehd dihf-ər-ens)–The estimated difference of an animal from that of its parents or offspring.

predisposition (prē-dihs-pō-sih-shuhn)–(1) Stress or anything that renders an animal liable to an attack of disease without actually producing it. (2) The effect of one or more environmental factors that makes a plant vulnerable to infection by a pathogen.

preen (prēn)–Of a bird, to arrange or dress its feathers with its beak.

preen gland (prēn glahnd)–An oil-secreting gland at the root of the tail in most birds whose secretion dresses their feathers. Also called rump gland.

preference (prehf-rehns)–The selection of palatable plants over others by grazing animals.

pregnancy mare serum (prehg-nahn-sē mär sē-ruhm)–A gonadotrophic hormone, secreted by the uterus, obtained from the blood of pregnant mares.

pregnancy (prehg-nahn-sē)–The condition of having a developing fetus in the uterus. See gestation.

prehensile (prē-hehn-sihl)–Adapted for grasp and seizing.

prehension (prē-hehn-shuhn)–The act of seizing or grasping.

preictal (prē-ihck-tahl)–The period before a seizure; also called aura (aw-rah).

premix (prē-mihcks)–A ration mixed with various feedstuffs at the feedmill.

premolar (prē-mō-lahr)–The cheek teeth found between the canine teeth and molars; abbreviated P in dental formula.

prenatal (prē-nāt-ahl)–Occurring or existing before birth.

preparturient (pre-pahr-tōō-rē-ahnt)–Occurring before birth.

prepotency (prē-pō-tehn-sē)–The ability to transmit characteristics to offspring to an unusual degree.

prepotent (prē-pō-tehnt)–Designating an animal that transmits its character to its progeny to a marked or highly uniform degree.

preprandial (prē-prahn-dē-ahl)–Before eating.

prepuce (prē-pyoos)–The retractable fold of skin covering the glans penis; also called foreskin.

prerenal (prē-rē-nahl)–Before the kidney.

prescription drug (pər-skrihp-shuhn druhg)–Medication that may be purchased by prescription or from a licensed professional.

presentation (prē-sehn-tā-shuhn)–The portion of fetus that can be touched during labor.

pressure necrosis (prehs-shuhr neh-krō-sihs)–An area of dead tissue resulting from excessive pressure.

prevalence (preh-vah-lehns)–The number of cases of disease in a population at a certain time.

prevention (prē-vehn-shuhn)–To avoid; also called prophylaxis (prō-fih-lahck-sihs).

priapism (prī-ahp-ihzm)–A persistent penile erection not associated with sexual excitement.

prick (prihck)–To pierce or cut a muscle in the tail of a horse so that the tail will be carried higher; an unlawful act in many states.

primapara (prī-mihp-ah-rah)–A female that has borne one offspring.

primary feathers (prī-mār-ē fehth-ərz)–The outermost group of major wingfeathers (usually ten) located on the third joint of a bird's wing, hidden when the wing is folded. Also called flight feathers, flights.

primary SPF pig (prī-mār-ē ehs pē chf pihg)–A pig removed from its mother just before the normal birth date by a surgical process (cesarean section) and raised in laboratory isolation to prevent exposure to specific antigens; SPF stands for specific pathogen free.

primary (prī-mār-ē)–The first in order or the main cause.

primigravida (prih-mih-grahv-ih-dah)–The first pregnancy.

proboscis (prō-bohs-ihs)–An elongated nose, such as the snout of a hog.

process (proh-cehs)–A projection of a bone.

produce (prō-doos)–In animal breeding, a female's offspring.

proestrus (prō-ehs-truhs)–The phase of the estrous cycle just before estrus; characterized by the development of the ovarian follicle.

profile (prō-fī-uhl)–(1) A group of laboratory tests; also called screen or panel. (2) The general outline of an animal's body.

progenitor (prō-jehn-ih-tōr)–A parent or ancestor; an individual animal or plant that is recognized as the source of a certain type or character in its offspring.

progeny (proh-jehn-ē)–Offspring or descendants.

progesterone (prō-jehs-tər-ōn)–A hormone produced by the corpus luteum of the ovary that functions in preparing the uterus for pregnancy and maintaining it if pregnancy occurs.

proglottids (prō-gloht-ihdz)–The segments or parts of a tapeworm other than the head (scolex) and neck region.

prognosis (prohg-nō-sihs)–A prediction of the outcome of disease; to know before.

progressive (prō-grehs-ihv)–Of a disease, developing through successive stages, usually in a certain direction, whether improving or deteriorating.

projection (prō-jehck-shuhn)–The path of the x-ray beam.

prolactin (prō-lahck-tihn)–A hormone of the anterior pituitary gland that functions in stimulating the secretion of milk.

prolapse (prō-lahpz)–The protrusion of viscera; "to fall forward."

prolapse retainer (prō-lahpz rē-tā-nər)–A device used to support the prolapsed uterus of a ewe. The uterus is held in place until healing occurs.

prolapsed uterus (prō-lahpsd yoo-tər-uhs)–A condition in which the uterus has exteriorized through the lips of the vulva; also called uterine prolapse.

prolapsed vagina (prō-lahpsd vah-jī-nah)–The protrusion of the vagina through the vulva usually due to pressure from the fetus during parturition.

prolific (prō-lihf-ihck)–Having the ability to produce many offspring.

pronation (prō-nā-shuhn)–Rotating a limb or body so that the ventral surface is turned downward.

propagate (proh-pah-gāt)–To reproduce.

properties (prohp-ər-tēz)–Characteristics by which a substance can be identified.

prophase (prō-fās)–The first phase of cell division wherein many of the preparatory steps take place, such as shortening and thickening of the chromosomes, disappearance of the nuclear membrane, and formation of the spindle.

prophylactic (prō-fihl-ahck-tihck)–Preventive or protective treatment against disease.

prophylaxis (prō-fihl-ahcks-sihs)–Prevention.

proprioceptive (prō-prē-ō-sehp-tihv)–Spatial orientation or perception of movement.

proptosis (prohp-tō-sihs)–Displacement of the eye from the orbit.

prostaglandins (prohs-tah-glahn-dihnz)–A large group of chemically related fatty acids that have various physiological effects in an animal's body. Artificial prostaglandins are used to synchronize estrus in cattle.

prostate gland (proh-stāt glahnd)–The gland surrounding the urethra that secretes a thick fluid to aid in the motility of sperm.

prostatitis (prohs-tah-tī-tihs)–Inflammation of the gland that surrounds the urethra (the prostate), which secretes a thick fluid to aid the motility of sperm.

prosthesis (prohs-thē-sihs)–An artificial substitute for a diseased or missing part of the body.

prostration (prohs-trā-shuhn)–A symptom of an animal's ailment; pronounced weakness and loss of strength, resulting in the animal lying in prone position unable to rise.

protein (prō-tēn)–Any of a large number of complex, organic compounds of amino acids that have a high molecular weight and are essential to all living organisms. Proteins consist largely of carbon, hydrogen, nitrogen, and oxygen.

proteinuria (prō-tēn-yoo-rē-ah)–The presence of large organic compounds made of amino acids (protein) in the urine.

proteolysis (prō-tē-ohl-ih-sihs)–The process by which casein or some insoluble casein derivative is broken down to water-soluble compounds through the acting of organisms.

proteolytic bacteria (prō-tē-ō-liht-ihk bahk-tē-rē-ah)–(1) Bacteria that produce a proteolytic action on the proteins in cheese and cause a strong odor and taste to develop. (2) Bacteria that act on proteins, breaking them down to simpler compounds.

prothrombin (prō-throhm-bihn)–One of the clotting proteins in plasma.

prothrombin time (prō-throhm-bihn tīm)– A diagnostic evaluation of the number of seconds required for thromboplastin to coagulate plasma.

protocol (prō-tō-kawl)–Written procedure for carrying out experiments.

protoplasm (prō-tō-plahzm)–A basic material that makes up the cell (includes cell membrane, cytoplasm, and nucleus).

protoplast (prō-tō-plahst)–A unit of protoplasm in one cell.

protozoa (prō-tō-zō-ah)–A group of one-celled organisms that generally do not contain chlorophyll, including amoebae, paramecia, flagellates, and certain spore-forming organisms; sometimes classified as one-celled animals.

proved (proovd)–An animal whose ability to pass on specific traits is known and predictable.

proven sire (proov-ehn sī-ər)–A bull with at least ten daughters that have completed DHIA lactation records and are out of dams with completed DHIA lactation records.

provender (prohv-ehn-dər)–All dry feed or fodder for domestic animals.

proventriculitis (prō-vehn-trihck-yoo-lī-tihs)–Inflammation of the glandular (or true) stomach that often occurs in growing chicks reared in confinement and occasionally in adult fowls.

proventriculus (prō-vehn-trihck-yoo-luhs)–The elongated, spindle-shaped, glandular stomach of birds.

provitamin (prō-vī-tah-mihn)–A precursor of a vitamin; a substance from which an animal organism can form a vitamin. Carotene is provitamin A and ergosterol is provitamin D.

proximal (prohck-sih-mahl)–Nearest the midline or beginning of a body structure.

proximal convoluted tubules (prohcks-ih-mahl kohn-vō-lūd-ehd too-byoolz)–The hollow tubes located between Bowman's capsule and the loops of Henle.

proximate analysis (prohck-sih-maht ah-nahl-ih-sihs)–A system of analysis used to determine the total composition of nutrients in feed.

pruritus (proo-rī-tuhs)–Itchiness.

pseudoalbino (soo-dō-ahl-bī-nō)–A coat color of some animals that is a very light cream to white.

pseudohermaphrodite (soo-dō-hər-mahf-rō-dīt)–A person or animal having internal genital organs of one sex, while its external genital organs and secondary sex characters resemble in whole or in part those of the opposite sex.

psittacine (siht-ah-sēn)–A group of parrot-like birds that includes parrots, macaws, budgerigars, and parakeets.

psomophagia (sō-mō-fā-jē-ah)–The rapid eating and swallowing (bolting) of food without thorough chewing. Often induced by nervousness and anxiety and usually results in obesity; also psomophagy (sō-mō-fā-jē).

pteryla (tehr-ih-lah)–The feather tract of birds; plural is pterylae (tehr-ih-lā).

ptosis (tō-sihs)–Prolapse or drooping.

ptyalin (tī-ah-lihn)–An enzyme in saliva that digests starch.

ptyalism (tī-uh-lihz-uhm)–Excessive salvation; also called hypersialosis (hī-pər-sī-ahl-ō-sihs).

puberty (pū-behr-tē)–The time when sexual maturity is reached. In the female, ova on the ovaries begin to develop. In the male, sperm production is initiated in the testicles.

pubescent (pū-behs-ehnt)–Strictly, this means covered with soft, short, fine hairs; as commonly used, however, the term means hairy, bearing hairs, in a generalized sense, without reference to the type of hair.

pubic symphysis (pū-bihck sihm-fih-sihs)–The cartilaginous connection between the pubic bones on the ventral surface of the pelvis.

pubis (pū-bihs)–One of the paired bones constituting the pelvis that make up the floor of the pelvis.

puller (puhl-ər)–A tool used to remove worn horseshoes.

pullet (puhl-eht)–An immature female chicken.

pulmonary (puhl-mah-nār-ē)–The lungs.

pulmonary circulation (puhl-mah-nār-ē sihr-kyoo-lā-shuhn)–Blood flow out of the heart to the lungs and from the lungs to the heart.

pulmonary edema (puhl-mohn-ār-ē eh-dē-mah)–The accumulation of fluid in lung tissue.

pulmonary fibrosis (puhl-mohn-ār-ē fī-brō-sihs)–An abnormal formation of fibers in the alveolar walls.

pulmonary semilunar valve (puhl-mohn-ār-ē sehm-ē-loo-nahr vahlv)–The membranous fold located between the right ventricle and the pulmonary artery; semilunar means half moon.

pulse (puhlz)–The rhythmic expansion and contraction of an artery produced by the pressure of blood flowing through it; vital sign that tells the number of times the heart beats per minute; also called pulse rate; pulse is taken by palpation of an artery.

punch (puhnch)–(1) To herd cattle; to take care of cattle. (2) To forcibly drive sheep up hills when they are tired (New Zealand).

punctum (puhnck-tuhm)–A point or small spot.

punk (puhnck)–A small, scrubby horse.

pupa (pū-pah)–The stage between the larva and the adult stage in insects with complete metamorphosis, a nonfeeding and usually inactive stage; the plural is pupae (pū-pē).

pupil (pū-puhl)–The circular opening in the center of the iris.

pupillary light reflex (pū-puhl-ār-ē līt rē-flehcks)–The iris' response to light; when light is shown in, pupil constriction should take place; abbreviated PLR.

pups (puhpz)–Young mice and rats.

pure line (puhr līn)–A strain of organism that is comparatively pure genetically (homozygous) because of continued inbreeding, etc.

pure strain (puhr strān)–An animal that is similar to a purebred but the breeding program usually involves a greater degree of inbreeding.

purebred (puhr-brehd)–Designating an animal belonging to one of the recognized breeds of livestock.

purebred breeder (puhr-brehd brēd-ər)–A person who raises purebred animals, that may or may not be pedigreed.

purebreeding (puhr-brēd-ihng)–The practice of breeding animals from within the same breed or line; the production of purebreds.

purgative (puhr-gah-tihv)–A medicinal agent that actively empties the bowels. Aloes, arecoline, and calomel are examples of laxatives, drastic purgatives, and cholagogue purgatives.

purpura (pər-pə-rah)–A skin bruising condition characterized by hemorrhage into the skin.

purring (pər-rihng)–The controlled cyclic respiration in feline that alternate between activity of the diaphragm and laryngeal muscles.

purulent (pər-ū-lehnt)–Producing or containing pus.

pus (puhs)–An inflammatory product made up of leukocytes, cell debris, and fluid.

pustule (puhs-tyuhl)–A small, circumscribed, pus-filled skin elevation.

put off (puht ohf)–To lack the desire to copulate, such as a bull.

putrefaction (pyoo-treh-fahck-shuhn)–Foul-smelling decay.

putrid (pū-trihd)–Decomposed, rotten; said of organic materials.

pyelitis (pī-eh-lī-tihs)–Inflammation of the renal pelvis.

pyelonephritis (pī-eh-lō-neh-frī-tihs)–Inflammation of the renal pelvis and kidney.

pyemia (pī-ē-mē-ah)–A generalized infection in the bloodstream caused by pyogenic organisms resulting in the occurrence of numerous abscesses in various parts of the body.

pygostyle (pihg-ō-stī-uhl)–A bony termination of the vertebral column in birds where tail feathers attach; also called rump post. Not found in rumpless fowls.

pylorus (pī-lōr-uhs)–A narrow passage; the bottom aperature that connects the stomach and duodenum.

pyoderma (pī-ō-dər-mah)–A skin disease containing pus.

pyogenic (pī-ō-jehn-ihck)–Pus-producing.

pyometra (pī-ō-mē-trah)–Pus in the uterus.

pyothorax (pī-ō-thō-rahcks)–The accumulation of pus in the pleural cavity.

pyridoxine (pehr-ih-dohck-sēn)–Vitamin B_6 which appears to be necessary for the normal growth of chickens and pigs and for the prevention of a type of nerve disorder.

pyriform (pēr-ih-fōrm)–Pear-shaped.

pyuria (pī-yoo-rē-ah)–Pus in the urine.

Q

quadrivalent (kwohd-rih-vā-lehnt)–In genetics, a group of four associated homologous chromosomes.

quadruped (kwohd-roo-pehd)–Four footed.

quail (kwāl)–Any of the various species of small upland, gallinaceous game birds belonging to the genus *Coturnix* and allied genera of the family odontophoridae.

qualitative traits (kwohl-ih-tā-tihv trāts)–Traits having a sharp distinction between phenotypes, and which are usually controlled by only a few genes, e.g., various coat colors and the horned trait in domestic animals.

quantitative traits (kwohn-tih-tā-tihv trāts)–Traits that do not have a sharp distinction between phenotypes, and usually require some kind of measuring tools to make the distinctions. These traits are normally controlled by many pairs of genes, e.g., growth rate, milk production, and carcass quality. See genotype, phenotype.

quarantine (kwahr-ehn-tēn)–A regulation under police power for the exclusion or isolation of an animal to control the spread of disease.

quarter (kwahr-tər)–(1) In slaughtering for meat, one-half of the side of beef, as the forequarter or hindquarter. (2) Pertaining to a horseshoe, the branch between the last nail hole and the heel. (3) A unit of weight: (a) one-quarter cwt. (25 pounds) (avoirdupois); (b) eight bushels, formerly one-quarter ton (especially of grain.) (4) A section of the bovine udder. One-fourth of the udder.

quarter boot (kwahr-tər boot)–A leather piece fitted around a horse's forefoot to prevent self-injury from striking with the hind foot.

quarter clip (kwahr-tər klihp)–A clip on the shoe for the hind hoof of a horse. The clips are placed on the quarters of the shoe on the outside or inside to prevent the shoe from shifting laterally on the foot.

quarter crack (kwahr-tər krahck)–A vertical split in the wall of the hoof of a horse that results from improper hoof growth or shoeing. Also called sand crack.

quarters (kwahr-tərz)–(1) The parts of the body of a horse or other quadruped above the legs, as the breast and hips. (2) A place used to house workers on plantations or ranches. (3) Farm or ranch housing for domestic animals.

quaternary (kwaht-eh-nār-ē)–Fourth in order.

queen (kwēn)–An intact female cat.

queening (kwē-nihng)–Giving birth to cats.

quick (kwihck)–The dermis under the nail or hoof that has a blood supply and is sensitive.

quick-release knot (kwihk rē-lēs noht)–A knot that breaks loose easily.

quidding (kwihd-ihng)–A condition in which a horse drops food from the mouth while chewing.

quill (kwihl)–The hollow, horny part of a feather.

quittor (kwihd-ər)–-Festering of the foot anywhere along the border of the cononet; necrosis of the lateral cartilage of the third phalanx.

R

rabbit (rahb-iht)–Any of certain small mammals of the family leporidae, genus *Sylvilagus*.

rabbit hutch (rahb-iht huhch)–An off-the-floor cage or box for raising rabbits.

rabbit test (rahb-iht tehst)–(1) A pregnancy test in which virgin female rabbits are used. (2) A test for sweet clover poisoning in which rabbits are fed sweet clover hay and if no injurious effects appear the hay is considered safe for cattle.

race (rās)–(1) Pathogens of the same species and variety that are structurally indistinguishable but which differ in their physiological behavior, especially in their ability to parasitize varieties of a given host. (2) A narrow passage or fenced land in a sheep yard for branding, dipping, etc. (3) An elongated white mark on the face of a horse or dog.

rachis (rā-kuhs)–The distal end of the feather shaft.

rack (rahck)–(1) A fast, flashy, unnatural four-beat gait in which each foot meets the ground separately at equal intervals; also called single-foot. (2) A metal device that supports caging units. (3) The rib portion of a sheep carcass. (4) A framework for holding feed for cattle, swine, sheep, etc., with upright partitions so that the animal can insert its head between the partitions and have access to the feed.

raddle (rahd-l)–The coloring smeared on the chest of a ram to mark ewes when they have mated.

radiation therapy (rā-dē-ā-shuhn thehr-ah-pē)–The treatment of a neoplasm through the use of x-rays.

radiculitis (rah-dihck-yoo-lī-tihs)–An inflammation of the root of a spinal nerve.

radiograph (rā-dē-ō-grahf)–A record of ionizing radiation used to visualize internal body structures; also called x-ray; note that the term *graph* is used to mean record, as opposed to the term *gram*.

radiographic contrast medium (rā-dē-ō-grahf-ihck kohn-trahst mē-dē-uhm)–A substance used to show structures on an x-ray that are otherwise difficult to see; also called radiographic contrast material.

radiography (rā-dē-ohg-rah-fē)–A procedure in which film is exposed as ionizing radiation passes through the patient and shows the internal body structures in profile.

radioimmunoassay (rā-dē-ō-ihm-yoo-nō-ahs-ā)–A laboratory technique in which a radioactively labeled substance is mixed with a blood specimen to determine the amount of a particular substance in the mixture; also called radioassay.

radiologist (rā-dē-ohl-ō-jihst)–A specialist who studies internal body structures after exposure to ionizing radiation.

radiology (rā-dē-ohl-ō-jē)–The study of internal body structures after exposure to ionizing radiation.

radiolucent (rā-dē-ō-loo-sehnt)–The quality of appearing black or dark gray on a radiograph.

radiopaque (rā-dē-ō-pāg)–The quality of appearing white or light gray on a radiograph.

radius (rā-dē-uhs)–One of the forelimb long bones that articulate proximally with the humerus to form the elbow joint and distally with the bones of the carpus.

ragged (rahg-ihd)–Of the fur or hair of animals, shaggy, rough, and hanging in tufts.

ragged hips (rahg-ihd hihpz)–An irregular or poorly conformed rear quarters of an animal.

raise (rāz)–To grow or produce, such as to raise corn or cattle.

rale (rahl)–An abnormal, discrete rattle, crackle, or other noise heard during inspiration; also called crackles.

ram (rahm)–An intact male sheep.

ram jacket (rahm jahck-eht)–A jacket, designed to prevent coitus (breeding), placed on a ram to permit it to run with the flock before the breeding season.

ram lamb (rahm lahm)–A male sheep under one year of age.

ramus (rā-muhs)–A branch from a larger structure.

ranch (rahnch)–An expression used mostly in the western United States to describe a tract of land, including land and facilities, used for the production of livestock.

rancid (rahn-sihd)–Designating an offensive smell or taste resulting from the chemical transformation or putrefaction of fat, butter, milk, ice cream, and other products.

random mating (rahn-dohm mā-tihng)–A mating system with no selection where every male has an equal chance of mating with every female.

random sample (rahn-dohm sahmp-l)–A sample taken without bias from an area or from a population in which every part has an equal chance of being taken, in contrast to systematic sampling.

rangeland (rānj-lahnd)–(1) Land on which the natural plant cover is composed principally of native grasses, forbs, or shrubs valuable for forage. (2) Land used for grazing by livestock and big game animals that is dominated by grasses, grasslike plants, forbs, and shrubs.

rangy (rān-jē)–Designating an animal that is long, lean, leggy, and not too muscular in appearance.

ranny (rā-nē)–A poor-quality calf.

ranting (rahnt-ihng)–The restless activity of a boar as it matures sexually and as the breeding season approaches. It is usually accompanied by loss of appetite and weight.

rasorial (rah-zōr-ē-ahl)–Referring to the animals that scratch the ground to obtain food, such as a fowl.

rasp (rahsp)–A tool used for leveling the foot.

rat (raht)–*Rattus*, family Muridae; a long-tailed rodent that is much larger than a mouse.

ratchet (rah-cheht)–A graded locking portion of an instrument located near the finger rings.

rate of genetic improvement (rāt ohf jeh-neh-tihck ihm-proov-mehnt)–The rate of improvement of an animal per unit of time (year). The rate of improvement is dependent on: (a) heritability of traits considered; (b) selection differentials; (c) genetic correlations among traits considered; (d) generation interval in the herd; and (e) the number of traits of which selections are made.

rate of growth (rāt ohf grōth)–The rate at which a young animal increases weight and height.

ration (rah-shuhn)–The amount of food consumed by an animal in a 24-hour period.

ration of maintenance (rah-shuhn ohf mān-tehn-ehns)–The feed necessary to maintain the body of an animal.

ratios (rā-shōz)–The performance of an animal compared with its contemporaries, with 100 being average.

rattail (raht-tāl)–A slim, hairless tail of a horse.

rawhide (raw-hīd)–The underdressed skin of cattle.

rawhide quirt (raw-hīd kwihrt)–A riding whip made of rawhide.

reach (rēch)–The difference between the average merit of a herd or flock, in one or several traits, and the average merit of those selected to be parents of the next generation.

reactor (rē-ahck-tər)–An animal that reacts positively to a foreign substance, especially in testing for a disease, e.g., a tuberculous animal would be a reactor to tuberculin.

rearing (rehr-ihng)–To care for and support up to maturity, such as to raise animals or fish to adults.

reagent (rē-ā-jehnt)–Any substance involved in a chemical reaction.

reata (rē-ah-tah)–A lariat; a kind of strong Mexican rope made by twisting thongs of hide together, used in roping cattle on western ranges (United States).

receptor (rē-sehp-tər)–A sensory organ or three-dimensional unit on a membrane that receives information.

recessive (rē-sehs-ihv)–In genetics, a gene or trait that is masked by a dominant gene. See dominant.

reconstitute (rē-kohn-stih-toot)–To restore to the original form or condition by adding water, such as reconstituted milk.

rectal (rehck-tahl)–Pertaining to the rectum.

rectal palpation (rehck-tahl pahl-pā-shuhn)–A method of determining pregnancy, the phase of the estrus cycle, or a disease process via insertion of a gloved arm into the rectum of the animal and feeling for a specific structure.

rectovaginal (rehck-tō-vah-jī-nahl)–Pertaining to the rectum and vagina.

rectrices (rehck-trih-sēz)–Tail flight feathers; singular is rectrix (rehck-trihcks).

rectum (rehck-tuhm)–The distal portion of the large intestine.

recumbent (rē-kuhm-behnt)–Lying down.

red (rehd)–A common coat color for several species of animals. It can range from dark red, deep rich-red, blood-red, golden-red, light red, or yellow-red. Shades of red are most common in cattle.

red cell count (rehd sehl kownt)–The number of erythrocytes per cubic millimeter of blood.

red roan (rehd rōn)–A coat color of horses and cattle that is bay, chestnut, or brown with a red background color.

reduction (rē-duhck-shuhn)–The theory of using the minimal number of animals for a project that will yield valid results; one of the three R principles of Russell and Burch.

refinement (rē-fīn-mehnt)–The theory of inflicting minimal stress and pain to animals in research; one of the three R principles of Russell and Burch.

reflex (rē-flehcks)–An autonomic, involuntary response to change.

reflex ovulation (rē-flehcks ohv-yoo-lā-shuhn)–Ovulation that is triggered by the sexual act. Only a few animals do this, including the rabbit, and in these animals ovulation generally will not take place without such stimulus. Also called induced ovulation.

refraction (rē-frahck-shuhn)–The process of the lens bending the light rays to help them focus on the retina.

refractometer (rē-frahck-tōh-mē-tər)–An instrument for determining the deviation of light through objects; used to measure solute concentration.

regimen (reh-geh-mehn)–Directions.

registered (rehj-ih-stehrd)–An animal whose name, along with the name and number of its sire (father) and dam (mother), has been recorded in the record books of its breed association. The association gives the animal a number, known as a registration number.

registration certificate (rehj-ih-strā-shuhn sər-tih-fih-kiht)–A paper issued by a breed association that shows that a particular animal has been registered.

registry number (rehj-ih-strē nuhm-bər)–A number assigned to a particular purebred animal that has been registered with a breed association.

regression (reh-grehsh-uhn)–The measure of the relationship between two variables. The value of one trait can be predicted by knowing the value of the other variable.

regurgitate (rē-guhr-jih-tāt)–The return of swallowed food into the oral cavity.

regurgitation (rē-gərg-ih-tā-shuhn)–Backflow; used to describe backflow of blood due to imperfect closure of heart valves.

rein (rān)–To control, check, stop, guide, or back up a horse or horses with the reins.

reins (rānz)–The part of a horse's harness fastened to each side of the bit or curb by which the rider or driver directs and controls it.

reinsemination (rē-ihn-sehm-ih-nā-shuhn)–To repeat the process of insemination.

relaxation (rē-lahck-sā-shuhn)–The lessening of tension.

relaxin (rē-lahck-sihn)–An ovarian hormone produced at the time of parturition that is thought to aid in the relaxation of the birth canal.

remiges (rehm-ih-jehz)–The primary wing feathers; singular is remix.

remission (rē-mihs-shuhn)–A partial or complete disappearance of disease signs and/or symptoms.

remnant teeth (rehm-nahnt tēth)–The first premolar teeth of the horse, which usually remain embedded under the gum but occasionally erupt. Also called wolf teeth.

remount (rē-mownt)–A fresh horse used to substitute for another worn out or otherwise incapacitated horse. The term was once used to denote horses raised by government purchase as calvary horses.

remuda (rih-moo-dah)–A collection or string of broken horses.

renal (rē-nahl)–Pertaining to the kidney.

renal failure (rē-nahl fāl-yər)–The inability of the kidney(s) to function.

renal infarction (rē-nahl ihn-fahrck-shuhn)–Decreased blood flow to the kidney(s).

renal pelvis (rē-nahl pehl-vihs)–An area of the kidney that collects urine from the nephrons before it enters the ureters.

render (rehn-dər)–To melt down fat by heat.

rennin (rehn-ihn)–A coagulant enzyme occurring particularly in the gastric juice of cows, and also in some plants and lower animals.

repeatability (rē-pēt-ah-bihl-ih-tē)–The tendency of animals to repeat themselves in certain performance traits in successive seasons, pregnancies, or lactations.

replacement (rē-plās-mehnt)–(1) An animal that is raised for addition to the herd (one that replaces a less desirable animal). (2) The theory of using cell or tissue culture or mathematical models instead of animals in research if possible; one of the three R principles of Russell and Burch.

replacement animal (rē-plās-mehnt ahn-ih-mahl)–A young animal that is being raised to take the place of an older animal that is being culled.

reproduce (rē-prō-doos)–(1) Of animals, to bring forth young. (2) Of plants, to bear fruit and seeds.

reproduction (rē-prō-duhck-shuhn)–(1) The production of offspring by organized bodies. (2) The creation of a similar object or situation; duplication; replication.

reproductive system (rē-prō-duhck-tihv sihs-tehm)–The organs of the body, either male or female, involved in producing offspring.

repulsion (reh-puhl-shuhn)–In genetics, the condition in which an individual heterozygous for two pairs of linked genes receives the dominant member of one pair from one parent and the dominant member of the second pair from the other parent, e.g., AAbb x aaBB.

research (reh-suhrch or rē-suhrch)–The process whereby all effort is directed toward increased knowledge of natural phenomena and the environment as well as toward solving problems in all fields of science.

reservoir host (rehz-ər-vwahr hōst)–An animal in which an infectious agent lives and multiplies and depends upon primarily for survival. The reservoir host is usually not greatly affected by the infectious agent that it harbors.

resistant (rē-zihs-tahnt)–Not susceptible; designating an animal capable of withstanding disease, inclement weather, or other adverse environmental conditions. Also not susceptible to disease.

resolution (rehs-ō-loo-shuhn)–The ability to separately identify different structures on a radiograph or ultrasound.

respiration (rehs-pihr-ā-shuhn)–(1) An exchange of gases. (2) The vital sign that tells the numbers of respirations (one total inhale and one total exhale) per minute; also called respiration rate and abbreviated RR.

respirations (rehs-pihr-ā-shuhnz)–One inhalation and one exhalation.

respiratory (rehs-pih-tōr-e)–The system that brings oxygen from the air into the body for delivery via the blood to the cells.

respiratory rate (rehs-pihr-ah-tōr-ē rāt)– The number of respirations (one inhalation and one exhalation) per minute; abbreviated RR.

responsive mouth (reh-spohn-sihv mowth)–Designating a horse so trained that it easily obeys commands given by means of the reins.

rested pasture (rehst-ehd pahs-chər)–A pasture ungrazed for an entire growing season.

restricted feeding (reh-strihck-tehd fēd-ihng)–A system of feeding whereby feed is provided only during certain periods of the day. See free-choice feeding.

restriction enzymes (reh-strihck-shuhn ehn-zīmz)–Enzymes used in genetic engineering to remove a gene from a piece of DNA.

rest-rotation grazing (rehst rō-tā-shuhn grā-zihng)–An intensive system of management whereby grazing is deferred on various parts of the range during succeeding years, allowing the deferred part complete rest for one year.

resuscitation (reh-suhs-ih-tā-shuhn)–To restore life.

retained placenta (rē-tānd plah-sehn-tah)–The fetal membranes or afterbirth that a mammal mother fails to expel normally within a few hours after her young is born.

retained teeth (rē-tānd tēth)–The deciduous teeth that are not shed at the appropriate time.

retaining pen (rē-tān-ihng pehn)–A pen for holding cattle, sheep, and hogs at the time of dehorning, shearing, dipping, or weighing.

rete testis (rē-tē tehs-tihs)–A network of tubules located inside the testis in the mediastinum connecting the seminiferous tubules to the efferent ducts.

reticulocyte (reh-tihck-yoo-lō-sīt)–An immature erythrocyte characterized by polychromasia (Wright's stain) or a meshlike pattern of threads (new methylene blue stain).

reticuloendothelial system (reh-tihck-ū-lō-ehn-dō-theh-lē-ahl sihs-tehm)–A widely spread network of cells in the body concerned with blood cell formation, bile formation, and engulfing or trapping of foreign materials, which includes cells of bone marrow, lymph, spleen, and liver.

reticulum (rē-tihck-yoo-luhm)–The most cranial compartment of the ruminant forestomach. Has a honeycomb-textured lining, so is often called the honeycomb.

retina (reht-ih-nah)–The nervous layer of the eye that receives images located in the posterior chamber.

retinal detachment (reht-ih-nahl dē-tahch-mehnt)–Separation of the nervous layer of the eye from the choroid; also called detached retina.

retinol (reht-ih-nohl)–Form of vitamin A found in animals.

retinopathy (reht-ih-nohp-ah-thē)–Any disorder of the retina.

retired to pasture (rē-tī-ərd too pahs-chər)–Designating an animal whose productive life has ended and which is turned out to graze.

retired to the stud (rē-tī-ərd too theh stuhd)–Designating a horse that no longer races but is retained for breeding purposes.

retractile (rē-trahck-tīl)–The ability to draw back; feline claws can be drawn back.

retractor (rē-trahck-tər)–An instrument used to hold back tissue.

retractor muscle (rē-trahck-tohr muhs-uhl)–The part of the male reproductive system that helps extend the penis from the sheath and draws it back after copulation.

retrograde (reh-trō-grād)–Going backward.

retrograde pyelogram (reh-trō-grād pī-lō-grahm)–A radiographic study of the kidney and ureters in which a contrast material is placed directly into the urinary bladder.

retroperitoneal (reh-trō-pehr-ih-tō-nē-ahl)–Behind or underneath the peritoneum.

rheum (room)–See catarrh.

rhinitis (rī-nī-tihs)–Inflammation of the nasal mucous membranes.

rhinopneumonitis (rī-nō-nū-moh-nī-tihs)–Inflammation of the nasal cavity and lungs.

rhinorrhea (rī-nō-rē-ah)–Nasal discharge.

rhoncus (rohn-kuhs)–An abnormal, continuous, musical, whistling sound heard during inspiration or expiration; also called wheeze; plural is rhonchi (rohn-kī).

riata (rē-ah-tah)–See lariat.

rib (rihb)–(1) The flat, curved bone that forms the thoracic wall. (2) (a) A cut of meat made from five to eight ribs, depending upon the method of cutting the forequarter.

(b) Any cut of meat that comes from the rib section, as a rib roast. (3) In judging of livestock, "spring of rib" indicates conformation and vigor. (4) To divide a side of beef into a fore and hind quarter. (5) A prominent vein.

ribbed-up (rihbd uhp)–Said of a horse on which the back ribs are well arched and incline well backwards, bringing the ends closer to the point of the hip and making the horse shorter in coupling.

riboflavin (rī-bō-flā-vihn)–Vitamin B$_2$; lactoflavin; C$_{17}$H$_2$N$_4$O$_6$; an essential nutrient for people and animals, riboflavin functions as a coenzyme concerned with oxidative processes.

ribonucleic acid (RNA) (rī-bō-nū-klē-ihck ah-sihd)–The substance in the living cells of all organisms that carries genetic information needed to form protein in the cell.

rickettsia (rih-keht-sē-ah)–The bacterium transmitted by lice, fleas, ticks, or mites.

ride (rīd)–(1) To mount a horse for going from one place to another. (2) To push down a fence, as a cow. (3) To mount a cow, as another cow indicating heat.

ridgeling (rihdg-lihng)–A cryptorchid male horse; also called rig.

riding stable (rīd-ihng stā-buhl)–A stable where saddle horses are maintained by the individual owners or are offered for hire by the management.

rigging ring (rihg-ihng rihng)–The ring attached to the saddle tree for a cinch.

rigor mortis (rihg-ōr mōr-tihs)–A physiological process following the death of an animal in which the muscles stiffen and lock into place.

ring (rihng)–(1) (a) A circular band of metal or wood, such as the metal ring in the nose of a bull. (b) To place a ring through the cartilage of the nose of an animal, e.g., to prevent a hog from harmful rooting or to control a bull, etc. (2) (a) A circular, metal or plastic band placed on the leg of a fowl for identification purposes. (b) To place a ring on the leg of a fowl. Also called ringing birds. (3) A ridge that encircles the horns of a cow, the number increasing with age. (4) A circular exhibition place for the showing or sale of livestock or the racing of horses.

ring bone (rihng bōn)–A bony enlargement, involving the pastern bones just above the hoof, that interferes with the action of the joints and tendons, thus causing lameness to an animal. Usually only seen in horses.

ring test (rihng tehst)–A milk test for brucellosis.

ringing (rihng-ihng)–(1) The act of implanting a wire ring through a pig's nose to discourage rooting. (2) Clipping the wool from a breeding ram around the neck, belly, and penis region in order to facilitate proper mating.

ringworm (rihng-wehrm)–A common name for a skin disease of humans and animals caused by parasitic fungi, usually marked by distinct, circular patches with a scaly appearance.

ripper (rihp-pər)–An unusually strong, large horse.

rising two (rī-zihng too)–Describing a horse 1½ to 2 years old. Also called coming two.

roan (rōn)–(1) A coat color of a horse that is chestnut or bay or may be red or strawberry roan, blue roan, or chestnut roan depending upon the intermingling of the background colors. (2) Designating the red-white color phase of shorthorn cattle.

roarer (rōr-ər)–A wind-broken animal that makes a loud noise in drawing air into the lungs.

roaring (rōr-ihng)–A defect in the air passage of a horse that causes him to roar or whistle when respiration is forced.

roaster (rōst-ər)–A chicken of either sex that weighs between 3½ and 5 pounds and is less than eight months old. See broiler, fryer.

robust (rō-buhst)–Designating a strong vigorous animal or plant.

rodent (rōd-nt)–A classification of mammals, mostly vegetarians, characterized by their single pair of chisel-shaped, upper incisors. Rodents are members of the orders Rodentia (rats, mice, squirrels, etc.) and Lagomorpha (rabbits, etc.).

rodenticide (rō-dehn-tih-sīd)–Any poison that is lethal to rodents.

rodeo (rō-dē-ō)–(1) The rounding up of cattle. (2) A public performance presenting such features as cattle round-up, lariat throwing, bronco riding, and bulldogging.

rods (rohdz)–The specialized cells of the retina that react to light.

roentgen (rehnt-kihn)–The international unit of radiation.

rogue (rōg)–A horse without inclination to work or cooperate with its handler.

rolling (rō-lihng)–An excessive side-to-side shoulder motion. Common in horses with abnormally wide fronts or chests.

Roman-nosed (rō-mahn nōzd)–Refers to a horse or other animal having a profile that is convex from poll to muzzle.

ronguers (rohn-jūrz)–Forceps with cupped jaws used to break large bone pieces into smaller ones.

roost (roost)–(1) A resting or lodging place for fowls. (2) A group of roosting fowls. (3) To rest upon a roost or perch.

rooster (roo-stər)–An intact male chicken; also called cock.

rooting (root-ihng)–Digging the earth with the snout, as hogs.

rope burn (rōp buhrn)–A skin irritation caused by contact pressure and speedy movement of a rope upon the surface of the skin of a human or animal.

rope halter (rōp hahl-tər)–A halter made of rope instead of leather.

rope walking (rōp wahlk-ihng)–The leg motion in which the horse swings the striding leg around and in front of the supporting leg in walking or running.

roped (rōpd)–Designating an animal that has been lassoed or tied.

rosette (rō-seht)–A swirl hairgrowth pattern in the Abyssinian guinea pig.

Ross test (rohs tehst)–A urine test for the presence of ketone bodies, used in detection of acetonemia in cattle and sheep. See acetonemia.

rostral (rohs-trahl)–The nose end of the head.

rot (roht)–A state of decay caused by bacteria or fungi. See decay.

rotated pasture (rō-tā-tehd pahs-chər)–(1) A pasture in the regular crop rotation that is grazed for a few years, usually two or three, and then plowed for other crops. (2) A pasture that is divided into segments by use of fences; the livestock being confined to one segment at a time in a definite rotation pattern.

rotation (rō-tā-shuhn)–A circular movement around an axis.

rotation grazing (rō-tā-shuhn grā-zihng)– Grazing forage plants on well-managed pastures in such a manner as to allow for a definite recovery period following each grazing period.

rotational crossbreeding (rō-tā-shuhn-ahl krohs-brēd-ihng)–A system of crossing two or more breeds where the crossbred females are bred to bulls of the breed contributing the least genes to that female's genotype.

rotten (roht-ehn)–(1) Designating decomposed or putrid organic matter. (2) Designating ground or soil extremely soft and yielding because of decay, or rocks partially decomposed. (3) Designating sheep attacked by rot.

rough (ruhf)–(1) To remove the major part of the plumage of a fowl leaving scattered feathers for the finisher. (2) In meat judging, designating uneven contour or uneven distribution of fat on a carcass. (3) Designating food of low quality. (4) Designating a horse not properly trained. (5) A calk for the horse's shoe.

rough coat (ruhf kōt)–A coarse, tangled, unkempt animal's coat which may indicate a lack of thrift or be the result of the animal remaining in the open during the winter.

roughage (ruhf-ahj)–A type of feed that is high in fiber and low in total digestible nutrients; examples include pasture and hay; also called forage.

round (rownd)–The center portion of the hindquarter of a beef animal; the cut of beef taken from the round.

round-up (rownd-uhp)–(1) The deliberate gathering of domestic animals, usually range cattle, for branding, fastening ear tags, injections, pesticide applications, inventory, and potential sales.

round window (rownd wihn-dō)–A membrane that receives sound waves through fluid after it has passed through the cochlea.

roundworms (rownd wehrmz)–A parasitic, unsegmented cylindrical nematode worm, of the class Nematoda, that primarily invades the gastrointestinal tract.

rowel (row-ehl)–In horseriding, the wheel of a spur with blunt to sharp projecting points.

rubber sleeve (ruhb-ər slēv)–A protective, rubber covering attached to a rubber glove for the hand, arm, and shoulder which is used by veterinarians and artificial inseminators in various phases of their work.

rubbing post (ruhb-ihng pōst)–Any of several devices consisting mainly of a post set in the ground with a can of oil attached to the top, so arranged that the oil or medication is allowed to slowly drip or run down the sides of the post so that when hogs or other livestock rub against it a small amount of oil will be deposited on the rubbed or infected area of the skin. Also called hog oilers, livestock oilers.

rubefaciant (roo-beh-fā-shehnt)–Liniment, plaster, or any substance that produces redness of the skin when applied to it.

ruddy (ruhd-dē)–An orange-brown color with ticking of dark brown or black.

rudimentary (roo-dih-mehn-tār-ē)– Incompletely developed.

rudimentary copulatory organ (roo-dih-mehn-tār-ē koph-ū-lah-tōr-ē ōr-gahn)–A very small, shiny or glistening eminence in chicks, which is located in the median portion of the fold between the urodaeum and the proteodaeum. By examination of this organ the sex of day-old chicks can be determined.

rudimentary teat (roo-dih-mehn-tār-ē tēt)–An undeveloped teat that may or may not be connected with milk-secreting tissue.

ruga (roo-gah)–An irregular fold in the mucous membranes; plural is rugae (roo-gā).

rugged (ruhg-ihd)–Designating size, strength, and vigor in the body build or conformation of an animal.

rumen (roo-mehn)–The largest compartment that serves as a fermentation vat of the ruminant fore-stomach. Also called the paunch.

rumen fistula (roo-mehn fihs-tyoo-lah)–A fistula of the rumen or first stomach of a cud-chewing animal.

rumen magnet (roo-mehn mahg-neht)–A smooth oblong magnet that is placed in the rumen to collect small metal objects that are swallowed by the animal during grazing.

rumenology (roo-mehn-oh-lō-jē)–A branch of animal science concerned with the study of the rumen.

rumenotomy (roo-mehn-oh-tō-mē)–The operation of cutting into the rumen to remove foreign bodies or to observe activity.

ruminant (roo-mihn-ehnt)–A cud-chewing animal that has a fore-stomach that allows for fermentation of ingesta.

ruminating (roo-mehn-ā-tihng)–Chewing cud.

rumination (roo-mehn-ā-shuhn)–A series of digestive functions in ruminants that includes regurgitation, remastication, and reswallowing of regurgitated food.

rump (ruhmp)–That part of the rear end of an animal that includes the buttocks or fleshy part of the rear quarters.

rumping (ruhm-pihng)–A method of restraining sheep by placing them in a sitting position with the front legs elevated; also called tipping.

run (ruhn)–(1) Unrestricted movement, as the colts have the run of the pasture. (2) A fenced-in pen used for the exercise of animals or poultry. (3) To feed; to graze, as steers run on the open range. (4) To maintain animals, as he runs sheep. (5) To work a dog with sheep or cattle. (6) To discharge pus, as a sore runs. (7) To move rapidly, as a horse.

run free (ruhn frē)–To be loose, as animals that are not restrained and are allowed to move about freely in the barn, yard, or pasture.

run on (ruhn ohn)–To graze or pasture on, as to run on the range.

running out (ruhn-ihng owt)–The condition in which an improved variety of animal is reverting to a former and inferior type or is losing some of its desired qualities.

running through (ruhn-ihng throo)–Lactation extended beyond 365 days.

running walk (ruhn-ihng wawlk)–A slow, four-beat gait intermediate in speed between a walk and rack.

runt (ruhnt)–Any animal smaller than normal that is culled and not used for breeding. Smallness may be due to genetics, injury, or disease. Also called puny.

rupture (ruhp-chər)–Forcible tearing.

rustler (ruhs-lehr)–A cattle thief.

rut (ruht)–The season of heightened sexual activity in male mammals that coincides with the season of estrus in the female. See estrus.

rutter (ruht-ər)–A female mammal, such as a cow, that remains in heat constantly as the result of some abnormality. Also called a buller.

rutting season (ruh-tihng sē-zohn)–The recurring, usually annual, period when deer, cattle, etc., are in heat. See estrous cycle.

S

sable (sā-buhl)–A color pattern with a cream-colored undercoat and black guard hairs on the feet, tail, and mask.

saccule (sahck-yoo-uhl)–One of the small, hairlike sacs of the inner ear that are partially responsible for equilibrium.

sacculectomy (sahck-yoo-lehck-tō-mē)–The surgical removal of a saclike part; usually refers to surgical removal of the anal sacs.

sack (sahck)–To stroke or rub a horse. (A highly nervous horse is frequently sacked by the buster who, from a safe position, soothes the animal by rubbing it with a burlap sack or broom.)

sacrum (sā-kruhm)–The fourth set of fused vertebrae that are found where the pelvis attaches to the spinal column; abbreviated S in vertebral formula.

saddle (sah-duhl)–(1) A seat designed to fit a horse's back and to make riding easy, comfortable, and safe. (2) The rear part of the back of a male chicken extending to the tail, covered by saddle feathers.

saddle gait (sah-duhl gāt)–The foot movement or gait of a saddle horse; a definite manner of walking, running, etc., such as the walk, running walk, trot, canter, rack.

saddle gall (sah-duhl gahl)–A sore or wound on the back of a horse that results from abrasion by a saddle.

saddle girth (sah-duhl gihrth)–In horses, a strong band or strap that encircles the body, usually just behind the front legs, and helps fasten a saddle to the back. Also called cinch.

saddle horn (sah-duhl hōrn)–A knoblike elevation at the front center of a western saddle, used in roping as an anchor for the lariat.

saddle skirt (sah-duhl skərt)–The leather part of a western-type saddle that extends beyond the sides and to the rear of the saddle seat.

saddle sore (sah-duhl sōr)–(1) A sore on the horse's back caused by an improperly fitted saddle. (2) A sore on the rider caused by the chaffing of the saddle.

saddlebred (sah-duhl-brehd)–Of, pertaining to, or designating a type of horse that has the breeding and qualities desired in a saddle horse.

saddle-broken (sah-duhl-brō-kehn)–Designating a horse trained to the saddle.

sagittal plane (sahdj-ih-tahl plān)–An imaginary line dividing the body into unequal right and left halves.

saline (sā-lēn)–A common term for 0.9% NaCl (normal saline); abbreviated NS.

saliva (sah-lī-vah)–A secretion from glands in the oral cavity that moistens food, aids in bolus formation, and contains small amounts of digestive enzymes.

salivary gland (sahl-ih-vahr-ē glahnd)–The oral cavity gland that secretes saliva.

salivation (sahl-ih-vā-shuhn)–A discharge of saliva.

salpingitis (sahl-pihn-jī-tihs)–(1) Inflammation of a fallopian tube (oviduct). (2) An inflammation of the oviduct of poultry characterized by a discharge from the duct that causes irritation of the vent and a smeary appearance of the feathers below it.

salt (sahlt)–(1) Sodium chloride (common salt), NaCl; a white crystalline compound occurring abundantly in nature as a solid or in solution.

salve (sahv)–Usually a preparation, consisting of a fatty, greasy, or jellylike base with added medications, which has properties of soothing, softening, healing, etc.

sand crack (sahnd krahck)–A vertical split that can develop at any part of the wall of a horse's hoof but is commonly seen at the frog in the hind foot and at the inside quarter of the forefoot. Also called quartercrack.

sandy bay (sahn-dē bā)–A coat color of a horse that is a light bay.

sanitary trap (sahn-ih-tehr-ē trahp)–A part of a vacuum pump installation in a milking machine that is designed for collecting and holding unwanted materials. It can be easily emptied and cleaned.

sanitation (sahn-ih-tā-shuhn)–The developing and practical application of measures designed to maintain or restore healthful conditions, such as the treatment, removal, or destruction of contaminated or infested materials and possible sources of infection or infestation.

sanitize (sahn-ih-tīz)–To disinfect and make sanitary the utensils used in preparation of food products, as dairy utensils.

saprogenesis (sahp-rō-jehn-eh-sihs)–That period in the life cycle of a pathogen in which it is not directly associated with a living host and in which it may be either dormant or living as a saprophyte. See saprophyte.

saprophyte (sahp-rō-fīt)–Plants, including certain bacteria and molds, capable of obtaining nutrients and energy from dead organic matter.

sarcoma (sahr-kō-mah)–A malignant neoplasm arising from any type of connective tissue.

sarcous (sahr-kuhs)–Of, or pertaining to, flesh or muscle.

satiety (sah-tī-eh-tē)–The loss of desire to eat or drink (water).

saturated (sahch-eh-rāt-ehd)–Filled to the maximum capacity under existing conditions.

scab (skahb)–(1) The crust formed over a wound or sore. (2) A mangy disease of animals. (3) Any of the numerous fungal or bacterial diseases of plants characterized by dark or rough, scablike spots, as apple scab, cereal scab, cucumber scab, gladiolus scab, potato scab.

scald (skahld)–To dip in hot water or to pour on hot liquid, as to scald a killed bird to loosen its feathers.

scalding vat (skahld-ihng vaht)–A tank, vat, or barrel used to hold very hot water for the dipping and scalding of slaughtered hogs or poultry to cause the hair or feathers to loosen for easy removal.

scale (skāl)–(1) Any instrument used to determine weight. (2) Either of the pans of a balance. (3) An instrument or device on which graduated spaces for measurement have been stamped or attached. (4) Any thin flake that peels from a surface, such as from skin. (5) One of the thin, flat, membranous plates that forms the outer, protective covering of certain animals or parts thereof, as the shanks of birds, etc. (6) To form into scales. (7) To cover with scales or with an incrustation.

scalpel (skahl-puhl)–A small, straight knife with a thin, sharp blade used for surgery and dissection.

scalping (skahlp-ihng)–In horses, a striking of the front of the hind coronet, pastern, or cannon against the front foot when running.

scapula (skahp-yoo-lah)–The flat bone in the pectoral girdle that articulates distally with the humerus; also known as the shoulder blade.

scar (skahr)–A mark left by a healing lesion where excess collagen was produced to replace injured tissue; a wound that has healed.

scarious (skahr-ē-uhs)–Thin, dry, and membranous; often translucent.

scent gland (sehnt glahnd)–A saclike secretory organ located at the base of the tail in snakes.

Schiotz' tonometer (shē-ohtz toh-noh-meh-tər)–An instrument that measures intraocular pressure by direct application of weights to the cornea.

schooled horse (skoold hōrs)–A horse, usually of a light-weight breed, that is trained or drilled for some special performance. See high school horse.

Schwann cell (shwahn sehl)–A type of neuroglia cell found in the peripheral nervous system that wraps around the axons and forms the myelin sheath.

scirrhous cord (skihr-uhs kōrd)–In castrated animals, abnormal swelling of the cut end of the spermatic cord caused by infection. It is usually accompanied by discharge of thick, white pus from the castration wound, and may be chronic.

sclera (sklehr-ah)–The fibrous outer layer of the eye that maintains the shape of the eye; also known as the white of the eye.

scleral injection (skleh-rahl ihn-jehck-shuhn)–Dilation of blood vessels into the sclera.

scleritis (skleh-rī-tihs)–Inflammation of the sclera.

scolex (skō-lehcks)–The anchoring or holdfast organ of a tapeworm that serves for the attachment of the parasite to the intestinal lining. Also called head.

scour (skowr)–(1) To cleanse the bowels of an animal by purging. (2) To rub off the flesh sticking to a hide.

scouring (skow-ər-ihng)–Cleaning wool.

scours (skowrz)–Diarrhea in livestock.

scout film (skowt fihlm)–A plain radiograph without the use of contrast medium.

scrag end (skrahg ehnd)–The nape or back part of the neck, especially in a sheep. Also called scrag.

scratch feed (skrahch fēd)–The grain part of a ration for poultry that consists of cracked or whole grain (corn, wheat, oats, etc.) or a mixture of such grains.

scratches (skrahch-ihz)–A low grade infection or scab in the skin follicles around the fetlock; also called grease heel.

scrotal circumference (skrō-tahl sehr-kuhm-fehr-ehns)–A measurement of testes size obtained by measuring the distance around the testicles in the scrotum with a tape. Testicle size is related to semen-producing capacity.

scrotal hernia (skrō-tahl her-nē-ah)–A protrusion of loops of intestine or other viscera into the scrotum through an enlarged inguinal ring or through an accidental opening in the inguinal region, which results in the scrotum being greatly enlarged.

scrotal hydrocele (skrō-tahl hī-drō-sēl)–A fluid-filled cyst in the testes or along the spermatic cord.

scrotum (skrō-tuhm)–The external sac that enclosed and supports the testes; also called scrotal sac.

scud (skuhd)–In tanning, to scrape a dehaired and trimmed hide or pelt to remove any hairs, lime, or other materials remaining in the hair follicles.

scurfy (skuhrf-ē)–Covered with small scales.

scurs (skərz)–A hornlike substance that grows from the skin at the location of the horn pits on polled animals.

scurvy (scər-vē)–A common term for vitamin C deficiency.

season (sē-zohn)–(1) A portion of the year, e.g., a growing season, when plants grow; the fire season, when danger of fire is greatest; the grazing season, when grazing is possible. (2) An animal that is ready for breeding is said to be in season.

seasonal breeder (sē-zohn-ahl brē-dər)–An animal that breeds naturally during a certain season of the year and rarely at other seasons, e.g., sheep, goats, etc., are seasonal breeders in the fall.

seasonal distribution (sē-sohn-ahl dihs-trih-byoo-shuhn)–(1) In pasturage, the progressive grazing in a sequence of moves from one part of a range to another as vegetation develops.

seasonal grazing (sē-zohn-ahl grā-zihng)–The grazing of a range only during a certain period or periods of the year that roughly correspond to one or more of the four seasons.

sebaceous (seh-bā-shuhs)–Referring to anything that secretes or resembles fatty matter; oily, greasy, as the secretions of the sebaceous glands of the skin.

sebaceous cyst (seh-bā-shuhs sihst)–A closed sac of yellow, fatty material.

sebaceous gland (seh-bā-shuhs glahnd)–A dermally located gland that secretes sebum (oil).

seborrhea (sehb-ō-rē-ah)–A skin condition characterized by overproduction of sebum (oil).

sebum (sē-buhm)–An oily substance produced by dermally located glands.

secondaries (sehck-ohn-dahr-ēz)–The long stiff wing feathers growing from the middle wing segments of a fowl.

secondary (sehck-ohnd-ār-ē)–Second in order or resulting from another (primary) process.

secondary infection (sehck-ehn-dahr-ē ihn-fehck-shuhn)–An invasion by a second and different organism after a first organism has become established in a host animal or plant.

secretin (sē-krē-tihn)–A hormone produced in the intestine (duodenum) that controls the secretion of the pancreatic juices or enzymes.

secretion (sē-krē-shuhn)–The metabolic act or process of synthesis and liberation of substances from cells or glands of animals or plants, as saliva, milk, etc.

secundines (sē-kuhn-dīnz)–The afterbirth; the placenta.

sed rate (sehd rāt)–An abbreviation for sedimentation rate; also abbreviated SR.

sedative (sehd-ah-tihv)–A drug that calms an animal.

seed (sēd)–(1) Offspring or progeny. (2) Sperm or semen.

seed stock (sēd stohck)–(1) Pedigreed or well-bred livestock that are maintained for breeding purposes.

segmentation (sehg-mehn-tā-shuhn)–A constriction of intestinal parts without movement of the ingesta backwards or forwards; "mixing."

segregation of genes (sehg-reh-gā-shuhn ohf jēnz)–The separation of genes on two homologous chromosomes and their distribution to separate gametes during gametogenesis; genes that once existed in pairs in a cell will become separated and distributed to different gametes.

seizure (sē-zhər)–A sudden, violent, involuntary contraction of muscles caused by a brain disturbance; also called convulsion.

selection (seh-lehck-shuhn)–Choosing certain individuals for breeding purposes in order to propagate or improve some desired quality or characteristic in the offspring.

selection difference (seh-lehck-shuhn dihf-ər-ehns)–The difference between the average for a trait in selected cattle and the average for the group from which they came. Also called reach.

selection index (seh-lehck-shuhn ihn-dehcks)–A formula that combines performance records from several traits or different measurements of the same trait into a single value for each animal. Selection indexes weigh the traits for their relative net economic importance and their heritabilities plus the genetic associations among the traits.

selection threshold (seh-lehck-shuhn threhs-hōld)–A trait or standard established by a breeder, below which the breeder will reject potential breeding animals.

selective breeding (seh-lehck-tihv brē-dihng)–The breeding of selected plants or animals chosen because of certain desirable qualities or fitness, as contrasted to random or chance breeding.

selective grazing (seh-lehck-tihv grā-zihng)–The tendency for livestock and other ruminants to prefer certain plants and to feed on these while grazing little on other species.

selenium (seh-lē-nē-uhm)–A nonmetallic element, related to sulfur, rarely present as more than a trace in soil.

selenodont (sē-lēn-ō-dohnt)–Animals with teeth that have crescents on their grinding surfaces, i.e., ruminants.

self (sehlf)–One-color fur.

self-feeder (sehlf fēd-ər)–(1) In poultry or other livestock farming, any feeding device by means of which the animals can eat at will, e.g., a hopper that supplies feed by gravity to a boxlike trough. (2) Any automatic mechanism for supplying materials to a given point in an operation.

self-sucker (sehlf-suhck-ər)–A cow that has the habit of sucking its own teats.

semen (sē-mehn)–The ejaculatory fluid that contains sperm plus secretions from the seminal vesicles, prostate gland, and bulbourethral glands (if present in species).

semen collector (sē-mehn kohl-ehck-tər)–Any device or receptacle used for the collection of semen from males of domestic animals. It may consist of an artificial cloaca or vagina, a test tube, a small beaker, etc.

semen plasma (sē-mehn plahs-mah)–The fluid portion of semen.

semen tank (sē-mehn tahnck)–A portable tank used to transport or store frozen semen.

semicircular canals (sehm-ī-sihr-kyoo-lahr kahn-ahlz)–Three half-circle tubes that contain fluid and hairlike cells within the inner ear; the three canals are the vestibular, tympanic, and cochlear.

seminal fluid (sehm-ih-nahl flū-ihd)–Semen.

seminal vesicle (sehm-ih-nahl vehs-ih-kuhl)–A gland that opens into the vas deferens as it joins the urethra.

seminiferous tubules (seh-mih-nihf-ər-uhs too-byoolz)–Channels in the testes in which sperm develop and through which sperm leave.

semipermeable membrane (sehm-ē-pehr-mē-ah-bl mehm-brān)–A membrane that permits the diffusion of one component of a solution but not the other. In biology, a septum that permits the diffusion of water but not of the solute.

senility (seh-nihl-ih-tē)–A condition of organisms that in old age revert to development resembling younger stages.

sensitivity (sehn-sih-tihv-ih-tē)–(1) The degree of susceptibility of a bacterial isolate to individual antibiotics. (2) The probability that a test will correctly identify the patients that are infected or have a condition.

sensitization (sehn-sih-tī-zā-shuhn)–The condition of being allergic or sensitive to, for example, a vaccine.

sensorium (sehn-sō-rē-uhm)–The whole sensory apparatus of the body, principally sight, smell, hearing, taste, and touch (which includes pain).

sensory neurons (sehn-sōr-ē nū-rohnz)–Peripheral nerve cells that send impulses from sense organs to the spinal cord.

sensus (sehn-suhs)–Meaning.

separator (sehp-ah-rā-tohr)–A mechanical device for separating cream from milk by centrifugal force.

sepsis (sehp-sihs)–A state of contamination or poisoning by pathogenic bacteria.

septate (sehp-tāt)–Containing or divided by a septum or septa. (2) Fungus hypha or spores that have cross-walls.

septic (sehp-tihck)–(1) Relating to, or caused by, the presence of disease-producing organisms, or their toxins, in the blood or tissues. (2) Putrefactive; putrefying. (3) A term sometimes used to refer to conditions where dissolved oxygen is absent and decomposition is occurring anaerobically, as in a septic tank.

septicemia (sehp-tih-sē-mē-ah)–A blood condition in which pathogenic microorganisms or their toxins are present.

septum (sehp-tuhm)–A separating wall or partition.

sequel (sē-kwehl)–A condition following as a consequence of a disease.

sequela (sē-kwehl-ah)–A consequence of disease.

sequestrum (sē-kwehs-truhm)–A piece of dead bone that is partially or fully detached from the adjacent healthy bone.

serological (sē-rō-lohj-ih-kahl)–Designating certain laboratory tests used in detecting diseases that involve the use of the liquid portion of the blood.

serology (sē-rohl-ō-jē)–The laboratory study of serum and the reactions of antigens and antibodies.

seroma (sehr-ō-mah)–The accumulation of serum beneath the surgical incision.

serration (sihr-ā-shuhn)–A sawlike edge or border.

Sertoli cells (sihr-tō-lē sehlz)–Specialized cells in the testis that support and nourish sperm growth.

serum (sē-ruhm)–(1) The clear portion of any animal liquid that is separated from its cellular elements. It usually applies to the amber-colored liquid called blood serum, which separates in the clotting of blood from the clot and corpuscles. (2) The term is applied often erroneously to skimmed milk.

serum neutralization (sē-ruhm nū-trahl-ih-zā-shuhn)–A test employed with certain viral infections to determine the presence or absence of specific antibodies in the blood serum of animals by mixing varying amounts of serum with a constant amount of known virus and determining afterwards by animal inoculation if inactivation of the virus has taken place.

serve (sərv)–To copulate with a female; to cover, as a stallion, bull, or other male animal.

sesamoid (sehs-ah-moyd)–A small, flat bone embedded in a tendon or joint capsule.

set (seht)–(1) To put in a particular place, position, condition, direction, adjustment, etc. (2) To put eggs under a fowl or in an incubator to hatch them. (3) To put a price or value on something. (4) To fix a trap or a net to catch animals or fish. (5) To adjust into proper apposition the ends of a broken bone, as to set a fracture.

set stocking (seht stohck-ihng)–Keeping animals in a given area of pasture continuously at a predetermined level of stocking for a defined period.

setae (sē-tā)–Sensitive bristles that grow on the heads of many birds; also called bristles.

setting hen (seht-ihng hehn)–A broody hen in the act of incubating eggs.

settle (seht-uhl)–Breeding successfully; said of a mare when she becomes pregnant.

settled (seht-uhld)–Designating a pregnant cow or mare.

sex (sehcks)–The distinction between male and female plants and animals. Ova (macrogametes) are produced by the female and sperm (microgametes) by the male. The union of these distinctive germ cells results in a new individual.

sex character (sehcks kahr-ahck-tər)–The peculiarity of appearance, other than the presence of sex organs, that distinguishes one sex from another, such as the thick neck and bold, rugged head of a bull as contrasted to the more finely turned neck and head of a cow. Technically, these are secondary sex characteristics.

sex chromosomes (sehcks krō-mō-zōmz)–A pair of chromosomes in animals that determines the sex of the progeny depending upon which one is distributed; one sex usually has two of the same kind of sex chromosome in its cells and the other sex has two kinds of sex chromosomes; in mammals the female is XX and the male is XY; in birds, the male is ZZ (or XX) and the female is ZW (or XY).

sex feathers (sehcks fehth-ərz)–(1) The feathers of a chick that designate the sex. The male's feathers are different from the female's. (2) The two top tail-feathers that characterize the mature males of most breeds of duck. These two tail-feathers at their outer ends turn upward and forward making a pronounced curl. This characteristic feathering is useful in sex determination.

sex-limited traits (sehcks lihm-iht-ehd trātz)–The traits expressed only in one sex although the genes for the trait are carried by both sexes.

sex-linked (sehchs lihnckd)–The characteristic developed from genes located on the sex chromosomes. The characteristic may be used to determine the sex of an animal, e.g., barring of Barred Plymouth Rock chickens.

sex linkage (sehcks lihnck-ahj)–The association of character with sex due to the fact that the gene for that characteristic is in a sex chromosome.

sex ratio (sehcks rā-shō)–The relationship existing between the number of male and female animals within a given herd or band.

sexual (sehck-yoo-ahl)–A method of reproduction that involves the union of sperm and an egg.

sexual dimorphism (sehcks-yoo-ahl dī-mōrf-ihzm)–The physical or behavioral differences between females and males of a given species.

sexual infantilism (sehcks-yoo-ahl ihn-fahn-tih-lihzm)–The failure to develop sexually.

shaft (shahft)–The quill or central part of a contour feather; also called scapus (skā-puhs).

shank (shahnck)–(1) That part of the leg joining the knee to the ankle in humans, or the corresponding part in various animals. (2) The tarsus of birds: the part of the leg, below the hock joint, that is covered with scales.

shank feathering (shahnck feh-thər-ihng)–Those feathers that grow on the outer side of the shank of certain birds, as those on the Cochin, Brahma, and Langshan breeds of chicken.

shear (shēr)–To shave off wool, hair, or fur.

shearing (shēr-ihng)–The act or operation of removing wool from a sheep or goat by means of shearing machines.

shearing lamb (shēr-ihng lahm)–In marketing, a subclass of lambs that are in full fleece and sold primarily for their wool.

shearling (shēr-lihng)–(1) A sheep after being shorn for the first time. (2) Pelt of slaughtered sheep carrying one-fourth to one inch growth of wool.

sheath (shēth)–The enclosing and protecting tubular structure within which the penis of a stallion, bull, and certain other male animals is retracted.

shed (shehd)–(1) A simple structure to shelter livestock, store feeds, or both; it is open on one or more sides and may be separate from or attached to another structure. (2) To lose or cast off something, as animals shed hair; birds molt or shed feathers.

shedder (shehd-ehr)–A brushlike device used to remove dead hairs from an animal's coat.

shedding (shehd-ihng)–Normal hair loss.

sheen (shēn)–Shininess or luster.

sheep (shēp)–Any of a variety of cud-chewing mammals, genus *Ovis*, related to goats, especially *Ovis aries*, that includes many breeds domesticated for their heavy wool, edible flesh (mutton), or skin.

sheep bell (shēp behl)–A bell hung about the neck of a sheep to indicate the location of the flock.

shell (shehl)–(1) A hard outer covering of certain animals, such as turtle, mollusk, or crustacean, which when ground may be used as a poultry feed supplement. (2) The hard calcareous exterior covering of an egg.

shell membranes (shehl mehm-brānz)–Either of the two membranes, the inner and outer shell membranes, surrounding the egg substance and located immediately inside the shell of the egg. At the large end of the egg, the membranes are separated to form the air cell.

shell quality (shehl kwahl-ih-tē)–The various characteristics of the shell of an egg, including thickness, porosity, texture, etc., which may vary widely between hens and may be improved through selective breeding.

shepherd's crook (shehp-ehrdz kruhck)–The staff used by a shepherd; specifically a staff with a crook at one end which is used in catching and holding a sheep (used by Old World shepherds since ancient times).

shin (shihn)–The anterior edge of the tibia.

shin boot (shihn boot)–A shaped, leather piece fitted to the skins and cannon bone of horses, especially saddle and running horses, to protect against injury from striking with a hoof when in motion.

shives (shīvz)–Small particles of vegetable matter other than burrs present in wool.

shoat (shōt)–A pig of either sex, usually between 60 and 160 pounds. This term is used infrequently. Also spelled shote.

shock (shohck)–(1) Inadequate tissue perfusion. (2) A small pile of hay. (3) A state of collapse in animals, resulting from severe loss of blood following surgery, accident, or sometimes under conditions of undue stress.

shod (shohd)–An equine with horseshoes.

shoddy (shohd-ē)–(1) That leather made by grinding waste leather to a pulp and pressing it into solid sheets with or without the addition of a binding material. (2) Reworked wool that has been recovered from wool cloth trimmings, rags, etc. (3) Anything of an inferior quality.

shoebox (shū-bohcks)–A type of caging that has a solid bottom flooring.

shorn (shohrn or shōrn)–(1) Designating a sheep, lamb, or goat that has had its fleece removed by shearing. (2) Referring to wool that has been sheared.

short pastern bone (shōrt pahs-tərn bōn)–In ungulates, phalanx II.

short-coupled (shōrt kuhp-uhld)–The term used to describe a horse having a short distance (usually not more than four fingers width) between the last rib and the point of the hip.

shote (shōt)–A young hog of either sex that is weaned but which usually weighs less than 150 pounds. Also spelled shoat. This term is an old term that is infrequently used.

shoulder (shōl-dər)–(1) The region around the large joint between the humerus and scapula. (2) The upper foreleg and/or the adjacent part cut from the carcass of a hog, sheep, or other animal; the upper part of the carcass to which the foreleg is attached. (3) The part of a bird at which the wing is attached.

shunt (shuhnt)–To bypass or divert.

shy (shī)–(1) Of, or pertaining to, a horse that is easily frightened or startled; timid; skittish. (2) To act in a startled manner, as a frightened horse may break gait and jump violently to one side, or otherwise show fright until he has a chance to observe the cause of this fear.

shy breeder (shī brē-dər)–A male or female of any domesticated livestock that has a low reproductive efficiency. It is associated most often with males or females from highly developed breeds.

sib (sihb)–In genetics, a brother or sister.

sib testing (sihb tehst-ihng)–A method of livestock selection in which an animal is selected on the basis of the performance of its brothers and sisters.

sibling (sihb-lihng)–One of several sons and/or daughters of the same parents. See sib.

sib-mating (sihb mā-tihng)–Mating between siblings, two or more individuals having one or both parents in common.

sick (sihck)–Suffering from illness of any kind; not sound or in fit condition.

sickle-hocked (sihck-ehl hohckd)–A term applied to an animal when the hind legs set too far forward, giving the impression of a sickle when viewed from the side.

sickles (sihck-ehlz)–The longer, curved tail feathers of a male chicken as distinct from smaller, similarly curved tail-coverts, the lesser sickles.

side (sīd)–(1) In the body of an animal, either of the two lateral surfaces or parts extending from the shoulders to the hips. (2) An entire lateral half of a beef, veal, or pork carcass split lengthwise through the backbone and the corresponding midundersurface. (3) Flesh taken from the lateral surface of an animal carcass, as a slab taken from the side of a hog carcass and used for preparing a side of bacon.

side-bone (sīd bōn)–Ossification of the lateral cartilages of a horse's foot, noticeable as hard, bony projections immediately above and toward the rear quarter of the hoof head. It is an unsound condition occurring especially in draft horses, as a result of injury, conformation, excessive use, or perhaps old age. Also called shell bone.

side check (sīd chehck)–A leather strap attached at the side of a horse's bridle to limit side movement of the head. See check rein.

side step (sīd stehp)–Stepping sideways; a foot movement expected of a show horse and desirable in a well-trained pleasure horse, as it assists in lining up horses in a show ring. Also called transversal or sidepass.

sigmoid flexure (sihg-moyd flehck-shər)–An S-shaped fold in the penis of a bull, ram, or boar that straightens during erection and allows it to extend from the sheath for copulation.

sign (sīn)–A characteristic of disease that can be observed by others.

silage (sī-lahj)–A type of roughage feed that is produced by fermenting chopped corn, grasses, or plant parts under specific moisture conditions to ensure preservation of feed without spoilage.

silo (sī-lō)–A pit, trench, aboveground horizontal container, or vertical cylindrical structure of relatively air-tight construction into which green crops, such as corn, grass, legumes, or small grain and other feeds are placed and converted into silage for later use as a livestock feedstuff.

silver-gray (sihl-vər grā)–A coat color of a horse that is light grey with a white mane and tail.

singe (sihnj)–To burn lightly, as to subject the carcass of a pig, fowl, etc., to a flame in order to remove bristles or hair, or to sear the ends of stems of flowers (which have a copious flow of sticky sap) to prolong freshness.

single comb (sihng-gehl kōm)–In chickens, a comb consisting of a single, fleshy, serrated (usually five-pointed) formation which extends from the beak backward over the crown of the head.

single cross (sihng-gehl krohs)–The crossing of one strain, variety, inbred line, or breed of plants or animals with a different strain, variety, inbred line, or breed. The seed, plant, or offspring so produced is designated as a single cross or F_1. See double cross, three-way cross.

single-rigged saddle (sihng-gehl rihgd sah-duhl)–A saddle that has one cinch.

sinoatrial node (sī-nō-ā-trē-ahl nōd)–A small mass of tissue located in the right atrium that regulates heart rhythm.

sinus (sīn-uhs)–An air-filled or fluid-filled cavity within a bone; also means channel.

sinusitis (sī-nū-sī-tihs)–Inflammation of a sinus.

sinusotomy (sī-nuhs-oht-ō-mē)–Surgical incision into a sinus.

sire (sī-ər)–(1) The male parent. (2) To father or to beget; to become the sire of, as to sire a fine cow.

sire evaluation (sī-ər ē-vahl-yoo-ā-shuhn)–An objective program designed and conducted by a breed association to increase the effectiveness of sire selection and to evaluate a bull's true genetic merit. It measures differences in sires using progeny information. It is the best measure of a bull's true transmitting ability.

sire index (sī-ər ihn-dehcks)–A mathematical index for measuring a bull's transmitting ability in terms of milk and butterfat. It tends to measure what production may be expected of the daughter when a bull is mated to cows of known productivity.

sitomania (sī-tō-mā-nē-ah)–Excessive craving for food. (Closely related to cynorexia, bulimia, and polyphagia.)

sitting breed (sih-tihng brēd)–A breed of poultry that retains the natural tendency to become broody once or twice a year.

six-tooth sheep (sihcks tooth shēp)–A three-year-old sheep.

skeletal muscle (skehl-eh-tahl muhs-uhl)–A type of striated muscle that is responsible for movement; movement is voluntary.

skewbald (skyoo-bohld)–A color pattern in horses involving white spots on any color but black.

skin (skihn)–(1) The flexible integument that forms the external covering of an animal, especially of vertebrates. (2) The pelt of a small animal, such as calf, sheep, goat, fox, mink, etc., which is usually dressed, tanned, or intended for such treatment. (3) To strip the skin from.

skin tag (skihn tahg)–A small growth that hangs from the body by a stalk.

skirt (skərt)–(1) The diaphragm muscle, the muscular part of the membrane that separates the abdominal from the thoracic cavity. It is edible meat. (2) To remove inferior grade wool from fleece. (3) See saddle skirt.

slat (slaht)–A sheepskin after the wool has been pulled, but before other treatment is given.

slaughter (slaw-tər)–(1) The butchering of cattle, sheep, and other animals for food. (2) To kill cattle, sheep, and other livestock for food; to butcher. (3) Designating an animal suitable for slaughter.

sleeper (slē-pər)–An unbranded calf; a maverick.

slick (slihck)–Unmarked stock; unbranded calf. Also called a maverick.

slick ears (slihck ērz)–Unbranded range horses or animals having no ear-slits to mark ownership. Also called slicks.

sling (slihng)–(1) A bandage for supporting part of the body. (2) A swine restraint device with four leg holes and an additional hole under the neck for blood collection; the device looks like a hammock.

slink (slinck)–(1) The young of an animal brought forth prematurely or abortively, especially a calf. (2) The flesh or skin of such a calf or other animal.

slip (slihp)–(1) An incompletely castrated male. (2) Curdled milk. (3) To abort. (4) To palpate a membrane by allowing it to fall through one's fingers or hands.

slipped tendon (slihpd tehn-dohn)–A crippling condition in young growing birds that results from a nutritional disturbance.

slitlamp examination (sliht-lahmp ehcks-ahm-ih-nā-shuhn)–Visual testing of the cornea, lens, fluids, and membranes of the interior of the eye using a narrow beam of light.

slobber (slohb-ər)–The saliva dripping or drooling from the mouth of an animal, usually a condition found following extreme exertion or excitement, or after eating certain types of feed; sometimes a result of infection.

slop (slohp)–A homemade, wet feed mixture made from flour mill by-products, grains, whole or crushed, and other waste or surplus farm products, such as kitchen waste, skim milk, etc.

slow breeding (slō brē-dihng)–(1) Delayed conception or settling after insemination. Also called delayed settling. (2) Designating an animal that lacks sexual desire. See shy breeder.

slow gait (slō gāt)–One of the several forward movements or gaits of horses, faster than a walk, but slower than a canter. There are three slow gaits: the running walk, the fox trot, and the slow pace.

slow gallop (slō gahl-ohp)–See canter.

slow poke (slō pōk)–The last cow in a herd to come up; a trailer.

smear (smēr)–Material smeared on a surface, usually a small piece of glass (slide), and examined under a microscope.

smegma (smehg-mah)–A dried and hardened, fatty secretion found in the sheaths of stallions and geldings, bulls, etc. When not removed by washing, excessive accumulations can shut off the urinary passageway.

smoky (smō-kē)–(1) Of the color of smoke, a brownish or bluish shade of gray; dusky; cloudy. (2) An abnormality in the appearance of a horse's eye, which becomes cloudy, whitish, and pearly colored. It is indicative of impaired vision. The condition is referred to as smoky eye.

smooth going (smooth gō-ihng)–(1) Designating an easy gait or forward movement of a horse used for riding and driving. (2) Designating any practice that proceeds at a brisk, even pace without particular hindrance.

smooth mouth (smooth mowth)–A condition where no cups are present in the permanent teeth.

smooth-mouthed sheep (smooth mowthd shēp)–An aged sheep, usually a ewe, whose incisor teeth are missing or so badly worn out that it cannot eat normally; a gummer. Also called broken-mouthed.

smooth muscle (smooth muhs-uhl)–A type of unstriated muscle; movement is involuntary.

smother (smuhth-ər)–To kill by suffocation, as panicky chickens or sheep are smothered when they pile up in a heap. (2) Of plants, to kill by covering thickly with dirt, straw, ice sheet, black sheet plastic, etc. (3) Of plants, to retard in growth or to kill by a dense or more vigorously growing crop.

snaffle (snahf-ehl)–Any type of mouth control used in handling horses, especially in breaking.

snap (snahp)–(1) A spring fastening or clasp that closes with a clicking sound, used as a fastener on harnesses, gate chains, dog chains, etc. (2) In horses, especially in harness or saddle horses, to move briskly with animation; to pick up feet quickly and sharply.

snare (snār)–A restraint method in which the pig's snout is secured via a loop tie that is attached to a long handle; also called hog snare.

snip (snihp)–A white marking between the nostrils of a horse.

snood (snowd)–The long, fleshy extension at the base of a turkey's beak. Also known as dew bill.

snout (snowt)–The projecting part of an animal's head that contains the nose and jaws, as the snout of a hog.

snub (snuhb)–To control an animal by roping it and tying the rope to an adjacent post or other solid object to restrict its movement.

snub down (snuhb down)–To place the hand on or over the nose of an animal to stop its breathing in order to control it.

snubbing post (shuh-bihng pōst)–Any smooth, sturdy post to which animals are fastened to restrict their movement.

snubbing ring (shuh-bihng rihng)–A heavy metal ring attached to the floor, or any other solid object, through which a control rope is passed and by means of which an animal's movement is restricted as in slaughtering or treating for an ailment.

soilage (soy-lahj)–Freshly cut green fodder fed to confined animals.

soiling (soyl-ihng)–In livestock management, the cutting and bringing of green forage to livestock in place of allowing the animals to eat the green feed where it grows.

sole (sōl)–The ventral surface of the hoof.

solid manure (sohl-ihd mah-noo-ər)–Manure in solid form as contrasted to liquid manure.

solitary (sohl-ih-tehr-ē)–Occurring singly or in pairs, not in colonies.

soluble (sohl-ū-bl)–Capable of changing form or changing into a solution.

soma (sō-mah)–The body, in contrast to the germ or germ plasm.

somatic (sō-maht-ihck)–(1) Designating body tissues. (2) Having chromosomes in pairs, one of each pair normally coming from the female parent and one from the male, as contrasted with terminal tissue which gives rise to germ cells.

somatic cell count (sō-mah-tihck sehl cownt)–A determination of the number of cells (leukocytes, epithelial cells, etc.) in milk to test for mastitis; abbreviated SCC.

sonogram (soh-nō-grahm)–A record of the internal body structures by recording echoes of pulses of sound waves above the range of human hearing.

sorrel (sohr-ehl)–A light shade of chestnut coat color in horses.

sound (sownd)–(1) In good condition, not damaged; acceptable. (2) Designating an animal free from blemishes of any kind. (3) Designating wool that has good strength.

soundness (sownd-nehs)–Freedom from defect or blemishes as: (a) a wool fleece that shows strong wool fiber and no weak spots; (b) an animal that is free from any noticeable defect; (c) an animal that shows good feet and legs.

sour (sowr)–(1) A term used to denote a horse that has been overworked or trained to the point that he refused to perform. (2) A flavor defect most frequently associated with dairy products, such as milk, butter, and ice cream, which is characterized by a sour taste and odor resulting from the reduction of milk sugar to lactic acid by lactic acid-producing bacteria.

sow (sow)–An intact female pig or guinea pig.

sow production breeding value (SPBV) (sow prō-duhck-shuhn brē-dihng vahl-ū)–The value assigned to a sow based on records on all the litters produced by the sow as well as the estimates of heritability and repeatability of the traits. SPBV is an estimate of the ability of a sow to pass her productivity on to her offspring. See sow productivity index.

sow productiviy index (sow prō-duhck-tihv-ih-tē ihn-dehcks)–An index used to compare and identify the top producing sows in a herd.

sp. (spp.)–The abbreviation, singular (plural) for species, e.g., when one species of a genus of plant or animal is referred to, the name may be written *Canis* sp. meaning one species of dog; or *Canis* spp. meaning more than one species of the genus *Canis*.

span (spahn)–A pair of animals usually harnessed together as a team. Also spelled spann.

spasm (spahzm)–An involuntary, sudden, painful, and violent muscular contraction.

spasticity (spahs-tihs-ih-tē)–Uncontrolled contractions of skeletal muscles.

spavin (spahv-ihn)–A disease affecting the hock joint of a horse's hind leg, usually a bony growth on the inner, lower part of the hock that causes lameness. See bog spavin, bone spavin, occult spavin.

spawn (spawn)–A common term for eggs and sperm.

spay (spā)–To remove the ovaries and uterus of a female animal.

spayed heifer (spād hehf-ər)–A heifer that has had both ovaries removed.

species (spē-shēz)–In the naming of plants and animals, Latin is used. Each kind of plant or animal can be identified by genus (plural, genera) and species (both singular and plural), e.g., the generic name (genus) of the domestic cat is *Felis* and the species name is *catus*.

specific (speh-sihf-ihck)–(1) A medicine that cures a particular disease. (2) Pertaining to a species. (3) Produced by a particular microorganism. (4) Restricted by nature to a special animal, thing, etc. (5) Exerting a peculiar influence over any part of the body.

specific gravity (speh-sihf-ihck grah-vah-tē)–A measurement that reflects the amount of wastes, minerals, and solids in a substance.

specific pathogen free (speh-sihf-ihck pahth-ō-jehn frē)–A management system in which animals are obtained from cesarean section and raised in isolation to prevent certain infectious diseases; abbreviated SPF; not disease free.

specificity (spehs-ih-fih-sih-tē)–The quality of having a certain action or reacting only with certain substances.

speckle (spehck-ehl)–A small patch or dot of color.

spectacle (spehck-tah-kuhl)–A transparent, highly vascular, unshed, abnormal covering over the cornea of some reptiles; also called an eyecap.

speculum (spehck-yoo-luhm)–An instrument to enlarge the opening of a canal or cavity.

spent (spehnt)–Of animals, especially horses, completely exhausted.

sperm (spərm)–The male sex cell, produced by the testicles.

spermatic cord (spər-mah-tihck kohrd)–A fibrous tissue that suspends the testes. It contains the vas deferens, the blood vessels, the nerves of the testicle, and a small muscle.

spermatid (spər-mah-tihd)–A haploid cell produced from the second division of meiosis in spermatogenesis that has not yet undergone the changes to form a sperm cell.

spermatocyte (spər-mah-tō-sīt)–One of the various kinds of cells produced in the development of male gametes in animals.

spermatogenesis (spər-mah-tō-jehn-eh-sihs)–The production of male gametes (sperm cells).

spermatogonium (spər-mah-tō-gō-nē-uhm)–A primary germ cell in the testis that will undergo spermatogenesis to produce spermatozoa. The plural is spermatogonia (spər-mah-tō-gō-nē-ah).

spermatozoon (spər-mah-tō-zō-uhn)–The male sex cell or gamete; plural is spermatozoa (spər-mah-tō-zō-ah).

spermiogenesis (spər-mē-ō-jehn-eh-sihs)–The part of the process of spermatogenesis involving the changes that permit spermatids to become spermatozoa.

sphincter (sfingk-tər)–A ringlike muscle that constricts an opening.

sphygmomanometer (sfihg-mō-mah-nohm-eh-tər)–An instrument used to measure blood pressure.

spinal (spī-nahl)–Referring to the spine or vertebral column.

spinal cavity (spī-nahl kahv-ih-tē)–A hollow space that contains the spinal cord within the spinal column.

spindle (spihn-dl)–The fine threads of achromatic protoplasm arranged in a fusiform mass within the cell during mitosis.

spine (spīn)–(1) A stiff, sharp-pointed outgrowth on an animal. (2) The vertebral column; the backbone.

spirochetes (spī-rō-kētz)–Spiral-shaped bacteria.

spirometer (spī-rohm-eh-tehr)–An instrument that measures the flow of air in and out of the lungs.

spit (spiht)–The common term for saliva.

splayed (splād)–A common fault found in horses or other animals where the front hooves are turned out and the heels turned in; also called splay-footed.

spleen (splēn)–A mass of lymphatic tissue located in the cranial abdomen that produces lymphocytes and monocytes, filters foreign material from the blood, stores red blood cells, and maintains the appropriate balance of cells and plasma in the blood.

splenectomy (splehn-ehck-tō-mē)–Surgical removal of the spleen.

splenomegaly (splehn-ō-mehg-ah-lē)–Enlargement of the spleen.

splint (splihnt)–A rigid or flexible appliance for fixation of movable or displaced parts.

splint bones (splihnt bōnz)–The common name for metacarpals II and IV or metatarsals II and IV in equine.

splints (splihntz)–A hard, bony enlargement on the splint bone that is located on the inside of the fore cannon bone of the leg of a horse.

split (splihnt)–To separate thick hides into layers in the preparation of leather.

split litter (splihnt lih-tər)–A rabbit litter in which some of the rabbits are born, and are followed some time later by the remaining rabbits.

spoilage (spoy-lihj)–(1) Hay or forage that has been improperly cured or stored. (2) Any objectionable change that has occurred in a food, feed, or material.

spondylitis (spohn-dih-lī-tihs)– Inflammation of the vertebrae.

spondylosis (spohn-dih-lō-sihs)–Any degenerative condition of the vertebrae, usually involving bony bridging of adjacent vertebrae.

sporadic disease (spō-rahd-ihck dih-zēz)–A disease that occurs in scattered or isolated instances.

spore (spōr)–A resistant, oval body formed within some bacteria.

sporocyst (spō-rō-sihst)–In flukes, or trematodes, the intermediate asexual generation or second larval stage and, in some cases, the third larval stage.

sporozoa (spō-rō-zō-ah)–A class of protozoa that consists of many parasitic species with complicated life cycles.

sporozoite (spō-rō-zō-īt)–A small, usually elongated, infective stage of sporozoan parasites, such as *coccidia*, *plasmodia* (malaria), etc.

sport (spōrt)–A random mutation.

sporulation (spōr-ū-lā-shuhn)–(1) In bacteria, the formation of spores within the body of the bacterium. The spores represent the inactive resting, or resistant, forms. (2) In coccidia, a kind of reproduction by which the fertilized cell within the oocyst wall splits up into new individuals, called sporozoites. Sporulation of coccidial oocysts usually occurs after the oocyst has been discharged from the body of the host. (3) The process of spore formation.

spot treatment (spoht trēt-mehnt)–The application of a pesticide or other material to a restricted or small area of heavier infestation.

spotter bull (spoh-tər buhl)–A vasectomized male bovine used to find and mark female bovines in estrus.

sprain (sprān)–A severe wrench or strain of the parts around a joint that causes pain, swelling, and difficulty in moving.

spraying (sprā-ihng)–Urination on objects to mark territory.

spread (sprehd)–(1) A straddle; the difference in price between two delivery months in the same or different markets, or the sale of one thing against a simultaneous purchase of another. Straddling between a foreign and domestic market is often referred to as arbitrage. (2) A ranch, including the buildings and the extent of land grazed by cattle or sheep (western United States). (3) The distribution of a disease. (4) To disseminate disease.

spreader (sprehd-ər)–(1) An instrument provided with hooks or flanges used to spread open the incision made in the body wall of a bird. (2) An animal, etc., capable of acting as a parasite or disease vector. (3) A species of microorganism that tends to grow profusely over the entire surface of the culture medium.

spring chicken (sprihng chihck-ehn)–A young chicken, usually only a few months old.

spring lamb (sprihng lahm)–A lamb that is marketed in the spring of the year and prior to July 1; they are usually born in the fall.

springing (sprihng-ihng)–Anatomic changes in a ruminant that indicate parturition is near.

springing cow (sprihng-ihng kow)–A pregnant cow or heifer that shows signs of the approaching birth of its young, as evidenced by a relaxation of the ligaments and muscles on either side of the tail-head and also by a slight elevation of the tail-head. Also called springing heifer.

springing heifer (sprihng-ihng hehf-ər)–A young female bovine that is pregnant with her first calf.

sprite (sprīt)–A spayed female ferret.

spur rowel (spər rowl)–A small wheel with radiating points, attached at the end of a horseman's spur as an added goad.

spurs (spərz)–An abnormal projection (as from a bone) or a sharp, horn-covered, bony projection from the shank of male birds of some species.

sputum (spyoo-tuhm)–Phlegm ejected through the mouth; lower respiratory tract secretion.

squab (skwohb)–A nestling pigeon raised for its flesh, which may be marketed when fully feathered under the wings and just before it is ready to leave the nest, usually at twenty-five to thirty-five days of age when it may weigh from 12 to 24 ounces.

square-gaited (skwahr gā-tehd)–Designating the gait of a horse in which the action is straight on all four feet.

squeeze chute (skwēz shoot)–A narrow stall with a hinged side that is used for restraining animals. The animal's head is caught in a head-catch chute, and the sides of the chute are moved against the animal to restrict movements.

stag (stahg)–(1) In animals, a male castrated after reaching sexual maturity and showing pronounced sexual development. (2) A horse that is thick and coarse in the throat latch and crest from late castration. (3) An imperfectly or recently castrated sheep or steer. (4) In marketing, an uncastrated male chicken with flesh slightly darkened and toughened and with comb and spur development showing the bird to be in a state of development between a roasting chicken and a cock. (5) A boar hog usually castrated after having passed breeding usefulness. (6) A wild mature male deer.

stake (stāk)–(1) To tether an animal to a stake.

stall (stawl)–The space in a barn that is occupied by a single animal, such as a dairy cow or horse, for feeding and handling.

stall barn (stawl bahrn)–A barn used for sheltering dairy cattle and/or young stock where the adult animals are confined to stalls by means of stanchions, straps, halters, or chains during part of the year, such as in the winter and for milking. Roughages and concentrates may be fed at the individual stalls. None, part, or all of feeds and bedding may be stored in the structure. Also called stanchion barn.

stall-feeding (stawl fē-dihng)–A system of management where animals are housed more-or-less continuously (except for exercise), and forage crops are cut and carried to them.

stallion (stahl-yuhn)–An intact male horse four years old and over.

stampede (stahm-pēd)–A wild rush of cattle or horses as a result of fright.

stanchion (stahn-shuhn)–A restraint device that secures cattle around the neck to allow accessibility for milking, feeding, and examining.

standard cow day (stahnd-ahrd kow dā)–A feeding index for a cow that is 16 pounds total digestible nutrients per day.

stand (stahnd)–(1) The density of game per acre, area, etc. (2) A stallion's court. (3) To cease walking or moving; to take or keep a certain position. (4) To rise to the feet. (5) Of a stallion, to be available for breeding purposes.

standing heat (stahnd-ihng hēt)–A female that freely accepts the male.

stapes (stā-pēz)–The auditory ossicle known as the stirrup.

staphylococci (stahf-ih-lō-kohck-sī)–A grapelike cluster of round bacteria.

stapling (stā-plihng)–A method of suturing that involves the use of stainless steel staples to close a wound.

star (stahr)–A small white spot in the center of an animal's forehead, especially of a horse. Also called star-faced.

star and stripe (stahr ahnd strīp)–A marking on the forehead of a horse with a stripe to the nasal peak.

star, stripe, and snip (stahr, strīp, ahnd snihp)–A marking on the forehead of a horse with a narrow extension of the nasal peak and opening up again between the nostrils. These may not be connected.

staring coat (stahr-ihng kōt)–Hairs standing on end, indicative of illness or poor condition.

starter (stahr-tər)–(1) The first food provided for young animals. (2) A prepared culture of desired organisms used for inoculating milk and cream in the making of various dairy products, such as butter, cheese, etc., chiefly to enhance the flavor.

starter pig (stahr-tər pihg)–Swine from about 10–40 pounds.

staunchness (stawnch-nehs)–Strong and steady while on point.

stay apparatus (stā-ahp-ahr-ah-tuhs)–An anatomical mechanism of the equine limb that allows the animal to stand with little muscular effort; includes many muscles, ligaments, and tendons.

steady (stehd-ē)–(1) A command to a horse to calm him or to make him go more slowly. (2) Regular, constant, uniform; as a steady breeze.

steaming up (stēm-ihng uhp)–Increased feeding, particularly of concentrates during the latter part of pregnancy.

steapsin (stē-ahp-sihn)–Pancreatic lipase, a lipolytic enzyme secreted by the pancreas that has the power of hydrolizing fats to fatty acids and glycerol.

stearic acid (stē-ahr-ihck ah-sihd)–One of the fatty acids occurring in combined form in animal and vegetable oils; one of the major fatty acids in butter and most fats. Commercial stearic acid is used in large quantities, in rubber compounding and in the preparation of soaps, greases, and chemicals.

steatitis (stē-aht-ī-tihs)–Inflammation of fat, usually caused by feeding cats too much oily fish; also called yellow fat disease.

steel gray (stēl grā)–The coat color of a horse that is an even shade of gray over the body with a black or dark gray mane and tail.

steer (stēr)–A male bovine castrated while young.

stenosis (stehn-ō-sihs)–The narrowing of an opening.

stenotic nares (stehn-oh-tihck nār-ēz)–Narrowed nostrils.

stent (stehnt)–A small expander implanted in a blood vessel to prevent it from collapsing; also a device to hold tissue in place or to provide support for a graft.

sterile (stehr-ihl)–(1) In animals, incapable of reproduction; unable to produce normal living young. (2) In biological products, etc., free from contamination with living bacterial, fungal, or viral organisms; or designating an organism not capable of growing or multiplying.

sterility (stər-ihl-ih-tē)–The inability to reproduce.

sterilization (stehr-ih-lī-zā-shuhn)–(1) The destruction of all living organisms. In contrast, disinfection is the destruction of most of the living organism. (2) To make animals infertile.

sterilize (stehr-ih-līz)–To destroy all organisms.

sternebra (stər-nah-brah)–An unpaired bone on the ventral chest that together comprise the sternum; plural is sternebrae (stər-nah-brā).

sternum (stər-nuhm)–The bone that forms the ventral portion of the rib cage, also known as the breastbone.

steroid (stehr-oyd)–A common term for a hormone or medication that controls metabolism, inflammation, immune function, salt and water balance, development of sexual characteristics, and the ability to withstand illness.

sterol (stē-rohl)–Any of a group of solid cyclic alcohols, such as cholesterol and phytosterol, with wide distribution among animals and plants.

stethoscope (stehth-ō-skōp)–An instrument used to listen to body sounds.

stick (stihck)–(1) To stab or cut with a knife, as in animal slaughter, to sever the jugular veins to cause free bleeding; also with poultry, to pierce the brain with a narrow-bladed knife to cause feather follicles to relax for ease in plucking feathers.

stifle (stī-fuhl)–The synovial joint located between the femur and tibia.

stifled (stī-fuhld)–A horse is said to be stifled when the patella (or kneecap) slips out of place and temporarily locks in a location above and to the inside of its normal location. A bull is said to be stifled when the stifle muscle is torn.

still-born (stihl bōrn)–Born lifeless; dead at birth.

stimulus (stihm-yoo-luhs)–An agent, act, or influence that produces a reaction; plural is stimuli (stim-yoo-lī).

stirk (stuhrk)–A six-to-twelve-month-old heifer.

stirrup (stihr-uhp)–An attachment to a riding saddle serving as an aid in mounting and as a footrest for the rider; made of metal, wood, or heavy leather in many different styles.

stock (stohck)–(1) Livestock; domesticated farm animals. (2) The original type from which a group of animals has been derived. (3) A small enclosure in which an animal is secured in a standing position during shoeing or an operation. (4) To provide or supply with livestock, as to stock a pasture or range with cattle, sheep, etc. (5) Of, or pertaining to, livestock, as stock barn, stock feed, etc.

stock cattle (stohck kah-tuhl)–Usually young steers or cows, light, thin, lacking in maturity and finish. Also called stockers.

stock solution (stohck sō-loo-shuhn)–A concentrated solution from which a portion is taken and diluted as needed, as a spray, etc.

stock tank (stohck tahnk)–Any structure holding water for livestock.

stocker (stohck-ər)–In marketing, a meat-producing animal capable of additional growth and finish and usually considered to be thinner than a feeder.

stockinet (stohck-ih-neht)–A knitted, coarse, elastic, netlike type of cotton cloth, similar to stocking material, which is used as a bandage material or wrap during surgical procedures.

stocking (stohck-ihng)–(1) The relative number of livestock per unit area for a specific time. In range management, the relative intensity of animal population, ordinarily expressed as the number of acres of range allowed for each animal for a specific period. (2) A white leg on an animal.

stocking rate (stohck-ihng rāt)–The actual number of animals, expressed in either animal units or animal unit months, on a specified area at a specific time.

stockyard (stohck-yahrd)–(1) A yard for keeping or holding livestock. (2) A series of pens or yards where market animals are collected for sale. It may be only a pen or two along a railroad siding in a small town, or it may be as extensive as the great stockyard systems of the city of Chicago.

stoma (stō-mah)–An opening or mouth.

stomach (stuhm-ahck)–The digestive pouch at the distal end of the esophagus that mechanically and chemically breaks down food.

stomatitis (stō-mah-tī-tihs)–Inflammation of the mouth.

stool (stool)–(1) Fecal material; evacuation from the digestive tract.

store (stōr)–An animal not yet ready for slaughter.

stot (stoht)–See steer.

stover (stō-vər)–(1) The stem and leafy parts of corn fodder after the ears have been removed. (2) Sorghum forage.

strabismus (strah-bihz-muhs)–A disorder in which the eyes are not directed in a parallel manner; deviation of one or both eyes.

straggler (strahg-glər)–An animal that wanders or strays from a flock or herd.

straightbred (strāt-brehd)–Designating an animal with a straight line of ancestry or pedigree within a recognized breed, such as a registered thoroughbred horse or a purebred Jersey cow.

strain (strān)–(1) A group of individuals within a breed that differ in one or more characters from the other members of the breed, e.g., the milking shorthorns. (2) An organism or group of organisms that differ in origin or in minor respects from other organisms of the same species or variety. (3) A virus entity whose properties and behavior indicate a relationship to a type virus and are sufficiently constant to enable the entity to be recognized whenever isolated. (4) A severe muscular effort on the part of a draft or other animal that may result in muscle, ligament, or other damage. (5) To filter a liquid and free it of impurities by passing it through some medium or fabric which can retain the solid matter and allow the liquid to pass.

strain cross (strān krohs)–A cross between members of two different strains.

stranguria (strahng-yoo-rē-ah)–Slow or painful urination.

straw (straw)–The mature, dry foilage of a crop after the grain has been removed.

strawberry roan (straw-beh-rē rōn)–The coat color of a horse, predominately red or reddish-bay with white hairs rather uniformly intermixed.

stray (strā)–An animal that has wandered away from the herd, flock, or farm to which it belongs.

streak canal (strēk kah-nahl)–A passageway that takes milk from the teat cistern to the outside; also called the papillary duct or teat canal.

strength (strenkth)–(1) The capacity to resist force; solidity or toughness; the quality of bodies by which they endure the application of force without breaking or yielding. (2) Bodily or muscular power; force; vigor. (3) The potency or power of a liquid or other substance; intensity of active properties. (4) The firmness of a market price for a given commodity; a tendency to rise or remain firm in price.

streptococci (strehp-tō-kohck-sī)–Round bacteria that form twisted chains.

Streptococcus (strehp-tō-kohck-uhs)–Any bacterium of the genus *Streptococcus*, family Coccaceae. Several species, such as *S. pyogenes*, *S. scarlatinae*, etc., are responsible for various diseases in humans and animals. Certain others, such as *S. defir* and *S. cremoris*, are used in making cheese, etc.; the plural is streptococci (strehp-tō-kohck-sī).

stress (strehs)–Abnormal or adverse conditions and factors to which an animal cannot adapt or adjust satisfactorily, resulting in physiological tension and possible disease; the factors may be physical, chemical, and/or psychological.

stretch (strehch)–The amount of body length of an animal; usually referring to the distance between the shoulders and the hips.

striate (strī-āt)–With fine grooves, ridges, or lines of color.

striated (strī-āt-ehd)–Striped.

stricture (strihck-shər)–An abnormal band of tissue narrowing a passage.

strictus (strihck-tuhs)–Erect.

stride (strīd)–The distance between successive imprints of the same foot.

stridor (strī-dōr)–An abnormal, harsh, high-pitched sound heard during inspiration or expiration.

strike (strīk)–A defensive or aggressive movement of a horse in which the front leg is moved quickly and cranially.

string (strihng)–(1) A group of partly broken horses that are assigned to a cowboy or horse trainer for his personal use or for further breaking or training. (2) A group of horses in a pack train. (3) To unroll wire prior to stretching and fastening to posts, etc.

strip (strihp)–(1) Each of the squirts or streams of milk as taken from a cow's teat in milking. (2) To take the last of the milk from a cow's udder by hand milking after the machine milker has been removed. (3) A narrow marking extending vertically between the forehead and nostrils of a horse.

strip cup (striph kuhp)–A metal cup with a lid that is used for detecting mastitis.

strobila (strō-bī-lah)–The entire adult tapeworm, which has a head (scolex), neck, and a chain of segments.

stroke (strōk)–(1) The linear distance traveled in one motion by a piston or ram, whether in an engine or pump. (2) A sudden and severe attack, as of paralysis.

stroma (strō-mah)–The supporting tissue of an organ.

strong (strohng)–(1) Having a specific quality to a high degree, as strong flavor, etc.

structural soundness (struhck-uhr-ahl sownd-nehs)–The physical condition of the skeletal structure (especially the feet and legs) of an animal.

strut (struht)–In an animal, an exaggerated step or gait.

stub (stuhb)–The quill portion of a short feather appearing on the shanks or toes of otherwise clean-shanked birds.

stubble pasture (stuh-buhl pahs-chər)–A field from which a crop of wheat or other grain has been previously harvested and on which animals are placed to consume the crop residue and weeds that may follow the grain crop.

stuck yolk (stuhck yōk)–The condition of an egg in which the yolk adheres to the inside of the shell. Stuck yolks may occur in eggs that have deteriorated from long holding. Such eggs are classed as inedible.

stud (stuhd)–(1) A male animal used for breeding purposes. (2) A unit of selected animals kept for breeding purposes, usually applied to horses. (3) The farm where stallions are kept.

stunt (stuhnt)–To check or hinder the growth or development of an animal or plant.

stupefacient (stoo-peh-fā-shehnt)–A drug used to cause birds or other animals to go into a state of stupor so they can be captured and removed.

stupor (stoo-pər)–Impaired consciousness with unresponsiveness to stimuli.

sty (stī)–A pen where swine are housed and fed. Usually it consists of a low shed and a dirt yard. Also called pigsty.

style (stīl)–(1) The manner in which an animal displays itself while at rest or in action. (2) The manifestation of those characteristics that contribute to the general beauty, pleasant appearance, and attractiveness of an animal.

subacute (suhb-ā-kūt)–A clinical condition intermediate between acute and chronic.

subarachnoid space (suhb-ah-rahck-noyd spās)–An area located below the arachnoid membrane and above the pia mater.

subbreed (suhb-brēd)–An offshoot or subdivision of a major breed of livestock, such as polled Jersey.

subclavian (suhb-klā-vē-ahn)–Under the collar bone.

subclinical (suhb-klihn-ih-kahl)–Without showing signs of disease.

subcutaneous (suhb-kyoo-tān-ē-uhs)–Under the dermis.

subcutis (suhb-kyoo-tihs)–Beneath the skin.

subdural hematoma (suhb-doo-rahl hē-mah-tō-mah)–A mass or collection of blood below the dura mater and above the arachnoid membrane.

subdural space (suhb-doo-rahl spās)–The area located below the dura mater and above the arachnoid membrane.

sublethal (suhb-lē-thahl)–Less than fatal in effect, as a sublethal dosage or application of a toxic substance, etc.

sublingual (suhg-lihn-gwahl)–Under the tongue.

subluxation (suhb-luhck-sā-shuhn)–Partial displacement of a bone from its joint.

submandibular (suhb-mahn-dihb-yoo-lahr)–Under the lower jaw.

submaxillary (suhb-mahcks-ihl-ār-ē)–Under the upper jaw.

subspecies (suhb-spē-shēz)–A major subdivision of a species, ranking between species and variety. It has somewhat varying connotations, depending on the user of the term, and often implies a distinct geographic distribution for the taxon.

succus (suhck-uhs)–Juices or fluids extracted from or secreted by an organism.

suck (suhck)–(1) Of an animal, to draw milk from the teat of the udder by application of suction by the mouth; to nurse. (2) To draw a liquid, air, etc., by vacuum-producing action.

suckle (suhck-l)–To obtain or provide milk from mammary glands.

suckling pig (suhck-lihng pihg)–A young pig still nursing its mother. When slaughtered at this stage it produces a small carcass for roasting whole.

sucklings (suhck-lihngz)–The young of mammals that are being nursed by the female; unweaned animals.

suction (suhck-shuhn)–Aspiration of gas or fluid by mechanical means.

suet (sū-eht)–Fat from the abdominal cavity of a ruminant animal, especially from cattle or sheep.

suint (swihnt)–Solid deposits from the perspiration of sheep found in the wool; a source of potassium.

sulcus (suhl-kuhs)–A depression or groove; also called fissure; plural is sulci (suhl-sī) or fissures.

sulking (suhlck-ihng)–In horses, the refusal to obey commands promptly.

sunburn (suhn-bərn)–A superficial inflammation of the skin of hogs (especially white hogs) on rape pasture, when they are wet from the dew on the plants and exposed to bright sunlight.

sunstroke (suhn-strōk)–A severe injury or killing of heat-sensitive people and animals by excessive heat during midsummer periods of cloudless skies with temperatures ranging above 100°F. Also called heatstroke.

superfetation (soo-pehr-fē-tā-shuhn)–The presence of two fetuses in the uterus that resulted from fertilization during two different estrous cycles.

superficial (soop-ər-fihsh-ahl)–Positioned at or near the surface.

superior (soop-pēr-ē-ər)–Uppermost, above, or toward the head.

supernumerary teats (soo-pər-nū-mahr-ē tētz)–More than the normal amount of nipples.

superovulation (soo-pehr-ohv-ū-lā-shuhn)–The stimulation of more than the usual number of ovulations during a single estrous cycle due to the injection of certain hormones.

superspecies (soo-pehr-spē-shēz)–A group of related species that are geographically isolated; without any implication of natural hybridization among them.

supination (soo-pih-nā-shuhn)–Rotating a limb or body so that the ventral surface is turned upward.

supine (soo-pīn)–Prostrate; lying with face upward.

supplemental pasture (suhp-leh-mehnt-ahl pahs-chər)–A pasture to augment range forage, particularly during emergency situations. Supplemental pasture may be provided by annual grasses and/or legumes, or by aftermath of meadows, grain fields, etc.

supplement (suhp-leh-mehnt)–A feed or feed mixture that is relatively richer in a specific nutrient than the basic feed ingredients in the ration to which it is added. It may be used to supply a single nutrient or may contain a mixture of vitamins, proteins, minerals, and other growth stimulants; also called additive.

suppuration (suhp-ū-rā-shuhn)–The formation of pus.

suppurative (suhp-yər-ah-tihv)–The formation or discharge of pus.

surcingle (suhr-sihng-gehl)–A strap or girth around the body of a beast of burden for holding the saddle or load in place or used for throwing the animal.

surfactant (sihr-fahck-tehnt)–A liquid that reduces surface tension of the lungs.

surgical clip (sihr-jih-kahl klihp)–A metal staplelike device used for vessel ligation.

susceptible (sah-sehp-tih-buhl)–Lacking resistance.

suspended cage (suh-spehnd-ehd kāj)–Caging that hangs from a metal rack and has wire flooring.

suspended waterers (suh-spehn-ehd wah-tər-ərz)–Poultry waterers that are suspended from the ceiling. They are raised up as the birds grow taller.

suspension (suh-spehn-shuhn)–The period of time in which none of the feet are in contact with the ground.

suture (soo-chuhr)–A jagged line where bones join and form a joint that does not move; also means to stitch or refers to the material used in closing a surgical or traumatic wound with stitches.

swaged needle (swehgd nēd-l)–A needle in which the needle and suture material are joined in a continuous unit; "eyeless" needle.

sward (swahrd)–A closely grazed or mowed area in which the grass and other plant species are close-growing, making an almost complete ground cover. See pasture.

sweat (sweht)–A substance that is comprised of water, salt, and waste produced by glands located in the dermis.

sweat collar (sweht kohl-ər)–A thick, padded leather collar that is attached to the neck of a beef animal to help shrink a dewlap that is excessive. This is done to make the animal look better in the show ring.

sweat gland (sweht glahnd)–The gland located dermally that produces and secretes sweat.

sweat scraper (sweht skrā-pər)–A smooth curved tool used to remove sweat from a horse's coat.

sweat shed (sweht shehd)–A properly ventilated shed or enclosure used to hold sheep so that their natural body heat causes them to sweat. Such a practice immediately preceding the shearing operation softens the yolk and makes shearing easier.

sweating process (sweht-ihng proh-sehs)–The practice of putting sheep skins in a warm, moist room to loosen the wool in preparation for pulling.

sweeny (swē-nē)–A wasting or atrophy of the shoulder muscles of a horse. Also called suprascapular paralysis.

sweetfeed (swēt-fēd)–Food that consists of grains and pellets mixed with molasses to increase palatability. Also called sweet feed.

swell (swehl)–(1) In a saddle, the part in front of the seat which rises to the horn. Also called fork, front.

swill (swihl)–Liquid food for domestic animals, especially swine, consisting of ground feed mixed with water or milk or liquid garbage from the kitchen. Also called slop.

swine (swīn)–Any mammal of the family Suidae. Domesticated species are grown for their edible flesh and fat, for their hides, and for their bristles. Swine are very important in the agricultural economy of the United States. Also called hog, pig.

swing (swihng)–The non-weight-bearing phase of a stride.

switch (swihch)–The distal part of a bovine tail that consists of long, coarse hairs.

swollen (swohl-ehn)–Enlargement of a body part due to fluid retention.

swollen joints (swō-lehn joyntz)–An acute infectious disease, commonly called navel ill, caused by certain bacteria which often gain entrance into the navel soon after birth, becoming septicemic and localizing in the joints.

symbiosis (sihm-bī-ō-sihs)–The close association of two dissimilar organisms, each known as symbiont. The associations may have five different characteristics as follows: mutualism: beneficial to both species; commensalism: beneficial to one but with no influence on the other; parasitism: beneficial to one and harmful to the other; amensalism: no influence on the

other; synnecrosis: detrimental to both species of organisms.

symbiotic relationship (sihm-bī-oht-ihck rē-lā-shuhn-shihp)–A relationship between two different types of organisms that is beneficial to both of them.

sympathetic nervous system (sihm-pah-theh-tihck nər-vuhs sihs-tehm)–The portion of the autonomic nervous system that prepares the body for stressful and emergency situations.

symphysis (sihm-fih-sihs)–An area where two bones join and are held firmly together so that they function as one bone; also known as a cartilaginous (kahr-tih-lah-jihn-uhs) joint.

symptom (sihmp-tuhm)–A characteristic of disease that can only be sensed by the patient; an incorrect term in veterinary medicine.

synapse (sihn-ahps)–A nerve impulse transfer point between two neurons or between a neuron and receptor.

syncope (sihn-kō-pē)–The temporary suspension of respiration and circulation; commonly called fainting.

syndrome (sihn-drōm)–A set of signs that occur together.

synechia (sī-nēk-ē-ah)–An adhesion that binds the iris to an adjacent structure; plural is synechiae (sī-nēk-ē-ā).

synergist (sihn-ər-gihst)–Things that work together.

syngamy (sihng-gah-mē)–The union of the gametes in fertilization.

synovial joint (sih-nō-vē-ahl joynt)–A movable joint.

synovial membrane (sihn-nō-vē-ahl mehm-brān)–The lining of the bursa or joint that secretes fluid to lubricate the joint.

synovitis (sihn-ō-vī-tihs)–Inflammation of the synovial membrane of joints.

synthetic (sihn-theh-tihck)–Artificial production; chemically manufactured.

syrinx (sehr-ihnks)–The voice organ of birds.

systemic (sihs-tehm-ihck)–Pertaining to the body as a whole and not confined to one organ or part of the body, as a systemic infection.

systemic circulation (sihs-tehm-ihck sihr-kyoo-lā-shuhn)–Blood flow to all parts of the body except the lungs.

systole (sihs-stohl-ē)–Contraction (of the heart ventricles).

T

tabby (tahb-bē)–Feline fur with two colors that may be either in stripes or spots.

tachycardia (tach-ē-kahr-dē-ah)–An abnormally rapid heartbeat.

tachypnea (tahck-ihp-nē-ah)–Abnormally rapid respiratory rates.

tack (tahck)–Equipment used in riding and driving horses.

tack room (tahck room)–A room in a stable in which saddles, bridles, spurs, boots, harness, etc., are kept.

tackling (tahck-lihng)–A draft horse harness.

tacky (tahck-ē)–An animal in poor condition.

tactile (tahck-tīl)–Touch.

tag (tahg)–(1) A dung-covered lock of wool. Also called dag, daglock. (2) A plastic or metal piece attached to an animal for identification, or a cardboard or cloth label attached to the container of a product, a feed or fertilizer, giving the content analysis, etc. (3) To place a tag on a product, animal, etc., for identification.

tagging (tahg-ihng)–(1) Clipping manured and dirty locks from sheep. (2) The process of attaching identifying tags to animals. See also brand, marking.

tail (tāl)–(1) The posterior part of the vertebral column of animals. It is usually covered with hair, some of which may be quite long. Also called brush. (2) A fanlike row of rather stiff feathers on the posterior part of a bird. (3) To remove the tail from a carcass. (4) To assist an undernourished cow to its feet by pulling on the tail.

tail autotomy (tāl aw-toh-tō-mē)–A lizard's ability to lose its tail.

tail band (tāl bahnd)–The crupper of a harness.

tail chewing (tāl choo-ihng)–The tendency of certain animals to gnaw at their own tails near the anus, sometimes caused by the infestation with worms or mange mites.

tail covert (tāl kō-vərt)–The feathers of a fowl that cover the base of the tail feathers in males and the larger portion of the tail in females. They are curved and pointed in males and oval in females.

tail end (tāl ehnd)–The poorest quality portion of a group of animals. To tail out is to remove animals from the bottom of the group.

tail head (tāl hehd)–The base of the tail where it connects to the body.

tail set (tāl seht)–(1) A device of leather and metal which, when attached to a horse, causes the horse to hold its tail high. (2) The position of the tail in relationship to the hips of an animal.

tail test (tāl tehst)–Intradermal test for Mycobacterium tuberculosis in which a small quantity of tuberculin is injected into the hairless skin on the undersurface of a bovine's tail. Reactions to the tuberculin injection are assessed 72 hours later. Also called tuberculin test.

tailed (tā-ld)–Designating an animal whose tail has been removed. See dock.

tailing (tā-lihng)–A restraint technique used in cattle in which the tailhead is grasped and raised vertically; also called a tail jack.

taint (tānt)–(1) A contamination or off-flavor of milk or other products. (2) To infect; contaminate.

take (tāk)–(1) To accept a male in coitus. (2) To result in a mild infection after vaccination.

tallow (tahl-ō)–(1) The fat extracted from the fat tissue of cattle and sheep. Used in candle making, soap manufacture, etc. See suet.

tally (tahl-ē)–(1) A label or tag attached to a product or animal for identification. (2) The number of products or animals handled or produced.

talon (tahl-ohn)–(1) The hind part or heel of the foot or hoof of such animals as deer, swine, horses, etc. (2) A claw of a bird, usually of a bird of prey.

tame (tām)–(1) Domesticated. (2) Cultivated. (3) Designating an animal that has been made docile or tractable, as a wild horse is tamed or broken. (4) To domesticate. (5) To make docile.

tan (tahn)–(1) A coat color of some animals, especially dogs. It is a yellowish-brown similar to the color of well-tanned leather, and varies in shade from light to dark. (3) To convert a hide into leather.

tankage (tahnk-ahj)–Animal residues left after rendering fat in a slaughter house that is used for feed or fertilizer.

tapadera (tahp-ah-dehr-ah)–A leather hood that covers the front of a stirrup on a saddle.

taper needle (tā-pər nē-dl)–A needle with a rounded tip that is sharp to allow the piercing of, but not the cutting of, tissue.

tapetum (tah-pē-duhm)–A layer of cells.

tapetum lucidum (tah-pē-duhm loo-sehd-uhm)–A brightly colored, iridescent reflecting layer of tissue of the choroid of some species; also called choroid tapetum.

tapetum nigrum (tah-pē-duhm nī-gruhm)–A black pigmented layer of tissue of the choroid in some species.

tapeworm (tāp-wehrm)–A parasitic intestinal worm of a flattened, tapelike form, order Cestoda, composed of separate parts or segments.

tar (tahr)–(1) A black, liquid mixture of hydrocarbons and their derivatives obtained by the distillation of wood, peat, coal, shale, etc. (2) Tar used as a disinfectant on a sheep accidentally cut by the shearer (Australia and New Zealand).

tarsectomy (tahr-sehck-tō-mē)–The surgical removal of all or part of the tarsal plate of the eyelid.

tarsorrhaphy (tahr-sōr-ah-fē)–The suturing together of the eyelids.

tarsus (tahr-suhs)–(1) The joint of the distal hindlimb located between the tibia and fibula and the metatarsals; also called the hock (hohck). (2) The platelike framework that provides stiffness and shape within the upper and lower eyelids.

taste (tāst)–(1) The flavor of a product as determined by placing the substance in the mouth. (2) A small amount or sample.

tattoo (taht-too)–The permanent identification of an animal using indelible ink that is injected under the skin.

tautonomy (taw-tohn-ō-mē)–Relations that exist if the same word is used for both the generic and specific name of a species.

taw (taw)–To dress and prepare, as the skins of sheep, lambs, goats, and kids, for gloves, etc., with alum, salt, and other softening and bleaching agents.

taxonomy (tahck-sohn-ō-mē)–The science of classifying organisms and other objects and their arrangement into systematic groups such as species, genus, family, and order.

team (tēm)–(1) Two or more horses, mules, etc., that are harnessed to the same vehicle, plow, etc. (2) Two or more specialists who are jointly investigating a problem. (3) To match two or more draft animals to serve as a team.

tease (tēz)–To stimulate an animal to accept coitus.

teased (tēzd)–The act of determining whether a mare is in heat (estrus) by presenting a stallion to her.

teaser (tē-zər)–(1) An animal that is used to sexually stimulate one of the opposite sex. (2) Designating a male animal that is used to locate females of the same species in heat.

teat (tēt)–A nipple, especially the large nipples of ruminants and equine.

teat cannula (tēt kahn-yoo-lah)–A short and narrow, rounded metal or plastic tube used to pass from the exterior through the teat canal and into the teat cistern to relieve obstructions.

teat cistern (tēt sihs-tehrn)–A cavity in the udder where milk is secreted before leaving the teat.

teat cup (tēt kuhp)–The part of a milking machine that is attached to the cow's teat. The milk is drawn out through suction.

teat dipping (tēt dihp-ihng)–Submerging or spraying a nipple with disinfectant to prevent the development of mastitis.

teat placement (tēt plās-mehnt)–The placement of teats on a sow's underline; they should be evenly spaced.

teat stripping (tēt strihp-ihng)–The removal of the last milk in the teat after milking by occluding the teat at the top between the thumb and forefinger and then pulling downward to express all the milk; also called stripping.

teg (tehg)–(1) A two-year-old sheep. (2) The fleece of a two-year-old sheep. Also spelled tegg.

telencephalon (tē-lehn-sehf-ah-lohn)–The portion of the brain that is the cerebral hemispheres.

telophase (tehl-ō-fāz)–The phase of cell division between anaphase and the complete separation of the two daughter cells; includes the formation of the nuclear membrane and the return of the

chromosomes to long, threadlike and indistinguishable structure.

temper (tehm-pər)–The relative mildness or viciousness of an animal.

temperament (tehm-pər-mehnt)–The emotional and mental qualities of an individual.

temperature (tehm-pər-ah-chər or tehm-pər-chər)–(1) The vital sign that tells about degree of heat or cold; recorded in degrees Fahrenheit or Celsius in animals. (2) The amount of heat or cold measured in degrees on different scales, as Fahrenheit or Centigrade. (3) The degree of heat in a living body. (4) Abnormal heat in a living body. Also called fever, running a temperature.

tendinitis (tehn-dih-nī-tihs)–Inflammation of a tendon.

tendon (tehn-dohn)–A fibrous connective tissue that connects muscle to bone.

tenectomy (teh-nehck-tō-mē)–Surgical removal of part of a tendon.

tenesmus (teh-nehz-muhs)–Painful, ineffective defecation.

teniae (tehn-ē-ā)–Smooth muscle bands.

tenotomy (teh-noht-ō-mē)–Surgical division of a tendon.

teratogen (tehr-ah-tō-jehn)–An agent or factor that causes the production of physical defects in the developing embryo of animals and people.

teratogenic (tər-ah-tō-jehn-ihck)–Producing defects in the fetus.

teratology (tehr-aht-ohl-ō-jē)–The study of embryo development.

term (tərm)–At term; the end of the normal period of gestation of pregnancy when birth is due to occur.

terminal (tər-mihn-ahl)–The end.

terminal sires (tər-mih-nahl sī-ərz)–Sires used in a crossbreeding system where all progeny, both male and female are marketed.

terrestrial (theh-rehs-trē-ahl)–(1) Referring to the earth. (2) Designating a ground bird such as a pheasant, partridge, or chicken, as contrasted to an aerial bird.

tertiary (tər-shē-ār-ē)–Third in order.

test group (tehst groop)–A collection of animals used for experimental manipulation.

test cross (tehst krohs)–A kind of genetic cross involving one individual expressing a dominant trait and one expressing the recessive trait, the purpose of the cross being to determine whether the individual expressing the dominant trait is heterozygous or homozygous.

testicle (tehs-tih-kuhl)–The male sex organ that produces sperm after sexual maturity; plural is testes or testicles.

testis (tehs-tihs)–The male gonad that produces spermatozoa; also called testicle; plural is testes (tehs-tēz).

testosterone (tehs-tohs-teh-rōn)–A hormone produced by the interstitial cells of the testes that functions in stimulating male sex drive, masculine characteristics, development of the male reproductive tract, and spermatogenesis.

tetanic (teht-ahn-ihck)–Pertaining to severe muscular contractions.

tetanus (teht-ah-nuhs or teht-nuhs)–A disease of the nervous system caused by an anaerobic bacillus that results in an inability to control certain muscles, particularly those in the region of the neck and jaw; also called lockjaw.

tetany (teht-ahn-ē)–Spasms of muscle.

tethering (tehth-ər-ihng)–Restraining an animal by tying it to a post, etc., with a long rope, chain, etc., to restrict grazing.

tetraplegia (teht-rah-plē-jē-ah)–Paralysis of all four limbs; also called quadriplegia (kwohd-rih-plē-jē-ah).

tetraploid (teht-rah-ployd)–An organism whose cells contain four haploid (monoploid) sets of chromosomes.

thalamus (thahl-ah-muhs)–The portion of the diencephalon that receives and

transmits sensory impulses to the cerebral cortex.

theave (thēv)–A ewe that has not borne a lamb.

theca (theh-kah)–The membranes that form the wall of the ovarian follicle.

therapeutic (thehr-ah-pū-tihck)–Pertaining to the treatment of disease; curative.

therapy (thehr-ah-pē)–The sum total of the treatment given to cure disease in plants, animals, and humans.

theriogenologist (thēr-ē-ō-jehn-ohl-ō-jihst)–A specialist who studies animal reproduction.

theriogenology (thēr-ē-ō-jehn-ohl-ō-jē)–Animal reproduction.

thiamin (thī-ah-mihn)–A member of the vitamin B complex; vitamin B_2.

thigh (thī)–(1) The part of an animal's hind leg between the hock and the trunk. (2) A piece or cut of fowl from the leg between the drumstick and the body.

thill (thihl)–Either of the shafts between which a horse or other animal is hitched to a vehicle.

thin (thihn)–(1) Designating an animal with little flesh. (2) Designating a pulse which is very feeble.

third ventricle (thərd vehn-trih-kuhl)–A cavity in the diencephalon that connects to the fourth ventricle.

thoracic (thō-rahs-ihck)–The chest.

thoracic cavity (thō-rahs-ihck kah-vih-tē)–A hollow space that contains the heart and lungs within the ribs; also called the chest cavity.

thoracic vertebrae (thō-rahs-ihck vər-teh-brā)–The second set of vertebrae found in the chest region; abbreviated T in vertebral formula.

thoracocentesis (thō-rah-kō-sehn-tē-sihs)–Puncture of the chest wall with a needle to obtain fluid or air from the pleural cavity.

thoracotomy (thō-rah-koht-tō-mē)–A surgical incision into the chest wall.

thorax (thō-rahcks)–The middle body region of an insect to which the wings and legs are attached.

thorough (thər-ō)–A dog that works every bit of ground and cover.

three-gaited horse (thrē-gā-tehd hōrs)–A saddle horse that has the following gaits: walk, trot, and canter.

three-way cross (thrē wā krohs)–The crossing of three different strains of breeds of plants or animals to produce a hybrid. See double cross, single cross.

thremmatology (threhm-ah-tohl-ō-jē)–The domestic breeding of plants and animals.

thriftiness (thrihf-tē-nehs)–The capacity to make good use of food or feed.

thrill (thrihl)–A vibration felt on palpation.

throatlatch (thrōt-lahch)–The brindle part that connects the bridle to the head located under the horse's throat.

thrombocyte (throhm-bō-sīt)–A nucleated clotting cell.

thrombocytopenia (throhm-bō-sīt-ō-pē-nē-ah)–An abnormal decrease in the number of clotting cells.

thrombocytosis (throhm-bō-sī-tō-sihs)–An abnormal increase in the number of clotting cells.

thrombosis (throhm-bō-sihs)–An abnormal condition in which a thrombus forms within a blood vessel.

thrombus (throhm-buhs)–An aggregation of blood attached to the interior wall of a vein or artery; plural is thrombi (throhm-bī).

throw (thrō)–(1) In horseback riding, the violent ejection of the rider from the horse's back by action of the horse. (2) To cause an animal, as a horse or cow, to fall to the ground before branding, treating, etc. (3) To wield a lasso.

throw up (thrō uhp)–To vomit.

throwbacks (thrō-bahcks)–Offspring that show an ancestor's characteristics that had not appeared in previous generations.

thumps (thuhmps)–An animal ailment resembling hiccoughs in humans; it is seen in baby pigs with anemia, swine influenza, and with verminous pneumonia.

thunk (thuhnck)–A wartlike growth in a cow's teat canal.

thurl (thuhrl)–The thigh of a horse.

thymectomy (thī-mehck-tō-mē)–Surgical removal of the thymus.

thymoma (thī-mō-mah)–A neoplasm of the thymus.

thymus (thī-muhs)–An immunologic functioning gland found predominantly in young animals located near the midline in the cranioventral portion of the thoracic cavity.

thyroid gland (thī-royd glahnd)–A ductless gland situated in the neck of humans and animals that secretes thyroxin, and triiodo thyronine hormones that control the rate of metabolism, and calcitonin, a hormone that takes calcium from blood to bone.

thyroidectomy (thī-royd-ehck-tō-mē)–Surgical removal of all or part of the thyroid gland.

thyroiditis (thī-roy-dī-tihs)–Inflammation of the thyroid gland.

thyroid-stimulating hormone (TSH) (thī-royd stihm-yoo-la-tihng hŏr-mŏn)–A hormone produced by the anterior pituitary gland that stimulates the thyroid to produce thyroxin.

thyromegaly (thī-rō-mehg-ah-lē)–Enlargement of the thyroid gland.

thyrotoxicosis (thī-rō-tohck-sih-kō-sihs)–An abnormal, life-threatening condition of excessive poisonous quantities of thyroid hormone.

tibia (tihb-ē-ah)–The larger of the two hindlimb long bones that articulates proximally with the femur to form the stifle joint and distally with the tarsus.

tick (tihck)–Any of the various blood-sucking arachnids that fasten themselves to warm-blooded animals. Some are important vectors of diseases.

tickborne (tihck-bōrn)–Transmitted by ticks, as cattle tick fever.

ticked coat (tihckd kōt)–A fur color where darker colors are found on the tips of each guard hair.

tick-infested (tihck ihn-fehs-tehd)–Designating an animal or area of land that abounds in ticks.

ticking (tihck-ihng)–A fur coat that has guard hairs with darker tips mixed in.

tie (tī)–(1) The period of copulation between a male and female canine, during which the two animals are locked together due to erectile tissue enlargement. (2) A string, rope, metal band, or wire that encircles a bale of cotton, hay, etc., holding the material in a limited space and form and making for ease of handling, storage, and transportation. (3) A rope, twine, etc., for fastening purposes. (4) A stanchion for securing an animal. (5) Depression, usually in the middle of the back of an animal, caused by the skin adhering to the backbone. (6) To fasten or secure with a rope, string, wire, etc.

tie stall (tī-stahl)–(1) A stanchion to which an animal is fastened. (2) A stall in which an animal is fastened by a halter or chain.

tight side (tīt sihd)–The right side of a beef carcass. See loose side.

timid (tihm-ihd)–Showing a lack of confidence or shy.

tissue (tihsh-yoo)–Groups of specialized cells that are similar in structure and function.

tissue forceps (tihs-yoo fōr-sehpz)–Tweezerlike, nonlocking instruments used to grasp tissue.

titer (tī-tehr)–The minimum quantity of a substance required to produce a specific reaction with a given amount of another substance.

tobiano (tō-bī-ahn-ō)–Basically, a white horse in which large, smooth, and solid blocks of color originate on the back and rump and extend down. The face is usually marked the same as in other color patterns found in horses.

toe (tō)–Any one of the digits on the foot.

toe mark (tō mahrk)–To punch holes in the membrane between the toes of fowls for identification purposes.

toed-out (tōd-owt)–A term used to describe an animal that walks with the feet pointed outward; splay-footed; slew-footed.

tom (tohm)–An intact male cat or turkey.

tomentose (tō-mehn-tōs)–Coated with short, matted wooly hair.

tomography (tō-moh-grahf-ē)–A recording of the internal body structures at predetermined planes.

tongue (tuhng)–A movable muscular organ in the oral cavity used for tasting and processing of food, grooming, and sound articulation.

tonic (tohn-ihck)–(1) A medicinal preparation to increase the strength or tone of the body system. (2) Referring to muscular contraction, as tonic convulsions.

tonometer (toh-noh-meh-tər)–An instrument used to measure intraocular pressure.

tonometry (toh-nohm-eh-trē)–A procedure using an instrument to measure intraocular pressure.

tonsil (tohn-sihl)–Lymphatic tissue that protects the nasal cavity and upper throat.

tonsillitis (tohn-sih-lī-tihs)–Inflammation of the tonsils.

tonus (tō-nuhs)–Tension.

tooth (tooth)–(1) In animals, one of the hard, bonelike appendages of the jaws used to tear and to masticate food.

tooth float (tooth flōt)–A device for filling horses' teeth. See float.

top (tohp)–(1) Scoured, combed, long wool. (2) To place the best articles of produce, as eggs, or fruit, on the top layer, in a container so that the whole container appears to include articles of higher quality than is actually the case. (3) To sort out animals that have reached a certain stage of development or finish.

topcoat (tohp-kōt)–Primary hairs that are stiff and smooth; also called guard hairs.

topcrossbred (tohp-krohs-brehd)–The progeny from the mating of inbred sires with noninbred dams of a different breed.

topincross (tohp-ihn-krohs)–The progeny from the mating of an inbred sire with a noninbred dam of the same breed.

topknot (tohp-noht)–A crest of feathers, usually decorative, on the head of a bird or a tuft of hair on the head of an animal.

toro (tōr-ō)–A bull.

torsion (tōr-shuhn)–An axial twist.

torticollis (tōr-tih-kō-luhs)–The contracted state of the cervical muscles producing torsion of the neck; also called wry neck.

tortoise shell (tōr-tihs shehl)–Feline fur with two colors (orange and black) producing a spotted or blotched pattern.

total digestible nutrients (tō-tahl dī-jehst-ih-buhl nū-trē-ehntz)–All the nutrients consumed by an animal that are digested and used; generally applied to proteins, carbohydrates, and fats.

tough (tuhf)–Designating animal products that are not brittle, crisp, or tender.

tourniquet (tōr-nih-keht)–A constricting band applied to a limb to control bleeding or to assist in drawing blood.

tower silo (tow-ər sī-lō)–A cylindrical tower made of wood, concrete, tile, metal, etc., used for storage of silage. It is the most common type of silo in the United States. Also called upright silo.

toxemia (tohcks-ē-mē-ah)–Blood poisoning that results from absorption of poisons (poisons or toxins are found in the bloodstream, and the bacteria or other poison producers are confined to a wound or other affected area).

toxic (tohcks-ihck)–Poisonous; caused by poison.

toxic residue (tohcks-ihck rehs-ih-doo)–The poisonous residue left on plants, in the soil, or on animals by a spray or dust.

toxicant (tohcks-ih-kahnt)–A substance that injures or kills an organism by physical, chemical, or biological action; e.g., heavy metals, pesticides, and cyanides.

toxicity (tohck-sihs-ih-tē)–The state or degree of being poisonous.

toxicology (tohcks-sih-kohl-ō-jē)–The science that deals with poisons, antidotes, toxins, effects of poisons, and the recognition of poisons.

toxin (tohcks-sihn)–A poison.

toxoid (tohcks-oyd)–A toxin that has been chemically altered so that it is no longer toxic but is still capable of uniting with antitoxins and/or stimulating antitoxin formation.

toy (toy)–Designating any small or dwarf variety, as a toy spaniel dog.

trace (trās)–(1) To follow the course of nutrient elements in plants or animals, as by use of radioisotopes. (2) Designating an artificial flavoring material with little or no true flavoring.

trachea (trā-kē-ah)–The windpipe; structure that carries air from the oral cavity to the bronchi.

tracheitis (trā-kē-ī-tihs)–Inflammation of the windpipe.

tracheoplasty (trā-kē-ō-plahs-tē)–Surgical repair of the windpipe.

tracheostomy (trā-kē-ohs-tō-mē)–The surgical creation of an opening into the windpipe (usually involves insertion and placement of a tube).

tracheotomy (trā-kē-oht-ō-mē)–A surgical incision into the windpipe.

tracks (trahcks)–(1) A term used to describe the leg movements of a horse while it is walking. (2) Prints of animal feet, as in tracks in the snow.

tract (trahckt)–(1) An area of land of any size, but bigger than a lot. (2) An anatomical structure such as the digestive tract, cerebellospinal tract.

trail (trāl)–(1) A pathway made either by repeated passage of people or animals or constructed for easier passage. It is not usable for vehicles. (2) To move cattle or sheep over a long distance to a pasture, a market, or shipping point.

trail drive (trāl drīhv)–To drive a herd along the trail to pasture or market.

trail herding (trāl hehrd-ihng)–Directing and controlling the movement of a group of livestock on restricted overland routes.

trailing (trā-lihng)–(1) The driving of livestock from place to place. (2) The voluntary wandering of livestock about a range, usually in search of forage, water, or salt.

trait (trāt)–Any observable feature or characteristic of an animal.

trait ratio (trāt rā-shō)–An expression of an animal's performance for a particular trait relative to the herd or contemporary group average.

trampling (trahmp-lihng)–(1) Treading under feet; the damage to plants or soil brought about by congested movements of livestock, including mechanical injury to tree reproduction and ground cover in woods. (2) Compacting soil in earthen dams and reservoirs by livestock to make the dam or reservoir impervious to water (now replaced by machine compaction).

transfix (trahnz-fihcks)–To pierce through and through.

transfusion (trahnz-fū-shuhn)–The introduction of whole blood or blood components into the bloodstream of the recipient.

transhumance (trahns-hyoo-mehns)–The seasonal nomadic movement of people and grazing livestock for part of the year combined with some form of permanent lowland farming in the wet season.

transitory range (trahn-sih-tōr-ē rānj)–Land that is suitable for temporary grazing, e.g., on disturbed lands, grass may cover the area for a period of time before being replaced by trees or shrubs.

translocation (trahns-lō-kā-shuhn)–In genetics, a change in position of a segment of a chromosome to another part of the same chromosome or to a different chromosome.

transluminal (trahnz-loo-mehn-ahl)–Through an opening within a vessel.

transmissible (trahnz-mihs-ih-buhl)–The ability to transfer from one animal to the next.

transmission (trahnz-mihs-shuhn)–Transfer from one animal to the next.

transovarially (trahnz-ō-vah-rē-ah-lē)–Through or across the ovary.

transplacentally (trahnz-plah-sehn-tah-lē)–Through or across the placenta.

transplant (trahnz-plahnt)–To transfer tissue from one part to another part.

transudate (trahnz-yoo-dāt)–Material that has passed through a membrane and is high in fluidity and low in protein, cells, or solid materials derived from cells.

transverse (trahnz-vərs)–Extending from side to side or at right angles to the long axis.

transverse plane (trahnz-vərs plān)–An imaginary line dividing the body into cranial and caudal portions; also called the horizontal plane.

trappy (trahp-pē)–A short, quick, choppy stride.

trauma (traw-mah)–A wound or injury; a deep emotional shock.

traumatic (traw-mah-tihck)–Pertaining to, resulting from, or causing injury.

travis (trah-vihs)–The partition or half-wall dividing individual stalls in a barn. Also called traverse.

tread (trehd)–(1) Any injury to the front legs of a horse caused by its overreaching with a hind foot as it runs.

treat (trēt)–(1) To care for a sick animal or diseased plant by giving it proper attention and medication. (2) To subject a product to an action or process to improve it in some manner. (3) To subject animals to various chemicals, practices, etc., in order to learn which are beneficial or harmful.

trematode (trehm-ah-tōd)–Parasitic flatworm of the class Trematoda, such as the sheep liver fluke.

tremor (treh-mər)–Involuntary trembling.

tremulous air cell (trehm-ū-luhs ahr sehl)–An enlarged air cell in poultry eggs that causes the membranes to split apart beyond the edges of the cell.

trench silo (trehnch sī-lō)–A trench excavated in a hillside or on firm ground, usually lined with wood or concrete retaining walls.

trephine (tre-fīn)–A circular, sawlike instrument used to remove pieces of bone or tissue.

tribe (trīb)–In animal breeding, a group or combination of animals descended through the feline line.

triceps (trī-sehpz)–A muscle with three heads.

trichobezoar (trī-kō-hē-zōr)–A hairball.

trick horse (trihck hōrs)–A horse trained to perform different feats. See high school horse.

tricuspid valve (trī-kuhs-pihd vahlv)–The membranous fold that controls blood flow through the opening between the right atrium and right ventricle; tricuspid means having three points; also called the right atrioventricular valve (ā-trē-ō-vehn-trihck-yoo-lahr vahlv).

triglyceride (trī-glihs-ər-īd)–One type of lipid that circulates in the blood.

trihybrid (trī-hī-brihd)–An individual that is heterozygous for three pairs of genes.

trimester (trī-mehs-tehr)–Three months, or one-third of the nine months of pregnancy in cattle. The nine months of pregnancy are divided into the first, second, and third trimesters.

trim (trihm)–To remove unwanted or undesirable portions/body part, as to trim fat from a ham or to trim hooves.

trio (trē-ō)–(1) A group consisting of three objects or organisms. (2) A male and two female birds of the same variety that are shown as a unit in exhibitions.

triploidy (trihp-loy-dē)–A condition where the cells possess three sets of homologous chromosomes rather than two.

trisomic (trī-sō-mihck)–Referring to an organism having three chromosomes of one type (chromosome formula 2n/1).

trocar (trō-kahr)–A sharp, needlelike instrument that has a cannula (tube) that is used to puncture the wall of a body cavity and withdraw fluid or gas.

trochanter (trō-kahn-tehr)–A large, flat, broad projection on a bone.

trochlea (trōck-lē-ah)–A pulley-shaped structure in which other structures pass or articulate.

trophic (trohf-ihck)–Or or pertaining to nutrition.

trot (troht)–A natural, rapid, two-beat, diagonal gait in which the front foot and the opposite hind foot take off and hit the ground at the same time.

trots (trohtz)–A diarrheal or abnormally loose condition of the bowels. Sometimes called skitters.

trotter (troht-ər)–(1) A horse used for sulky racing. (2) Lamb shank.

true (trū)–Like the parental type, without change, as a variety that breeds true.

true breeding (trū brē-dihng)–Designating varieties that conform to the parental type with respect to certain characteristics, such as color, disease resistance, etc.

true stomach (trū stohm-ihck)–See abomasum.

try (trī)–To determine if a female is in heat by bringing her and a male in close proximity. They are usually kept separated by a fence, wall, or teasing pole.

tryer (trī-ər)–A grade stallion that is used for the purpose of determining if a mare is in heat. See try.

trypsin (trihp-sihn)–One of the principal proteolytic enzymes of the pancreatic secretion in people and animals.

tryptophan (trihp-tō-fān)–One of a group of essential animo acids that constitute fundamental units from which proteins are built and which are needed by people and animals for the building and repair of tissues.

tubbing (tuhb-bihng)–Standing a horse's foot in a bucket of hot water with washing soda dissolved in it to soften the hoof.

tube dehorner (toob dē-hōrn-ər)–A device consisting of a metal tube that is sharp on one end and has a round ball-like knob on the other end that is used in a twisting motion to remove the horn.

tubercle (too-behr-kuhl)–A small, rounded nodule.

tuberculin (too-behr-kū-lihn)–A biological agent, derived from the growth and further processing of the tubercle bacilli, that is used for the detection or diagnosis of tuberculosis in animals and people.

tuberosity (too-beh-rohs-ih-tē)–An elevation.

tubule (too-byool)–A small tube.

tumor (too-mər)–A distinct mass of tissue with no physiological use or function. It can be either benign or malignant.

tunica (too-nih-kah)–A membranous covering.

tunica albugenia (too-nih-kah ahl-bū-jihn-ē-ah)–A capsulelike covering of an organ, in particular, the testes.

tup (tuhp)–A ram.

tupping (tuhp-ihng)–Mating with a ram.

turbid (tər-bihd)–Cloudiness.

turbinates (tər-bih-nātz)–The scroll-like bones of the nasal cavity.

turgor (tər-gər)–The degree of fullness or rigidity due to fluid content.

turken (tuhr-kehn)–The female offspring of the turkey cock and domestic hen.

turkey (tuhr-kē)–*Meleagris gallopavo*; a large, native American bird, now largely domesticated, that is raised for its delicious meat.

turn (tərn)–(1) To change the position of an egg in an incubator or in the nest where the hen is incubating it. (2) To restrain an animal, as a fence turns a cow. (3) To direct an animal to pasture. (4) To change.

tusk (tuhsk)–The overgrown canine tooth of a boar.

twin (twihn)–(1) Either one of two young born at one birth. (2) To bear two young at birth when only one is usual.

twist (twihst)–The region between the hind legs of an animal where the thigh muscles join. See twitch.

twitch (twihch)—A mode of restraint in which a device is twisted on the upper lip or muzzle. Also called twist.

tympanic bulla (tihm-pahn-ihck buhl-ah)–The osseous chambers that house the middle ear at the base of the skull.

tympanic membrane (tihm-pahn-ihck mehm-brān)–The tissue that separates the external ear from the middle ear and transmits sounds to the ossicles; also called eardrum.

tympany (tihm-pahn-ē)–The detection of intestinal or free gas via resonance.

typical (tihp-ih-kahl)–In appraising, that which most frequently occurs or exists in the particular situation under consideration.

U

ubiquitous (yoo-bihck-wih-tehs)–Occurring everywhere, e.g., house flies, house sparrows, weeds.

udder (uh-dər)–The milk production organ in ruminants and equine. Also called bag.

udder cannula (uh-dər kahn-yoo-lah)–A small plastic, metal, or hard rubber tube designed for the injection of medicines into the udder through the teat canal. Also called teat cannula.

udder felon (uh-dər fehl-ehn)–See mastitis.

ulcer (uhl-sihr)–Focal loss of superficial tissue, exposing the tissue beneath.

ulna (uhl-nah)–One of the forelimb long bones that articulate proximally with the humerus to form the elbow joint and distally with the bones of the carpus.

ultrasound (uhl-trah-sownd)–Imaging of internal body structures by recording echoes of high-frequency waves; also called ultrasonography (uhl-trah-soh-noh-grah-fē).

umbilical cord (uhm-bihl-ih-kuhl kōrd)–The structure that forms where the fetus communicates with the placenta.

umbilicus (uhm-bihl-ih-kuhs)–A pit in the abdominal wall marking the point where the umbilical cord entered the fetus; also known as the navel (nā-vahl).

unbroken (uhn-brō-kehn)–(1) In marketing, designating an egg free from cracks or breaks in the shell. (2) Designating an untrained horse.

undercoat (uhn-dər kōt)–Secondary hairs that are soft, thin, and wavy.

underfeeding (uhn-dər-fē-dihng)–Feeding animals less than is recommended.

underfur (uhn-dər-fər)–The very dense, soft, short hair that is found beneath the longer, coarser guard hairs.

undergrazing (uhn-dər-grā-zihng)–An intensity of grazing that fails to fully use the forage available for consumption in a given area under a system of conservation range management.

undershot jaw (uhn-dər-shoht jaw)–A condition where the mandible is longer than the maxilla.

understocking (uhn-dər-stohck-ihng)–Pasturing or grazing a number of livestock less than the carrying capacity of a particular pasture or range.

ungulate (uhng-yoo-lāt)–Animals with hooves.

unicellular (yoo-nih-schl-ū-lahr)–One-celled; refers to an organism that consists of a single cell.

unilateral (yoo-nih-lah-tər-ahl)–One side.

unipera (yoo-nih-pehr-ah)–An animal that regularly produces only one offspring at a parturition.

unipolar (yoo-nē-pō-lahr)–One end.

unit (yoo-niht)–(1) A single thing or item of produce. (2) A recognized measure of weight, volume, or distance, such as a bushel, gallon, or mile.

univalent (yoo-nih-vā-lehnt)–Designating a chromosome unpaired at meiosis.

unsex (uhn-sehcks)–To remove the ovaries of a female or the testes of a male animal or bird; to castrate; caponize.

unshod (uhn-shohd)–Designating a horse that has no horseshoes on its hooves.

unsound (uhn-sownd)–A term designating an animal with a defect that interferes with the complete utilization of the animal's ability to perform a service.

unthrifty (uhn-thrihf-tē)–Not thriving. The animal may be skinny, weak, sickly, or all three.

upbreeding (uhp-brē-dihng)–The use of a superior breed upon an inferior or mongrel one in order to improve the offspring of the inferior breed.

upper GI (uhp-pər gē-ī)–A contrast radiograph to visualize the structures of the upper gastrointestinal tract after a contrast material is swallowed; also called a barium swallow (bār-ē-uhm swahl-ō).

upper respiratory tract (uhp-pər rehs-pih-tōr-ē trahck)–The part of the respiratory system that consists of the nose, mouth, pharynx, epiglottis, larynx, and trachea.

upright silo (uhp-rīt sī-lō)–A vertical structure used for preserving chopped, green feed silage. Usually it is made of wooden staves, concrete or tile blocks, or metal sheets held together by bands or reinforcements. Diameter and height vary with the capacity desired.

upstanding (uhp-stahnd-ihng)–Refers to an animal that is long legged, rangy, and tall.

urea (yoo-rē-ah)–The end product of protein metabolism found in urine, blood, and lymph.

uremia (yoo-rē-mē-ah)–Urinary waste products in the blood. Also called uremic poisoning.

uremic (yoo-rē-mihck)–An excessive buildup of waste products in the bloodstream.

ureter (yoo-rē-tər)–The narrow tube located between each kidney and the urinary bladder.

ureterectasis (yoo-rē-tər-ehck-tah-sihs)–Distention of the ureter.

ureterectomy (yoo-rē-tər-ehck-tō-mē)–Surgical removal of the ureter.

ureterolith (yoo-rē-tər-ō-lihth)–A stone in the ureter.

ureteroplasty (yoo-rē-tər-ō-plahs-tē)–Surgical repair of the ureter.

ureterostenosis (yoo-rē-tər-ō-steh-nō-sihs)–A stricture or narrowing of the ureter.

urethra (yoo-rē-thrah)–A single tube connecting the urinary bladder to the outside of the body; in males the urethra passes through the penis to the outside of the body and serves both the reproductive and urinary systems.

urethritis (yoo-rē-thrī-tihs)–Inflammation of the urethra.

urethroplasty (yoo-rē-thrō-plahs-tē)–Surgical repair of the urethra.

urethrostenosis (yoo-rē-thrō-steh-nō-sihs)–A stricture or narrowing of the urethra.

urethrostomy (yoo-rē-throh-tō-mē)–Surgical creation of a permanent opening between the urethra and the skin.

urinalysis (yoo-rih-nahl-ih-sihs)–Examination of the physical and chemical properties of urine.

urinary bladder (yoo-rih-ār-ē blah-dər)–A hollow muscular organ that holds urine.

urinary calculi (urolithiasis) (yoo-rihn-ār-ē kahl-kyoo-lī [yoo-rō-lihth-ī-ah-sihs])–Stones of phosphates, urates, oxalates, and lime salts that lodge in some part of the urinary tract, such as the kidney, ureter, bladder, or urethra. Symptoms of the presence of these stones are obstruction to the flow of urine, passage of a few drops of blood, and convulsive pain when urinating.

urinary calculus (yoo-rihn-ār-ē kahl-kyoo-luhs)–Abnormal mineral deposit anywhere in the urinary system

urinary catheterization (yoo-rihn-ār-ē kahth-eh-tər-ih-zā-shuhn)–Insertion of a tube through the urethra and into the urinary bladder.

urinary incontinence (yoo-rihn-ār-ē ihn-kohn-tih-nehns)–The inability to control urine excretion.

urinary retention (yoo-rih-nār-ē rē-tehn-shuhn)–The inability to completely empty the urinary bladder.

urinary tract infection (yoo-rihn-ār-ē trahck ihn-fehck-shuhn)–An invasion of microorganisms into the urinary system which results in local cellular injury; abbreviated UTI.

urination (yoo-rih-nā-shuhn)–The process of excreting urine.

urine (yoo-rihn)–A fluid containing water and waste products filtered through the kidney(s).

urodeum (yoo-rō-dē-uhm)–An area of the cloaca in which the ureters and vagina open.

urogenital (yoor-ō-jehn-ih-tahl)–The urinary and reproductive systems.

urolith (yoo-rō-lihth)–A urinary bladder stone; also called a cystolith (sihs-to-lihth).

urolithiasis (yoo-rō-lihth-ī-ah-sihs)–A disorder characterized by the presence of urinary bladder stones.

urologist (yoo-rohl-ō-jihst)–A specialist who studies the urinary system.

urology (yoo-rohl-ō-jē)–The study of the urinary system.

uropoiesis (yoo-rō-pō-ē-sihs)–The process of urine production.

uropygial gland (yoor-ō-pihj-ih-ahl glahnd)–A gland located laterally to the tail feather attachment that secretes oil used to water-proof or preen feathers; also called the preen gland (prēn).

urostyle (yoor-ō-stīl)–A long caudal vertebra in some amphibians.

urticaria (ər-tih-kā-rē-ah)–A localized area of swelling that itches; also called hives; plural is urticariae (ər-tih-kā-rē-ā).

uterine involution (yoo-tər-ihn ihn-vō-loo-shuhn)–The return to normal size of the muscular-walled, hollow organ that houses and nourishes the fetus.

uterus (yoo-tər-uhs)–The muscular-walled, hollow organ that houses and nourishes the embryo and fetus of the female reproductive tract; the womb.

utricle (yoo-trih-kuhl)–One of the small, hairlike sacs of the inner ear that is partially responsible for equilibrium.

uvea (yoo-vē-ah)–A term used to describe the iris, ciliary body, and choroid.

uveitis (yoo-vē-ī-tihs)–Inflammation of the uvea.

V

V muscle (vē muhs-uhl)–The muscles of the chest of a horse that give the appearance of an inverted V.

vaccination (vahck-sihn-ā-shuhn)–The administration of an antigen (vaccine) to stimulate a protective immune response against a specific infectious agent; also called immunization (ihm-yoo-nih-zā-shuhn).

vaccine (vahck-sēn)–A preparation of pathogen (live, weakened, or killed) or portion of pathogen that is administered to stimulate a protective immune response against that pathogen.

vacreator (vahck-rē-ā-tōr)–A multiple unit consisting of steam jets and vacuum chambers designed to remove an undesirable odor from milk and/or cream.

vacuolated cytoplasm (vahck-ū-ōl-ā-tehd sī-tō-plahzm)–The living substance of cells, exclusive of the nucleus, when filled with or containing bubblelike structures.

vagina (vah-jī-nah)–A muscular tube lined with mucosa that extends from the cervix to outside the body.

vaginitis (vahj-ih-nī-tihs)–Inflammation of the vagina.

valgus (vahl-guhs)–Bent outward.

valine (vahl-ihn)–One of the naturally occurring amino acids.

valvotomy (vahl-voh-tō-mē)–A surgical incision into a valve or membranous flap.

vane (vān)–The thin web part of a bird's feather.

variance (vār-ē-ahns)–A statistical measure of the amount of variation that is observed within or among a group of animals or plants.

variant (vār-ē-ahnt)–A recognized entity different from normal.

variation (vahr-ē-ā-shuhn)–One of the laws of organic nature; organisms vary in time, from place to place, and also in one locality with time; they vary also in their appearance (morphology).

variety (vah-rī-eh-tē)–(1) A group of related animals that differs from other similar groups by characters too trivial or inconstant to be recognized as a species; often any category of lower rank than a species. (2) In domesticated animals, a subdivision of a breed based on some minor character such as color, etc.

variety hybrid (vah-rī-eh-tē hī-brihd)– A cross between two varieties of the same species.

varus (vār-uhs)–Bent inward.

vas deferens (vahs dehf-ər-ehnz)–The tube that carries sperm into the pelvic region toward the urethra.

vasculitis (vahs-kyoo-lī-tihs)–Inflammation of a blood or lymph vessel.

vasectomized (vahs-ehck-tō-mīzd)–Designating a male whose vas deferens or a portion of it has been surgically removed to make the male sterile or incapable of producing live sperm.

vasectomy (vah-sehck-tō-mē)–Sterilization of a male in which a portion of the vas deferens is surgically removed yet the animal may retain its libido.

vasoconstrictor (vahs-ō-kohn-strihck-tər)–A substance that narrows the blood vessels.

vasodilator (vahs-ō-dī-lāt-ər)–A substance that widens the blood vessels.

vasotocin (vahs-ō-tō-sihn)–An oxytocinlike hormone produced in birds that stimulates oviposition.

vat (vaht)–Any large vessel, tub, cistern, etc., used for holding liquids, as a vessel used in large-scale brewing or dipping of domestic animals.

veal (vēl)–A confined young dairy calf that is fed only milk or milk replacer to produce pale, soft, and tender meat.

vealer (vēl-ər)–A veal calf less than three months old.

vection (vehck-shuhn)–The passing of a disease from one plant or animal to another.

vector (vehck-tər)–Something that carries disease from one animal to another.

vegetarian (vehj-ē-tār-ē-ahn)–(1) One who, because of cultural reasons or personal conviction, abstains from eating meat (in the strictest sense, also milk, butter, and eggs). (2) An herbivorous animal or person.

vein (vān)–A blood vessel that returns blood to the heart.

vena cava (vē-nah kā-vah)–The large veins that carry blood into the right atrium of the heart.

velocity (vehl-oh-sih-tē)–The speed at which something travels through an object.

venereal disease (vē-nē-rē-ahl dih-zēz)–A contagious disease that is usually contracted by animals having sexual intercourse.

venipuncture (vehn-ih-puhnck-chər)–Withdrawing blood from a vein (usually with a needle and syringe).

venous (vē-nuhs)–The blood vessels that return blood to the heart.

vent (vehnt)–(1) A small opening or passage, as an opening for ventilation. (2) The opening of the cloaca; the anus of a bird. (3) A brand mark made to indicate that the animal no longer belongs to the original owner. (4) To cancel an old brand on an animal to show change in ownership.

vent shield (vehnt shē-uhld)–A shield used to cover a chicken's vent to prevent picking by other chickens.

ventral (vehn-trahl)–The belly or underside.

ventral midline incision (vehn-trahl mihd-līn ihn-sihz-zhuhn)–A surgical cut along the midsagittal plane of the abdomen along the linea alba.

ventral punctum (vehn-trahl puhnck-tuhm)–A small spot near the lower medial canthus where the nasolacrimal duct begins.

ventral recumbency (vehn-trahl rē-kuhm-behn-sē)–Lying on the underside or belly; also termed sternal recumbency (stər-nahl) or prone (prōn).

ventral root (vehn-trahl root)–The portion of the spinal nerves that enters the lowermost or ventral portion of the spinal cord and carries efferent motor impulses.

ventricle (vehn-trih-cuhl) (1) A small cavity; animals have a third, fourth, and lateral ventricle. (2) The inferior chamber of the heart; also a cavity of the brain.

ventriculus (vehn-trihck-yoo-luhs)–The muscular stomach of birds; also called the gizzard.

ventrodorsal projection (vehn-trō-dōr-sahl prō-jehck-shuhn)–An x-ray beam that passes from the belly to the back; abbreviated V/D.

venules (vehn-yoolz)–The smaller branches of veins.

vermicide (vehr-mih-sīd)–Any substance that kills internal, parasitic worms; an anthelmintic. See vermifuge.

vermiform (vehr-mih-fōrm)–Worm-shaped.

vermifuge (vehr-mih-fūj)–A drug or chemical that expels worms from animals; an anthelmintic. See vermicide.

vermin (vehr-mihn)–Any noxious animal; an insect, acarid, rodent, etc.

verminous (vehr-mih-nuhs)–Pertaining to or due to worms.

vermis (vər-mihs)–A worm; the term used to describe the central body of the cerebellum.

vernal (vehr-nahl)–Appearing in spring.

vernis (vehr-nihs)–Of spring.

verrucae (veh-roo-sē)–Warts.

vertebra (vər-teh-brah)–A single spinal column bone; plural is vertebrae (vər-teh-brā).

vertebral column (vər-teh-brahl koh-luhm)–The bones that support the head and body and provides bony protection for the spinal cord.

vertebrates (vehr-teh-brātz)–Animals with a spinal column or backbone, such as fishes, birds, mammals, etc.

vertical transmission (vər-tih-kahl trahnz-mihs-shuhn)–Disease transfer from mother to fetus.

vertigo (vər-tih-gō)–A sense of dizziness.

vesicle (vehs-ih-kuhl)–A circumscribed skin elevation that is less than 0.5 cm in diameter filled with fluid; also called a blister.

vestibular disease (vehs-tihb-yoo-lahr dih-zēz)–A neurologic disorder characterized by heat tilt, nystagmus, rolling, falling, and circling.

vestibule (vehs-tih-buhl)–The front part of the nostrils and nasal cavity.

vestigial (vehs-tihj-ē-ahl)–A structure that has lost a function it previously had.

veterinarian (veht-ehr-ih-nār-ē-ahn)–An authorized practitioner of veterinary science.

viability (vī-ah-bihl-ih-tē)–The ability to live (immediately after birth or hatching). (2) Pertaining to sperm cells in the semen, capable of living and successfully fertilizing the female gamete.

viable (vī-ah-buhl)–Capable of living outside the mother.

vibrissa (vī-brihs-ah)–Large tactile hair.

vice (vīs)–Any seriously undesirable habit of an animal, such as vent picking and cannibalism in poultry; balking, kicking, shying, etc., in horses.

vigor (vihg-ər)–Healthy growth; also means having high energy.

vilitis (vihl-ih-tihs)–A condition of a horse's hoof in which the soft part of the wall becomes inflamed.

villus (vihl-uhs)–A tiny, hairlike projection; plural is villi (vihl-ī).

viral (vī-rahl)–Having the nature of a virus; pertaining to a virus; like a virus.

viremia (vī-rē-mē-ah)–The condition where viral organisms are found in an animal's blood system.

virgin (vihr-jehn)–Any female that has not had coitus.

virile (vī-rihl)–A male capable of functioning in copulation.

virology (vī-rohl-ō-jē)–The study of viruses and viral diseases.

virucide (vī-rū-sīd)–A chemical or physical agent that kills or inactivates viruses; a disinfectant.

virulence (vihr-yoo-lehns)–The strength of an organism to cause disease.

virulent (vihr-ū-lehnt)–Highly pathogenic; having great disease-producing capacity; deadly; very poisonous or harmful.

virus (vī-ruhs)–A self-reproducing, eucaryotic agent that is considerably smaller than a bacterium and can multiply only within the living cells of a suitable host; plural is viruses.

virustatic (vihr-ū-staht-ihck)–A substance that prevents the multiplication of a virus.

visceral (vihs-ər-ahl)–An organ.

viscosity (vihs-koh-siht-ē)–Resistance to flow; degree of fluid thickness.

vitamin (vī-tah-mihn)–An organic substance, classified as either fat or water soluble, that performs specific and necessary functions in relatively small concentrations in an organism.

vitelline membrane (vī-tehl-ihn mehm-brān)–The thin membrane located inside the zona pellucida of an ovum that contains the cytoplasm or vitellus.

vitellus (vī-tehl-uhs)–The cytoplasm of a mammalian ovum; it contains relatively large amounts of nutritive material called yolk or deutoplasm.

vitreous chamber (viht-rē-uhs chām-bər)–The cavity in the posterior two-thirds of the eyeball.

vitreous humor (viht-rē-uhs hū-mər)–A soft, clear, jellylike mass that fills the vitreous chamber.

vivarium (vī-vahr-ē-uhm)–A glass box resembling an aquarium used to keep or raise animals or plants.

viviparous (vī-vihp-ahr-uhs)–Bearing live young.

vocal cords (vō-kahl kōrdz)–Two pairs of membranous bands in the larynx.

void (voyd)–To evacuate feces and/or urine.

voiding (voy-dihng)–The process of eliminating.

volatile fatty acids (vohl-ah-tihl faht-ē ah-sihdz)–A group of low-molecular weight acids that form gases rather easily, and are produced by microbial action in the rumen, e.g., acetic, propionic, and butyric acid.

volvulus (vohl-vū-luhs)–An end-to-end twist.

vomit (vohm-iht)–(1) Material ejected from the stomach through the mouth or nostrils. (2) To eject matter from the stomach through the mouth or nostrils; to throw up.

vomiting (voh-mih-tihng)–The forcible expulsion of the stomach contents through the mouth.

vomitus (voh-mah-tuhs)–Vomited material.

vulva (vuhl-vah)–The external female genitalia.

W

waif (wāf)–A stray animal, specifically a lost sheep.

walk (waulck)–A natural, slow, flat footed, four-beat gait in which each foot takes off and strikes the ground at separate intervals.

walk-through (waulck throo)–Designating a building or milking room so designed that the traffic moves in one door or gate and out another, rather than in and out the same door.

wallow (wahl-ō)–(1) A natural or artificial wading area to cool swine. (2) To roll, lie, move lazily in a mud hole, as a hog. (3) A land feature resembling a large hog wallow.

warm-blooded animal (wahrm bluhd-ehd ahn-ih-mahl)–Any animal, such as a bird or mammal, whose body temperature is warmer than its surrounding medium, as contrasted with the cold-blooded animals, such as reptiles. Also called homeothermic.

warren (wahr-ehn)–An area, place, or enclosure in which rabbits are bred and raised. Also called rabbit warren.

wart (wahrt)–(1) A tumor on the skin or mucous membrane of an animal composed of fibrous tissue covered over with epithelial cells similar to those of the part of the body on which they are located.

wasty (wahs-tē)–A term used to describe a carcass or live animal that is too fat.

water (wah-tər)–H$_2$O; hydrogen oxide; the liquid may contain associated molecules.

watering (waht-tər-ihng)–Furnishing water for the consumption by animals.

wattle (waht-tuhl)–(1) Appendages suspended from the head (usually the chin) in chickens, turkeys, and goats. (2) Fold of skin cut into the dewlap of cattle for identification purposes.

wavelength (wāv-lehngth)–The length that a wave must travel in one cycle.

waxing (wahcks-ihng)–An accumulation of colostrum on the nipples of mares usually prior to foaling; also called waxed teats.

wax-picking (wahcks pihck-ihng)–A method of picking poultry in which the bird carcasses are coated with hot, melted wax. When cold, the wax is removed, the feathers and pin feathers being carried with it.

wean (wēn)–(1) To make a young animal cease to depend on its mother's milk. (2) To accustom partly grown birds to do without artificial heat.

weaner (wē-nər)–(1) Any of many devices used to prevent a young animal from suckling its mother. (2) A weaned lamb, five to twelve months old, before it grows its permanent teeth (Australia). (3) Any lamb from the time it is weaned until it is sheared. (4) A weaned range calf.

weaning ring (wē-nihng rihng)–A ring with spikes that fits on a calf's muzzle to prevent nursing. When the calf approaches the mother, the projections on the ring make the mother move.

weanling (wēn-lihng)–An animal that has recently been weaned or taken from its mother.

web (wehb)–(1) The cotton disk used in filtering milk. (2) The fat that surrounds the paunch and intestines of sheep. (3) The membrane that unites the fingers and toes, especially of amphibians and water birds.

weed (wēd)–(1) A horse or other domestic animal that is undersized or a misfit. (2) Designating a flavor of dairy products, especially milk, that resembles that of certain vegetation (onion, wild garlic, leeks, etc.) (3) To take out, as in to weed the population.

weedy (wē-dē)–(1) Designating an off-flavor of milk or its products to suggest the odor or taste of certain weeds. (2) Designating a thin or undesirable animal.

well-bred (wehl-brehd)–Designating a plant or animal that has a good pedigree.

wen (wehn)–An encysted skin tumor.

western saddle (wehs-tərn sah-duhl)–A heavy, deep-seated saddle with a distinctly raised horn and cantle, as contrasted to the English saddle, which is light and comparatively flat. It is used for working livestock on the ranches in the western United States. Also called stock saddle.

wet (weht)–(1) To dampen; to sprinkle; to furnish water. (2) Moist, covered, soaked, or saturated with water.

wet band (weht bahnd)–A group of ewes or does accompanied by nursing young. Also called wet herd.

wet mare (weht mār)–An intact female horse that has foaled during the current breeding season and is nursing a foal.

wet sheep (weht shēp)–Ewes with suckling lambs. Wet band.

wet veal (weht vēl)–Slaughter calves under three weeks of-age.

wether (whehthər)–A castrated male sheep or goat.

whang (wahng)–Tough leather adapted for strings, thongs, and belt laces that is commonly made from calf skin.

wheal (whēl)–A smooth, slightly raised, swollen area that itches.

whelp (wehlp)–An unweaned canine.

whelping (wehl-pihng)–Giving birth to canines.

whinny (wihn-ē)–The gentle, soft cry of a horse.

whip (whihp)–(1) Any of several kinds of instruments for lashing animals, usually consisting of a handle and a lash. (2) To lash an animal. (3) To beat cream, egg white, etc., so that it becomes a froth and holds shape.

whip break (whihp brăck)–To train a horse by use of a whip. (The whip is used sparingly, and never out of anger, so that the horse learns that when it does certain things it is punished and when it does certain other things it is rewarded.)

whipworm (whihp-wehrm)–Any worm of the genus *Trichuris* and related genera that live internally in the intestines of humans and various domestic animals.

white cell count (whīt sehl cownt)–The number of leukocytes per cubic millimeter of blood.

white face (whīt fās)–A coat color characteristic of certain breeds of animals, such as Hereford cattle, which have a red body, a white face, and white markings on the brisket, underline, and switch of the tail.

white line (whīt līn)–Fusion between the wall and sole of the hoof in equine.

whorl (wehrl)–A swirl or cowlick in the hair on an animal's coat.

wicking (wihck-ihng)–The act of providing material that moves liquid from one area to another (thus providing a potential infection source).

wildling (wihld-lihng)–(1) Any wild animal. (2) In forestry, a seedling produced naturally outside of a nursery.

wiltshire side (wihlt-shehr sīd)–The entire half of a dressed pig, minus the head, shank, shoulder bone, and hip bone. All of the side except the ham and shoulder is sold as bacon.

wind (wihnd)–(1) The blast of air from a blower or fan. (2) The breathing of a horse.

windbreak (wihnd brāk)–A shelter in which an animal can stand and be protected from the wind.

wing (wihng)–(1) An organ of flight for birds, insects, bats, etc. (2) A piece of bird meat consisting of the organ of flight with the feathers removed. (3) The outside corner of the cutting edge of a plow.

winged web vaccination (wihngehd wehb vahck-sihn-ā-shuhn)–In poultry, the process of injecting a vaccine into the skin on the underside of the wing at the elbow.

winking (wihnk-ihng)–The opening of the labia to expose the clitoris while the female assumes a mating position.

winter pause (wihn-tər paws)–In poultry, the cessation of egg production in winter.

winter range (wihn-tər rānj)–The range, usually at low altitudes, used for pasture in the winter months in the western United States.

wintering (wihn-tər-ihng)–The care, feeding, and maintenance of livestock and tender plants through the winter months.

with calf (wihth kahlf)–Designating a cow that is pregnant.

withdrawal period (wihth-draw-uhl pehr-ē-ohd)–The length of time a feed additive or drug must not be fed or administered to an animal prior to slaughter.

withers (wih-thərz)–The dorsal region of the horse located between both scapula.

wolf teeth (wuhlf tēth)–The rudimentary first premolar in equine that is usually shed in maturity.

womb (woom)–The uterus.

woodland pasture (wood-lahnd pahs-chər)–(1) Farm woodlands used for grazing. (2) Wooded areas with grass and other grazing plants growing in open spaces among trees.

wool (wool)–The fibers constituting the soft, curly coats of sheep, camel, yak, etc. After cotton, sheep wool is the most important source of natural fiber used in apparel, upholstery, carpets, etc.

wool-ball (wool bahl)–Wool ingested by nursing lambs that have been sucking at tags instead of at the teats. It forms a ball in the stomach and may cause serious digestive disturbance. See hair ball.

wool carding (wool kahrd-ihng)–The process of disentangling and separating the fibers of wool and delivering them in a parallel condition suitable for spinning.

wool clip (wool klihp)–The annual wool crop.

wool combing (wool kōm-ihng)–The process of dealing with longstapled wools that are to be used in worsted manufacture. The operations are like carding and are effected with a gilling machine.

wool fat (wool faht)–The waxy, greasy substance that exudes from the sheep's skin and clings to the wool fibers. After refining, it is sold in the drug trade as lanolin. In the crude form it is often known in commerce as degras. Also called yolk.

wool washing (wool wahsh-ihng)–The process necessary to remove the suet (fat), or natural grease, and dirt from the fleece. It is treated first in alkaline liquor which saponifies the grease and then is washed in fresh water.

woolly (wool-ē)–Clothed with long, matted hairs.

worm (wehrm)–(1) Any small, soft-bodied, usually limbless animal, such as a larva, grub, maggot, earthworm, silkworm, etc. (2) To rid an animal of internal parasitic worms; the proper term is to deworm.

worm nest (wehrm nehst)–In cattle, a protuberance or swollen area on the flank or lower portion of the chest that contains filarial worms.

wormed (wehrmd)–Designating an animal that has been given a vermifuge or anthelmintic to kill worms; the proper term is deworm.

wound (wownd)–Any violently caused disruption of the continuity of internal or external tissue.

wrangler (rahng-glər)–A herder or handler of wild or range horses on ranches.

wry neck (rī nehck)–(1) A twisting of the neck and head of fowls into an unnatural position, possibly as a symptom of paralysis and other diseases. (2) An animal with its neck distorted as the result of having been carried in a faulty position in the uterus.

wry tail (rī tāl)–The tail of a fowl or mammal (usually cattle), when it is permanently carried to one side.

X

xanthosis (zahn-thō-sihs)–A yellowish discoloration.

xiphoid process (zī-foyd proh-sehs)–The cartilaginous caudal part of the sternum.

x-ray (ehcks-rā)–A roentgen ray.

Y

yard (yahrd)–(1) A unit of linear measurement equal to 3 feet or 36 inches. (2) The grounds that immediately surround a dwelling. (3) An enclosed area in which stock, fowls, etc., are kept, as a chicken yard. It may be used as a suffix, as in stockyard.

yean (yēn)–To give birth to young, especially by goats and sheep.

yeanling (yēn-lihng)–The newborn or new young of sheep or goats.

yearling (yēr-lihng)–A young horse between one and two years of age.

yearling weight (365 day adjusted weight) (yēr-lihng wāt)–Weight taken as a yearling or long yearling and adjusted within breed and sex for the age of the calf and the age of the dam to a standard mature-cow basis of the age group.

yeast (yēst)–A budding form of fungus.

yeld mare (yehld mār)–A mare that is not lactating; a dry mare.

yellow dun (yehl-ō duhn)–A coat color of a horse that is light, lemon-yellow.

yield (yēld)–(1) The grade in meat animals, referring to the amount of lean meat produced in a carcass. (2) The percentage of clean wool remaining in a lot after scouring. (3) The ratio of carcass weight to live weight.

yoke (yōk)–(1) A variously shaped wooden frame that is placed on the necks of two oxen to work them together as a team. (2) A clamp that unites two pieces (3) A team, as a yoke of oxen. (4) A wood or steel bar placed at the end of a tongue connected by rings and leather straps to the hames of the harness of a pair of horses.

yolk (yōk)–(1) The yellow part of a fowl's egg that has a germinal disk located on its outer edge from which the embryo develops. The remainder of the yolk, together with the white, provides the nourishment to sustain the embryo's growth. (2) In wool scouring, all the substances, such as wool grease and perspiration, present in the fleece of a sheep when it is sheared.

Z

zero pasture (zĕr-ō pahs-chər)–Putting green forage and hauling it to stock in corrals or in dairy barns in lieu of pasturing. Sometimes called zero grazing.

zinc (zihngk)–Zn; a metallic chemical element; one of the micronutrient elements in soils, essential for both plant and animal growth.

zona pellucida (zō-nah pehl-lū-sihd-ah)–The relatively thick covering or membrane that forms the outer surface of a mammalian ovum.

zoned heating (zōnd hē-tihng)–A system of heating a large building, such as a greenhouse or chicken house, by means of controls in different zones of the building.

zoonosis (zō-ō-nō-sihs)–A disease that can be transmitted between animals and humans.

zoophagous parasite (zō-ohf-ah-guhs pahr-ah-sīt)–Any parasite that thrives in or on animals.

zooplankton (zō-ō-plahnck-tohn)–(1) The animal constituents of plankton. (2) Tiny animals that drift with the currents. (3) The animal portion of the planktonic organisms.

zootechny (zō-ō-tehck-nē)–The scientific breeding and/or domestication of animals.

zygote (zī-gōt)–A fertilized ovum or egg, it is the diploid cell formed from the union of the sperm with an ovum.

Word Parts

a-	Prefix for without, no, not, away from
ab-	Prefix for away from
abdomin/o	Combining form for abdomen
-able	Suffix for capable of
abort/o	Combining form for abortion
abrad/o	Combining form for scrape off
abras/o	Combining form for scrape off
abrupt/o	Combining form for broken away from
abs-	Prefix for away from
absorpt/o	Combining form for suck up or in
ac-	Prefix for toward
-ac	Suffix for pertaining to
acanth/o	Combining form for spiny
acetabul/o	Combining form for hip socket
-acious	Suffix for characterized by
acne/o	Combining form for point
acoust/o	Combining form for sound or hearing
acous/o	Combining form for sound or hearing
acr/o	Combining form for extremities or top
acromi/o	Combining form for point of the shoulder blade
acu/o	Combining form for sharp, severe, sudden, needle
acuit/o or acut/o	Combining form for sharp
-acusia	Suffix for hearing
ad-	Prefix for toward or in the direction of
-ad	Suffix for toward, to, in the direction of
aden/o	Combining form for gland
adhes/o	Combining form for stickiness or clinginess
adip/o	Combining form for fat
adnex/o	Combining form for bound to
adren/o	Combining form for adrenal glands
adrenal/o	Combining form for adrenal glands

aer-	Prefix for air or gas
aer/o	Combining form for air or gas
aesthet/o	Combining form for sensation or feeling
af-	Prefix for toward
affect/o	Combining form for exert influence on
ag-	Prefix for toward
agglutin/o	Combining form for clumping
aggress/o	Combining form for attack or step forward
-ago	Suffix for attack, diseased condition
-agra	Suffix for excessive pain or seizure
-aise	Suffix for ease or comfort
-al	Suffix for pertaining to
al-	Prefix for similar
alb/i	Combining form for white
alb/o	Combining form for white
albin/o	Combining form for white
albumin/o	Combining form for albumin (one type of protein)
alg/e or alg/o	Combining form for pain
algesi/o	Combining form for pain or suffering
-algesia	Suffix for pain or suffering
algi-	Prefix for pain or suffering
-algia	Suffix for pain
align/o	Combining form for to bring in line or the correct position
aliment/o	Combining form for nourishment
all-	Prefix for other or different
all/o	Combining form for other or different
alopec/o	Combining form for baldness
alveol/o	Combining form for alveoli or small sac
amb-	Prefix for both sides or double
amb/i	Combining form for both sides or double
ambly/o	Combining form for dim
ambul/o	Combining form for to walk
ambulat/o	Combining form for to walk
ametr/o	Combining form for out of proportion
-amine	Suffix for nitrogen compound
amni/o	Combining form for fetal membrane
amph-	Prefix for around, on both sides, or doubly
amput/o	Combining form for cut away or cut off
amputat/o	Combining form for cut away or cut off
amyl/o	Combining form for starch

an-	Prefix for negative, no, not, or without
-an	Suffix for pertaining to
an/o	Combining form for anus or ring
ana-	Prefix for up, apart, backward, excessive
andr/o	Combining form for male
angi-	Prefix for vessel
angi/o	Combining form for blood or lymph vessels
angin/o	Combining form for strangling
anis/o	Combining form for unequal
ankyl/o	Combining form for crooked, bent, or stiff
anomal/o	Combining form for irregularity
ante-	Prefix for before or forward
anter/o	Combining form for front
anti-	Prefix for against
anxi/o or anxiet/o	Combining form for uneasy or distressed
aort/o	Combining form for aorta
ap-	Prefix for toward
aphth/o	Combining form for small ulcer or eruption
apic/o	Combining form for apex or point
aplast/o	Combining form for lack of or defective development
apo-	prefix for opposed or detached
aponeur/o	Combining form for aponeurosis
aqu/i	Combining form for water
aqu/o	Combining form for water
aque/o	Combining form for water
-ar	Suffix for pertaining to
arachn/o	Combining form for spider
arc/o	Combining form for bow or arch
-arche	Suffix for beginning
areat/o	Combining form for occurring in patches or circumscribed areas
arrect/o	Combining form for upright
arteri/o	Combining form for artery
arthr/o	Combining form for joint
articul/o	Combining form for joint
-ary	Suffix for pertaining to
as-	Prefix for toward
-ase	Suffix for enzyme
-ashenia	Suffix for weakness
asphyxi/o	Combining form for absence of a pulse
aspir/o	Combining form for breathe in

aspirat/o	Combining form for breathe in
asthen-	Prefix for weakness
-asthenia	Suffix for weakness
asthmat/o	Combining form for gasping
at-	Prefix for toward
atel/o	Combining form for incomplete
ather/o	Combining form for plaque or fatty substance
athet/o	Combining form for uncontrolled
-atonic	Suffix for lack of tone or strength
atop/o	Combining form for out of place
atres/i	Combining form for without an opening
atri/o	Combining form for atrium or chamber
attenuat/o	Combining form for diluted or weakened
aud-	Prefix for ear or hearing
aud/i	Combining form for ear or hearing
audi/o	Combining form for ear or hearing
audit/o	Combining form for ear or hearing
aur/i	Combining form for ear or hearing
aur/o	Combining form for ear or hearing
auscult/o	Combining form for listen
aut/o	Combining form for self
ax/o	Combining form for axis or mainstream
axill/o	Combining form for armpit
azot/o	Combining form for urea or nitrogen
bacill/o	Combining form for rod
bacteri/o	Combining form for bacteria
bar/o	Combining form for pressure or weight
bas/o	Combining form for base (not acid)
bi-	Prefix for two, double, or twice
bi/o	Combining form for life
bifid/o	Combining form for split or cleft
bifurcat/o	Combining form for divide into two branches
bil/i	Combining form for bile or gall
bilirubin/o	Combining form for bilirubin or breakdown product of blood
-bin	Suffix for twice, double, or two
bio-	Prefix for life
bis-	Prefix for twice, double, or two
-blast	Suffix for immature or embryonic or formative capability
blast/i	Combining form for embryonic
blast/o	Combining form for immature or embryonic

blephar/o	Combining form for eyelid
borborygm/o	Combining form for rumbling sound
brachi/o	Combining form for arm
brachy-	Prefix for short
brady-	Prefix for abnormally slow
brev/i	Combining form for short
brev/o	Combining form for short
bronch/i	Combining form for bronchial tube
bronch/o	Combining form for bronchial tube
bronchi/o	Combining form for bronchial tube
bronchiol/o	Combining form for bronchiole
brux/o	Combining form for grind
bucc/o	Combining form for cheek
bucca-	Prefix for cheek
burs/o	Combining form for bursa (sac of fluid near a joint)
cac-	Prefix for bad, diseased, or weak
cac/o	Combining form for bad, diseased, or weak
cadaver/o	Combining form for dead body
calc/i	Combining form for calcium or heel
calc/o	Combining form for calcium or heel
calcane/o	Combining form for heel bone
calcul/o	Combining form for stone
cali/o	Combining form for cup
calic/o	Combining form for cup
call/i	Combining form for hard
callos/o	Combining form for hard
calor/i	Combining form for heat
canalicul/o	Combining form for little duct
canth/o	Combining form for corner of the eye
capill/o	Combining form for hair
capit/o	Combining form for head
capn/o	Combining form for carbon dioxide
-capnia	Suffix for carbon dioxide
capsul/o	Combining form for little box
carb/o	Combining form for carbon
carcin/o	Combining form for cancerous or malignant growth of epithelial cells
cardi/o	Combining form for heart
cari/o	Combining form for decay
carot/o	Combining form for sleep or stupor

carp/o	Combining form for carpus or the joint of the front limbs between the radius/ulna and metacarpals
cartilag/o	Combining form for cartilage or gristle
caruncul/o	Combining form for bit of flesh
cat-	Prefix for down, lower, under, or downward
cata-	Prefix for down, lower, under, or downward
catabol/o	Combining form for breaking down
cath-	Prefix for down, lower, under, or downward
cathart/o	Combining form for cleansing or purging
cathet/o	Combining form for insert or send down
caud/o	Combining form for toward the tail
caus/o	Combining form for burning
caust/o	Combining form for burning
caut/o	Combining form for burning
cauter/o	Combining form for burning
cav/i	Combining form for hollow
cav/o	Combining form for hollow
cavern/o	Combining form for hollow
cec/o	Combining form for blind gut or cecum
-cele	Suffix for hernia, cyst, cavity, or tumor
celi/o	Combining form for abdomen
cement/o	Combining form for cementum
cent-	Prefix for hundred
-centesis	Suffix for surgical puncture to remove fluid or gas
centi	Prefix for one-hundredth
cephal-	Prefix for head
cephal/o	Combining form for head
-ceps	Suffix for head
cera-	Prefix for wax
cerebell/o	Combining form for cerebellum
cerebr/o	Combining form for cerebrum or largest part of brain
cerumin/o	Combining form for cerumen or earwax
cervic/o	Combining form for neck or necklike
chalas/o	Combining form for relaxation or loosening
-chalasis	Suffix for relaxation or loosening
chalaz/o	Combining form for hailstone
cheil/o	Combining form for lip
chem/i	Combining form for drug or chemical
chem/o	Combining form for drug or chemical
chemic/o	Combining form for drug or chemical

chol/e	Combining form for bile or gall
cholangi/o	Combining form for bile duct
cholecyst/o	Combining form for gallbladder
choledoch/o	Combining form for common bile duct
cholesterol/o	Combining form for cholesterol
chondr/o	Combining form for cartilage
chondri/o	Combining form for cartilage
chord/o	Combining form for cord or spinal cord
chori/o	Combining form for chorion or membrane
choroid/o	Combining form for choroid layer of the eye
chrom/o	Combining form for color
chromat/o	Combining form for color
-chrome	Suffix for color
chym/o	Combining form for juice
cib/o	Combining form for meal
-cidal	Suffix for death
-cide	Suffix for killing or destroying
cili/o	Combining form for microscopic hairlike projections or eyelashes
cine/o	Combining form for movement
circi/o	Combining form for ring or circle
circulat/o	Combining form for go around in a circle
circum-	Prefix for around
circumscrib/o	Combining form for confined or limited in space
cirrh/o	Combining form for tawny, orange-yellow
cis/o	Combining form for cut
-clasis	Suffix for break
-clast	Suffix for break
clav/i	Combining form for key
clavicul/o	Combining form for collarbone; clavicle
clitor/o	Combining form for female erectile tissue
clon/o	Combining form for violent action
clus/o	Combining form for to close or shut
-clysis	Suffix for irrigation
co-	Prefix for together, with
coagul/o	Combining form for clump or congeal
coagulat/o	Combining form for clump or congeal
cocc/i	Combining form for round
cocc/o	Combining form for round
-coccus	Suffix for round bacterium
coccyg/o	Combining form for tailbone

cochle/o	Combining form for snail-like or spiral
coher/o	Combining form for stick together
cohes/o	Combining form for stick together
coit/o	Combining form for coming together
col/o	Combining form for colon (part of large intestine)
coll/a	Combining form for glue
coll/i	Combining form for neck
colon/o	Combining form for colon (part of large intestine)
colp/o	Combining form for vagina
column/o	Combining form for pillar
com-	Prefix for together
comat/o	Combining form for deep sleep
comminut/o	Combining form for break into pieces
communic/o	Combining form for share
compatibil/o	Combining form for sympathize with
con-	Prefix for together or with
concav/o	Combining form for hollow
concentr/o	Combining form for remove excess water or condense
concept/o	Combining form for receive or become pregnant
conch/o	Combining form for shell
concuss/o	Combining form for shaken together violently
condyl/o	Combining form for knuckle or knob
confus/o	Combining form for disorder or confusion
conjunctiv/o	Combining form for conjunctiva or the pink mucous membrane of the eye
consci/o	Combining form for aware
consolid/o	Combining form for become solid
constipat/o	Combining form for pressed together
constrict/o	Combining form for draw tightly together or narrow
contact/o	Combining form for touched
contagi/o	Combining form for touching something or infection
contaminat/o	Combining form for pollute
contine/o	Combining form for keep in or restrain
continent/o	Combining form for keep in or restrain
contra-	Prefix for against or opposite
contracept/o	Combining form for prevention of fertilization of egg with sperm
contus/o	Combining form for bruise
convalesc/o	Combining form for become strong
convex/o	Combining form for arched
convolut/o	Combining form for twisted or coiled

convuls/o	Combining form for pull together
copi/o	Combining form for plentiful
copulat/o	Combining form for joining together
cor/o	Combining form for pupil
cord/o	Combining form for spinal cord
cordi/o	Combining form for heart
core-	Prefix for pupil
core/o	Combining form for pupil
cori/o	Combining form for skin or leather
corne/o	Combining form for transparent anterior portion of the sclera; cornea
coron/o	Combining form for crown
corp/u	Combining form for body
corpor/o	Combining form for body
corpuscul/o	Combining form for little body
cort-	Prefix for covering
cortic/o	Combining form for outer region or cortex
cost/o	Combining form for rib
cox/o	Combining form for hip or hip joint
crani/o	Combining form for skull
-crasia	Suffix for mixture
crepit/o	Combining form for crackling
crepitat/o	Combining form for crackling
crin/o	Combining form for secrete or separate
-crine	Suffix for secrete or separate
cris/o	Combining form for turning point
-crit	Suffix for separate
critic/o	Combining form for turning point
cry/o	Combining form for cold
crypt/o	Combining form for hidden
cubit/o	Combining form for elbow
cuboid/o	Combining form for cubelike
culd/o	Combining form for blind pouch
-cusis	Suffix for hearing
cusp/i	Combining form for pointed or pointed flap
cut-	Prefix for skin
cutane/o	Combining form for skin
cyan/o	Combining form for blue
cycl/o	Combining form for ciliary body of the eye or cycle
-cyst	Suffix for bag or bladder

cyst-	Prefix for bag or bladder
cyst/o	Combining form for urinary bladder, cyst, or sac or fluid
cyt/o	Combining form for cell
-cyte	Suffix for cell
-cytic	Suffix for pertaining to a cell
-cytosis	Suffix for condition of cells; implies increased cell numbers
dacry-	Prefix for tear or tear duct
dacry/o	Combining form for tear or tear duct
dacryocyst/o	Combining form for lacrimal sac
dactyl/o	Combining form for digits
dart/o	Combining form for skinned
de-	Prefix for from, not, down, or lack of
dec/i	Combining form for ten or tenth
deca-	Prefix for ten or tenth
decem-	Prefix for ten
deci	Prefix for one-tenth
decidu/o	Combining form for falling off or shedding
decubit/o	Combining form for lying down
defec/o	Combining form for free from waste
defecat/o	Combining form for free from waste
defer/o	Combining form for carrying down or out
degenerat/o	Combining form for breakdown
deglutit/o	Combining form for swallow
dehisc/o	Combining form for burst open
dek-	Prefix for ten
deka-	Prefix for ten
deliri/o	Combining form for wandering in the mind
dem/o	Combining form for population or people
-dema	Suffix for swelling (fluid)
demi-	Prefix for half
dendr/o	Combining form for resembling a tree or branching
dent-	Prefix for teeth
dent/i	Combining form for teeth
dent/o	Combining form for teeth
depilat/o	Combining form for hair removal
depress/o	Combining form for pressed or sunken down
derm/o	Combining form for skin
dermat/o	Combining form for skin
-derma	Suffix for skin
derma-	Prefix for skin

-desis	Suffix for surgical fixation of a bone or joint or to bind
deteriorat/o	Combining form for worsening
deut-	Prefix for second
deuto-	Prefix for second
deutero-	Prefix for second
dextr/o	Combining form for right side
di-	Prefix for twice
dia-	Prefix for through, between, apart, or complete
diaphragmat/o	Combining form for wall across or muscle separating the thoracic and abdominal cavities
diastol/o	Combining form for expansion
didym/o	Combining form for testes or double
diffus/o	Combining form for spread out
digest/o	Combining form for divided or distribute
digit/o	Combining form for digit
dilat/o	Combining form for spread out
dilatat/o	Combining form for spread out
dilut/o	Combining form for dissolve or separate
diophor/o	Combining form for sweat
diphther/o	Combining form for membrane
dipl/o	Combining form for double
dipla-	Prefix for double
dips/o	Combining form for thirst
-dipsia	Suffix for thirst
dis-	Prefix for duplication or twice; negative, apart, or absence of
dislocat/o	Combining form for displacement
dissect/o	Combining form for cutting apart
disseminat/o	Combining form for widely scattered
dist/o	Combining form for far
distend/o	Combining form for stretch apart
distent/o	Combining form for stretch apart
diur/o	Combining form for increasing urine output
diuret/o	Combining form for increasing urine output
divert/i	Combining form for turning aside
domin/o	Combining form for controlling
don/o	Combining form for give
dors/i	Combining form for back of body
dors/o	Combining form for back of body
drom/o	Combining form for running
-drome	Suffix for course, condition, run together, or running

du/o	Combining form for two
-duct	Suffix for opening
duct/o	Combining form for carry or lead
duoden/i	Combining form for duodenum
duoden/o	Combining form for duodenum
dur/o	Combining form for dura mater or tough, hard
dy-	Prefix for two
-dynia	Suffix for pain
dynovi/o	Combining form for synovial membrane or synovial fluid
dyo-	Prefix for two
dyph/o	Combining form for bent or hump
dys-	Prefix for difficult, bad, or painful
e-	Prefix or out of or from
-eal	Suffix for pertaining to
ec-	Prefix for out or outside
ecchym/o	Combining form for pouring out of juice
ech/o	Combining form for sound
eclamps/o	Combining form for flashing or shining forth
eclampt/o	Combining form for flashing or shining forth
-ectasia	Suffix for dilation or enlargement
-ectasis	Suffix for dilation or enlargement
ecto-	Prefix for out or outside
-ectomy	Suffix for surgical removal or excision
eczemat/o	Combining form for eruption
edem-	Prefix for swelling, fluid, or tumor
edemat/o	Combining form for swelling, fluid, or tumor
edentul/o	Combining form for toothless
ef-	Prefix for out
effect/o	Combining form for bring about a response
effus/o	Combining form for pouring out
ejaculat/o	Combining form for throw out
electr/o	Combining form for electricity
eliminat/o	Combining form for expel from the body
em-	Prefix for in
emanciat/o	Combining form for wasted away by disease or lean
embol/o	Combining form for something inserted or thrown in
embry/o	Combining form for fertilized egg
-emesis	Suffix for vomiting
emet/o	Combining form for vomit
-emia	Suffix for blood or blood condition

emmetr/o	Combining form for in proper measure
emolli/o	Combining form for soften
en-	Prefix for in, into, or within
encephal/o	Combining form for brain
end/o	Combining form for within, in, or inside
endo-	Combining form for within, in, or inside
endocrin/o	Combining form for secrete within
enem/o	Combining form for inject
ennea-	Prefix for nine
enter/o	Combining form for small intestine
ento-	Prefix for within
enzym/o	Combining form for leaven
eosin/o	Combining form for red or rosy
epi-	Prefix for upon, above, on, or upper
epidemi/o	Combining form for among the people
epididym/o	Combining form for tube at the upper part of each testis (epididymis)
epiglott/o	Combining form for lid covering the larynx (epiglottis)
episi/o	Combining form for vulva
epithel/i	Combining form for outer layer of skin or external surface covering
cquin/o	Combining form for horse or horselike
erect/o	Combining form for upright
erg/o	Combining form for work
eruct/o	Combining form for belch
eructat/o	Combining form for belch
erupt/o	Combining form for burst forth
erythem/o	Combining form for flushed or redness
erythemat/o	Combining form for flushed or redness
erythr/o	Combining form for red
es-	Prefix for out of, outside, or away from
-esis	Suffix for adnormal condition or state
eso-	Prefix for inward
esophag/o	Combining form for tube that connects the mouth to the stomach (esophagus)
esthesi/o	Combining form for sensation or feeling
-esthesia	Suffix for sensation or feeling
esthet/o	Combining form for feeling or sense of perception
estr/o	Combining form for female
ethm/o	Combining form for sieve
eti/o	Combining form for cause

eu-	Prefix for well, easy, or good
evacu/o	Combining form for empty out
evacuat/o	Combining form for empty out
eviscer/o	Combining form for disembowelment
eviscerat/o	Combining form for disembowelment
ex-	Prefix for without, out of, outside, or away from
exacerbat/o	Combining form for aggravate
excis/o	Combining form for cutting out
excori/o	Combining form for scratch
excoriat/o	Combining form for scratch
excret/o	Combining form for separate or discharge
excruciat/o	Combining form for intense pain
exhal/o	Combining form for breathe out
exhalat/o	Combining form for breathe out
-exis	Suffix for condition
exo-	Prefix for out of, outside, or away from
exocrin/o	Combining form for secrete out of
expector/o	Combining form for cough up
expir/o	Combining form for breathe out or die
expirat/o	Combining form for breathe out or die
exstroph/o	Combining form for twisted out
extern/o	Combining form for outside or outer
extra-	Prefix for outside or beyond
extrem/o	Combining form for outermost or extremity
extremit/o	Combining form for outermost or extremity
extrins/o	Combining form for contained outside
exud/o	Combining form for sweat out
exudat/o	Combining form for sweat out
faci-	Prefix for face or form
faci/o	Combining form for face or form
-facient	Suffix for producing
fasc/i	Combining form for fascia or fibrous band
fasci/o	Combining form for fascia or fibrous band
fascicul/o	Combining form for little bundle
fatal/o	Combining form for death or fate
fauc/i	Combining form for narrow pass
febr/i	Combining form for fever
fec/i	Combining form for sediment
fec/o	Combining form for sediment
femor/o	Combining form for thigh bone (femur)

fenestr/o	Combining form for window
fer/o	Combining form for carry or bear
-ferent	Suffix for carrying
-ferous	Suffix for carrying
fertil/o	Combining form for productive or fruitful
fet/i	Combining form for fetus
fet/o	Combining form for fetus
fibr/o	Combining form for fibrous tissue; tissue that is arranged in fibers
fibros/o	Combining form for fibrous tissue; tissue that is arranged in fibers
fibrill/o	Combining form for muscular twitching
fibrin/o	Combining form for threads of a clot (fibrin)
fibros/o	Combining form for fibrous connective tissue
fibul/o	Combining form for small bone or distal hind limb (fibula)
-fic	Suffix for producing or forming
fic/o	Combining form for producing or forming
filtr/o	Combining form for strain through
filtrat/o	Combining form for strain through
fimbri/o	Combining form for fringe
fiss/o	Combining form for crack, split, or cleft
fissur/o	Combining form for crack, split, or cleft
fistul/o	Combining form for tube or pipe
flamme/o	Combining form for flame colored
flat/o	Combining form for rectal gas
flex/o	Combining form for bend
flu/o	Combining form for flow
fluor/o	Combining form for luminous
foc/o	Combining form for point
foll/i	Combining form for bag or sac
follicul/o	Combining form for small sac (follicle)
foramin/o	Combining form for opening
fore-	Prefix for before or in front of
-form	Suffix for form, figure, or shape
form/o	Combining form for form, figure, or shape
fornic/o	Combining form for arch or vault
foss/o	Combining form for shallow depression
fove/o	Combining form for pit
fract/o	Combining form for break
fren/o	Combining form for bridle or device that limits movement
frigid/o	Combining form for cold
front/o	Combining form for forehead or front

-fuge	Suffix for drive away
funct/o	Combining form for perform
function/o	Combining form for perform
fund/o	Combining form for bottom, base, or ground
fung/i	Combining form for fungus
furc/o	Combining form for branching or forking
furuncul/o	Combining form for boil or infecion
-fusion	Suffix for pour
galact/o	Combining form for milk
gamet/o	Combining form for sperm or egg
gangli/o	Combining form for ganglion
ganglion/o	Combining form for ganglion
gangren/o	Combining form for gangrene
gastr/o	Combining form for stomach
gastrocnemi/o	Combining form for calf muscle of hind limb
gemin/o	Combining form for twin or double
gen/o	Combining form for producing or birth
-gene	Suffix for production, origin, or formation
-genesis	Suffix for producing
-genic	Suffix for producing
genit/o	Combining form for producing or birth
-genous	Suffix for producing
genti/o	Combining form for related to birth or the reproductive organs
ger/i	Combining form for old age
geront/o	Combining form for old age
germin/o	Combining form for bud or germ
gest/o	Combining form for bear or carry offspring
gestat/o	Combining form for bear or carry offspring
gigant/o	Combining form for giant or very large
gingiv/o	Combining form for gum
glauc/o	Combining form for gray
glen/o	Combining form for socket or pit
gli/o	Combining form for glue
globin/o	Combining form for protein
globul/o	Combining form for little ball
-globulin	Suffix for protein
glomerul/o	Combining form for glomerulus
gloss/o	Combining form for tongue
glosso-	Prefix for tongue
glott/i	Combining form for back of tongue

glott/o	Combining form for back of tongue
gluc/o	Combining form for sugar
glute/o	Combining form for buttocks
glyc/o	Combining form for sugar
glycer/o	Combining form for sweet
glycogen/o	Combining form for animal starch (glycogen)
gnath/o	Combining form for jaw
-gnosis	Suffix for knowledge
-gog	Suffix for make flow
-gogue	Suffix for make flow
goitr/o	Combining form for enlargement of the thyroid gland
gon/e	Combining form for seed or angle
gon/i	Combining form for seed or angle
gon/o	Combining form for seed or angle
gonad/o	Combining form for sex glands
goni/o	Combining form for seed or angle
gracil/o	Combining form for slender
grad/i	Combining form for move, go, or step
-grade	Suffix for go
-gram	Suffix for record or picture
granul/o	Combining form for granule(s)
-graph	Suffix for instrument that records; occasionally used to describe the record or picture
-graphy	Suffix for process of recording
grav/i	Combining form for heavy or severe
gravid/o	Combining form for pregnancy
gynec/o	Combining form for woman
gyr/o	Combining form for turning or folding
hal/o	Combining form for breath
halit/o	Combining form for breath
hallucin/o	Combining form for wander in the mind
hecato-	Prefix for hundred
hect-	Prefix for hundred
hecto-	Prefix for hundred
hem-	Prefix for blood
hem/e	Combining form for deep red iron-containing pigment
hem/o	Combining form for blood
hemangi/o	Combining form for blood vessel
hemat/o	Combining form for blood
hemi-	Prefix meaning half

hepa-	Prefix for liver
hepar-	Prefix for liver
hepat/o	Combining form for liver
hept-	Prefix for seven
hepta-	Prefix for seven
hered/o	Combining form for inherited
heredit/o	Combining form for inherited
herni/o	Combining form for protrusion of a part through tissues normally containing it
herpet/o	Combining form for creeping
heter/o	Combining form for different
hetero-	Prefix for different
hex-	Prefix for six
hexa-	Prefix for six
hiat/o	Combining form for opening
hidr/o	Combining form for sweat
hil/o	Combining form for notch or opening from a body part (hilus)
hirsut/o	Combining form for hairy or rough
hist/o	Combining form for tissue
histi/o	Combining form for tissue
holo-	Prefix for all
hom/o	Combining form for same
home/o	Combining form for unchanging
hormon/o	Combining form for urge on or excite
humer/o	Combining form for long bone of the proximal fore limb (humerus)
hydr/o	Combining form for water
hydra-	Prefix for water
hydro-	Prefix for water
hygien/o	Combining form for healthful
hymen/o	Combining form for membrane (hymen)
hyper-	Prefix for over, above, increased, excessive, or beyond
hyph-	Prefix for under
hypn/o	Combining form for sleep
hypo-	Prefix for under, decreased, deficient, or below
hyster/o	Combining form for uterus
-ia	Suffix for state or condition
-iac	Suffix for pertaining to
-iasis	Suffix for condition or abnormal condition
iatr/o	Combining form for relationship to treatment or doctor

-ible	Suffix for capable of being
-ic	Suffix for pertaining to
ichthy/o	Combining form for dry or scaly
icter/o	Combining form for jaundice
ictero-	Prefix for jaundice
idi/o	Combining form for peculiar to an individual or organ
idio-	Prefix for peculiar to an individual
-iferous	Suffix for bearing, carrying, or producing
-ific	Suffix for producing
-iform	Suffix for resembling or shaped like
-igo	Suffix for diseased condition
-ile	Suffix for capable of being or pertaining to
ile/o	Combining form for ileum
ili/o	Combining form for part of the pelvis (ilium)
illusi/o	Combining form for deception
immun/o	Combining form for protected, safe, or immune
impact/o	Combining form for pushed against
impress/o	Combining form for pressing into
impuls/o	Combining form for pressure or urging on
in-	Prefix for in, into, not, without
-in	Suffix for substance, usually with specialized function
incis/o	Combining form for cutting into
incubat/o	Combining form for hatching or incubation
indurat/o	Combining form for hardened
-ine	Suffix for pertaining to
infarct/o	Combining form for filled in or stuffed
infect/o	Combining form for tainted or infected
infer/o	Combining form for below or beneath
infest/o	Combining form for attack
inflammat/o	Combining form for flame within or to set on fire
infra-	Prefix for beneath, below, or inferior to
infundibul/o	Combining form for funnel
ingest/o	Combining form for carry or pour in
inguin/o	Combining form for groin
inhal/o	Combining form for breathe in
inhalat/o	Combining form for breathe in
inject/o	Combining form for to force or throw in
innominat/o	Combining form for nameless
inocul/o	Combining form for introduce or implant
insipid/o	Combining form for tasteless

inspir/o	Combining form for breathe in
inspirat/o	Combining form for breathe in
insul/o	Combining form for island
intact/o	Combining form for whole
inter-	Prefix for between or among
intermitt/o	Combining form for not continuous
intern/o	Combining form for within or inner
interstiti/o	Combining form for the space between things
intestin/o	Combining form for intestine
intim/o	Combining form for innermost
intoxic/o	Combining form for put poison in
intra-	Prefix for within, into, inside
intrins/o	Combining form for contained within
intro-	Prefix for within, into, or inside
introit/o	Combining form for entrance or passage
intussuscept/o	Combining form for to receive within
involut/o	Combining form for curled inward
ion/o	Combining form for charged particle
ipsi-	Prefix for same
ir/i	Combining form for iris
ir/o	Combining form for iris
irid/o	Combining form for iris
irit/o	Combining form for iris
is/o	Combining form for equal
isch/o	Combining form for holding back
ischi/o	Combining form for part of pelvis (ischium)
-ism	Suffix for state of or condition
iso-	Prefix for equal
-itis	Suffix for inflammation
-ium	Suffix for structure
jejun/o	Combining form for jejunum
jugul/o	Combining form for throat
juxta-	Prefix for nearby
kal/i	Combining form for potassium
kary/o	Combining form for nucleus
karyo-	Prefix for nucleus
kata-	Prefix for down
kath-	Prefix for down
kel/o	Combining form for growth or tumor
kera-	Prefix for horn or hardness

kerat/o	Combining form for hornlike, hard, or cornea
ket/o	Combining form for breakdown product of fat (ketones)
keton/o	Combining form for breakdown product of fat (ketones)
kilo-	Prefix for one thousand
kinesi/o	Combining form for movement
-kinesis	Suffix for motion
koil/o	Combining form for hollow or concave
kraur/o	Combining form for dry
labi/o	Combining form for lip
labyrinth/o	Combining form for maze (used to describe the inner ear)
lacer/o	Combining form for torn
lacerat/o	Combining form for torn
lacrim/o	Combining form for tear or tear duct
lact/i	Combining form for milk
lact/o	Combining form for milk
lactat/o	Combining form for secrete milk
lamin/o	Combining form for one part of the caudal portion of the vertebra (lamina)
lapar/o	Combining form for abdomen or flank
laps/o	Combining form for slip or fall downward
laryng/o	Combining form for voice box
lat/i	Combining form for broad
lat/o	Combining form for broad
later/o	Combining form for side
lax/o	Combining form for loosen or relax
laxat/o	Combining form for loosen or relax
lei/o	Combining form for smooth muscle
leiomy/o	Combining form for smooth muscle
lemm/o	Combining form for peel
lent/i	Combining form for lens of the eye
lenticul/o	Combining form for shaped like a lens
-lepsy	Suffix for seizure
lept/o	Combining form for thin
letharg/o	Combining form for drowsiness or indifference
leuco-	Prefix for white
leuk/o	Combining form for white
lev/o	Combining form for lift up
levat/o	Combining form for lift up
libid/o	Combining form for sexual drive
libidin/o	Combining form for sexual drive

ligament/o	Combining form for fibrous tissue connecting bone to bone (ligament)
ligat/o	Combining form for binding or tying off
lingu/o	Combining form for tongue
lip/o	Combining form for fat
-lite	Suffix for stone or calculus
-lith	Suffix for stone or calculus
lith/o	Combining form for stone or calculus
-lithiasis	Suffix for presence of stone
lob/i	Combining form for lobe or well-defined part of an organ
lob/o	Combining form for lobe or well-defined part of an organ
loc/o	Combining form for place
loch/i	Combining form for confinement or birthing process
-logist	Suffix for specialist in the study of
-logy	Suffix for the science or study of
longev/o	Combining form for long-lived
lord/o	Combining form for bent backward
lumb/o	Combining form for lower back or loin
lumin/o	Combining form for light
lun/o	Combining form for moon
lunul/o	Combining form for crescent
lup/i	Combining form for wolf
lup/o	Combining form for wolf
lute/o	Combining form for yellow
lymph/o	Combining form for lymphatic tissue
lymphaden/o	Combining form for lymph gland
lymphangi/o	Combining form for lymph vessel
-lysis	Suffix for break down, destruction, or separate
-lyst	Suffix for agent that breaks down or separates
-lytic	Suffix for reduce or destroy
macr/o	Combining form for large or abnormal size
macro-	Prefix for large or abnormal size
macul/o	Combining form for spot
magn/o	Combining form for great or large
major/o	Combining form for larger
mal-	Prefix for bad, poor, or evil
malac/o	Combining form for adnormal softening
-malacia	Suffix for adnormal softening
malign/o	Combining form for bad or evil

malle/o	Combining form for hammer
malleol/o	Combining form for little hammer
mamm/o	Combining form for breast
man/i	Combining form for rage or hand
man/o	Combining form for hand or pressure
mandibul/o	Combining form for lower jaw
-mania	Suffix for obsessive preoccupation
manipul/o	Combining form for handful or use of hands
manubri/o	Combining form for handle
masset/o	Combining form for chew
mast/o	Combining form for breast
mastic/o	Combining form for chew
masticat/o	Combining form for chew
mastoid/o	Combining form for mastoid process of temporal bone
matern/o	Combining form for mother
matur/o	Combining form for ripe
maxill/o	Combining form for upper jaw
maxim/o	Combining form for largest
meat/o	Combining form for opening or passageway
med-	Prefix for middle
medi/o	Combining form for middle
mediastin/o	Combining form for in the middle
medic/o	Combining form for medicine, healing, or doctor
medicat/o	Combining form for medication or healing
medull/o	Combining form for inner section, soft, or middle
mega-	Prefix for large
megal/o	Combining form for large
-megaly	Suffix for large or enlargement
mei/o	Combining form for less
melan/o	Combining form for black
mellit/o	Combining form for honey
membran/o	Combining form for thin skin (membrane)
men/o	Combining form for menstruation
mening/o	Combining form for meninges (membrane lining the central nervous system)
meningi/o	Combining form for meninges (membrane lining the central nervous system)
menisc/o	Combining form for crescent
mens-	Prefix for menstruation
mens/o	Combining form for menstruation

ment/o	Combining form for mind
mes-	Prefix for middle
mes/o	Combining form for middle
mesenter/o	Combining form for mesentery or middle intestine
mesi/o	Combining form for middle or median plane
meta-	Prefix for beyond, change, behind, after, or next
metabol/o	Combining form for change
metacarp/o	Combining form for beyond the carpus
metatars/o	Combining form for beyond the tarsus
-meter	Suffix for measuring device
metr/i	Combining form for uterus
metr/o	Combining form for uterus
metri/o	Combining form for uterus
mi/o	Combining form for smaller or less
micr/o	Combining form for small
micro-	Prefix for small; one-millionth of the unit it precedes
mictur/o	Combining form for urinate
micturit/o	Combining form for urinate
midsagitt/o	Combining form for dividing into equal left and right halves
milli-	Prefix for one thousandth
-mimetic	Suffix for mimic or copy
mineral/o stance	Combining form for naturally occurring nonorganic solid sub- (mineral)
minim/o	Combining form for smallest
minor/o	Combining form for smaller
mio-	Prefix for smaller or less
mit/o	Combining form for thread
mitr/o	Combining form for miter having two points on top
mobil/o	Combining form for capable of moving
mon/o	Combining form for single or one
monil/i	Combining form for string of beads
mono-	Prefix for single or one
morbid/o	Combining form for disease
moribund/o	Combining form for dying
morph/o	Combining form for shape or form
mort/i	Combining form for dead
mort/o	Combining form for dead
mort/u	Combining form for dead
mortal/i	Combining form for dead
mot/o	Combining form for movement

motil/o	Combining form for movement
mu/o	Combining form for close or shut
muc/o	Combining form for mucus
mucos/o	Combining form for mucus
multi-	Prefix for many
muscul/o	Combining form for muscle
mut/a	Combining form for genetic change
my/o	Combining form for muscle
myc/e	Combining form for fungus
myc/o	Combining form for fungus
mydri/o	Combining form for wide
myel/o	Combining form for spinal cord or bone marrow; "white substance"
myocardi/o	Combining form for heart muscle
myom/o	Combining form for muscle tumor
myos/o	Combining form for muscle
myring/o	Combining form for tympanic membrane; "eardrum"
myx-	Prefix for mucus or slime
myxa-	Prefix for mucus
myx/o	Combining form for mucus
myxo-	Prefix for mucus or slime
nar/i	Combining form for nostril
narc/o	Combining form for numbness or stupor
nas/i	Combining form for nose
nas/o	Combining form for nose
nat/i	Combining form for birth
natr/o	Combining form for sodium
nause/o	Combining form for sick feeling to stomach
ne/o	Combining form for new
necr/o	Combining form for death
-necrosis	Suffix for death of tissue
nect/o	Combining form for bind or tie
neo-	Prefix for new
nephr/o	Combining form for kidney
nephra-	Prefix for kidney
nerv/o	Combining form for nerve
neu-	Prefix for nerves
neur/i	Combining form for nerve
neur/o	Combining form for nerve
neutr/o	Combining form for neither or neutral
niter-	Prefix for nitrogen

nitro-	Prefix for nitrogen
noct-	Prefix for night
noct/i	Combining form for night
nod/o	Combining form for knot or swelling
nodul/o	Combining form for little knot
nom/o	Combining form for control
non-	Prefix for ninth or no
nona-	Prefix for ninth
nonus-	Prefix for nine
nor-	Prefix for chemical term meaning normal or parent compound
norm/o	Combining form for usual
novem-	Prefix for nine
nuch/o	Combining form for nape
nucle/o	Combining form for nucleus
nulli-	Prefix for none
numer/o	Combining form for count or number
nunci/o	Combining form for messenger
nutri/o	Combining form for nourishment or food
nutrit/o	Combining form for nourishment or food
nyct/o	Combining form for night
nyctal/o	Combining form for night
o-	Prefix for egg
ob-	Prefix for against
obes/o	Combining form for extremely fat
obliqu/o	Combining form for slanted or sideways
oblongat/o	Combining form for oblong or elongated
obstetr/i	Combining form for one who receives
obstetr/o	Combining form for one who receives
occipit/o	Combining form for back of the skull
occlud/o	Combining form for close up
occlus/o	Combining form for close up
occult/o	Combining form for hidden
oct-	Prefix for eight
octa-	Prefix for eight
octo-	Prefix for eight
ocul/o	Combining form for eye
oculo-	Prefix for eye
odont/o	Combining form for tooth
-oid	Suffix for like or resembling
-ole	Suffix for small

olecran/o	Combining form for proximal projection on the ulna
olfact/o	Combining form for smell
oligo-	Prefix for scanty or little
-ologist	Suffix for specialist
-ology	Suffix for the study of
-oma	Suffix for tumor, mass, or neoplasm (usually benign)
oment/o	Combining form for omentum or fat
onc/o	Combining form for tumor
ont/o	Combining form for existence
onych/o	Combining form for claw or nail
oo/o	Combining form for egg
oophor/o	Combining form for ovary
opac/o	Combining form for shaded or dark
opacit/o	Combining form for shaded or dark
oper/o	Combining form for work or perform
operat/o	Combining form for work or perform
opercul/o	Combining form for lid or cover
ophthalm/o	Combining form for eye
-opia	Suffix for vision
opisth/o	Combining form for backward
-opsia	Suffix for vision
-opsis	Suffix for vision
-opsy	Suffix for vision
opt/i	Combining form for eye
opti/o	Combining form for eye
optic/o	Combining form for vision
or/o	Combining form for mouth
orbit/o	Combining form for circle, bony cavity, or socket
orch/o	Combining form for testis
orchi/o	Combining form for testis
orchid/o	Combining form for testis
orect/i	Combining form for appetite
orex/i	Combining form for appetite
organ/o	Combining form for organ
orth/o	Combining form for straight, normal, or correct
ortho-	Prefix for straight, normal or correct
os-	Prefix for mouth or bone
-osis	Suffix for abnormal condition
osm/o	Combining form for smell or odor; pushing
-osmia	Suffix for smell or odor

oss/e	Combining form for bone
oss/i	Combining form for bone
oste/o	Combining form for bone
-ostomosis	Suffix for surgically creating an opening
-ostomy	Suffix for surgically creating an opening
ot/o	Combining form for ear or hearing
-otomy	Suffix for cutting into
-ous	Suffix for pertaining to
ov/i	Combining form for egg
ov/o	Combining form for egg
ovari/o	Combining form for small almond-shaped female reproductive organ that produce ova (ovary)
ox/i	Combining form for oxygen; also means quick or sharp
ox/o	Combining form for oxygen; also means quick or sharp
ox/y	Combining form for oxygen; also means quick or sharp
-oxia	Suffix for oxygen; oxygenation
oxid/o	Combining form for containing oxygen
oxy-	Prefix for sharp, acid, quick, or oxygen
pachy-	Prefix for thick
palat/o	Combining form for roof of the mouth
pall/o	Combining form for pale
palliat/o	Combining form for hidden
pallid/o	Combining form for pale
palm/o	Combining form for ventral surface of forepaw or hoof
palpat/o	Combining form for touch
palpebr/o	Combining form for eyelid
palpit/o	Combining form for throbbing
pan-	Prefix for all, entire, or every
pancreat/o	Combining form for pancreas
papill/i	Combining form for nipplelike
papill/o	Combining form for nipplelike
papul/o	Combining form for pimple
par/o	Combining form for apart from, beside, near or abnormal; also prefix for near, beside, abnormal, or apart from; reproductive labor
para-	In reference to neurologic conditions of quadrupeds para- implies hind or back portion
-para	Suffix for to bear or bring forth
paralys/o	Combining form for disable
paralyt/o	Combining form for disable
parasit/o	Combining form for near food

parathyr/o	Combining form for parathyroid gland
parathyroid/o	Combining form for parathyroid gland
pares/i	Combining form for disable
-paresis	Suffix for weakness
paret/o	Combining form for disable
pariet/o	Combining form for wall
parotid/o	Combining form for parotid gland
paroxysm/o	Combining form for sudden attack
part/o	Combining form for birth or labor
-partum	Suffix for birth or labor
parturit/o	Combining form for birth or labor
patell/a	Combining form for patella
patell/o	Combining form for patella
path/o	Combining form for disease
-pathic	Suffix for pertaining to disease
-pathy	Suffix for disease
paus/o	Combining form for stopping
pector/o	Combining form for chest
ped/o	Combining form for child or foot
pedi/a	Combining form for child
pedicul/o	Combining form for louse
pelv/i	Combining form for hip bone or pelvis
pelv/o	Combining form for hip bone or pelvis
pen/i	Combining form for penis
-penia	Suffix for deficiency or reduction in number
pent-	Prefix for five
penta-	Prefix for five
peps/i	Combining form for digestion
pept/o	Combining form for digestion
per-	Prefix for excessive or through
percept/o	Combining form for to become aware
percussi/o	Combining form for tap or beat
peri-	Prefix for around or surrounding
perine/o	Combining form for region between the scrotum or vulva and the anus
perme/o	Combining form for to pass through
pernici/o	Combining form for destructive or harmful
perone/o	Combining form for fibular
pertuss/i	Combining form for intense cough
petechi/o	Combining form for skin spot

-pexy	Suffix for surgical fixation
phac/o	Combining form for lens of the eye
phag/o	Combining form for eating or swallowing
-phage	Suffix for one that eats
-phagia	Suffix for eating or swallowing
phak/o	Combining form for lens of the eye
phalang/o	Combining form for phalanges
phall/o	Combining form for penis
pharmac/o	Combining form for drug
pharyng/o	Combining form for throat (pharynx)
phas/o	Combining form for speech
-phasia	Suffix for speech
phe/o	Combining form for dusky
pher/o	Combining form for bear or carry
-pheresis	Suffix for removal
phil/o	Combining form for love or attraction to
-philia	Suffix for attraction for or increase in numbers
phim/o	Combining form for muzzling or constriction of an orifice
phleb/o	Combining form for vein
phlegm/o	Combining form for thick mucus
phob/o	Combining form for fear
phon/o	Combining form for sound or voice
-phonia	Suffix for sound or voice
phor/o	Combining form for carry or movement
-phoresis	Suffix for carrying or transmission
-phoria	Suffix for to bear, feeling, or mental state
phot/o	Combining form for light
phren/o	Combining form for diaphragm
-phylactic	Suffix for protective or preventive
-phylaxis	Suffix for protection
physi/o	Combining form for nature
physic/o	Combining form for nature
-physis	Suffix for to grow
-phyte	Suffix for plant
pigment/o	Combining form for color
pil/i	Combining form for hair
pil/o	Combining form for hair
pineal/o	Combining form for pineal gland
pinn/i	Combining form for external ear or auricle
pituit/o	Combining form for pituitary gland

plac/o	Combining form for flat patch
placent/o	Combining form for round flat cake or placenta
-plakia	Suffix for thin flat layer or scale
plan/o	Combining form for flat
plant/i	Combining form for ventral surface or bottom of rear hoof, foot, or paw
plant/o	Combining form for ventral surface or bottom of rear hoof, foot, or paw
plas/i	Combining form for growth, development, formation, or mold
plas/o	Combining form for growth, development, formation, or mold
-plasia	Suffix for formation, development, or growth of number of tissue and cell
-plasm	Suffix for formative material of cells
plasm/o	Combining form for something formed
plast/o	Combining form for growth, development, or mold
-plasty	Suffix for surgical repair
-plegia	Suffix for paralysis
-plegic	Suffix for paralysis
pleur/o	Combining form for membrane lining the lungs (pleura)
plex/o	Combining form for network
plic/o	Combining form for fold or ridge
-pnea	Suffix for breathing
-pneic	Suffix for pertaining to breathing
pneu-	Prefix for relating to air or lungs
pneum/o	Combining form for lung or air
pneumon/o	Combining form for lung or air
pod/o	Combining form for foot
-poiesis	Suffix for formation
poikil/o	Combining form for irregular
poli/o	Combining form for gray metter of the brain and spinal cord
polio-	Prefix for gray matter of the brain and spinal cord
poly-	Prefix for many
polyp/o	Combining form for small growth
pont/o	Combining form for part of the brain (the pons)
poplit/o	Combining form for back of the patella
por/o	Combining form for small opening (pore)
-porosis	Suffix for passage or porous condition
port/i	Combining form for gate or door
post-	Prefix for after or behind
poster/o	Combining form for toward the tail

potent/o	Combining form for powerful
pract/i	Combining form for practice
practic/o	Combining form for practice
prandi/o	Combining form for meal
-prandial	Suffix for meal
-praxia	Suffix for action
pre-	Prefix for before or in front of
pregn/o	Combining form for pregnant or with offspring
prematur/o	Combining form for too early
preputi/o	Combining form for foreskin or prepuce
press/o	Combining form for press or draw
priap/o	Combining form for penis
primi-	Prefix for first
pro-	Prefix for before or in behalf of
process/o	Combining form for going forth
prococ/i	Combining form for early or premature
procreat/o	Combining form for reproduce
proct-	Prefix for first
proct/o	Combining form for anus and rectum
prodrom/o	Combining form for precursor
product/o	Combining form for lead forward or to yield
prolaps/o	Combining form for fall downward or slide forward
prolifer/o	Combining form for to reproduce
pron/o	Combining form for bent forward
pront/o	Combining form for bent forward
prostat/o	Combining form for prostate gland
prosth/o	Combining form for addition or appendage
prosthet/o	Combining form for addition or appendage
prot/o	Combining form for first, original, or protein
prote/o	Combining form for first, original, or protein
proto-	Prefix for first
proxim/o	Combining form for near
prurit/o	Combining form for itch
pseud/o	Combining form for false
pseudo-	Prefix for false
psor/i	Combining form for itch
psor/o	Combining form for itch
psych/o	Combining form for mind
ptomat/o	Combining form for a fall
-ptosis	Suffix for drooping or sagging

-ptyalo	Suffix for spit or saliva
-ptysis	Suffix for spitting
pub/o	Combining form for part of the hip bone (pubis)
pubert/o	Combining form for adult
pudend/o	Combining form for pudendum
puerper/i	Combining form for labor
pulm/o	Combining form for lung
pulmon/o	Combining form for lung
pulpos/o	Combining form for fleshy or pulpy
puls/o	Combining form for beating
punct/o	Combining form for sting, puncture, or little hole
pupill/o	Combining form for pupil of the eye
pur/o	Combining form for pus
purpur/o	Combining form for purple
pustul/o	Combining form for blister
py/o	Combining form for pus
pyel/o	Combining form for renal pelvis
pylor/o	Combining form for gate keeper or pylorus
pry/o	Combining form for fever or fire
pyret/o	Combining form for fever or fire
pyramid/o	Combining form for pyramid-shaped
quadr/i	Combining form for four
quadr/o	Combining form for four
quadri-	Prefix for four
quadro-	Prefix for four
quart-	Prefix for fourth
quinque-	Prefix for five
quint-	Prefix for fifth
rabi/o	Combining form for rage or madness
rachi/o	Combining form for spinal column or vertebrae
radi/o	Combining form for radiation or the radius (bone in distal fore limb)
radiat/o	Combining form for giving off radiant energy
radicul/o	Combining form for root or nerve root
raph/o	Combining form for seam or suture
-raphy	Suffix for suturing
re-	Prefix for back or again
recept/o	Combining form for receive
recipi/o	Combining form for receive
rect/o	Combining form for rectum or straight
recticul/o	Combining form for network

recuperat/o	Combining form for recover
reduct/o	Combining form for bring back together
refract/o	Combining form for bend back or turn aside
regurgit/o	Combining form for flood or gush back
remiss/o	Combining form for let go or give up
ren/o	Combining form for kidney
restor/o	Combining form for rebuild
resuscit/o	Combining form for revive
retent/o	Combining form for hold back
retin/o	Combining form for nervous tissue of the eye (retina) or net
retract/o	Combining form for draw back
retro-	Prefix for behind or backward
rhabdomy/o	Combining form for striated muscle
-rhagia	Suffix for bursting forth
-rhaphy	Suffix for suture
-rhea	Suffix for flow or discharge
rheum/o	Combining form for watery flow
rheumat/o	Combining form for watery flow
-rhexis	Suffix for rupture
rhin/o	Combining form for nose
rhiz/o	Combining form for root
rhonc/o	Combining form for snore
rhythm/o	Combining form for rhythm
rhytid/o	Combining form for wrinkle
rigid/o	Combining form for stiff
roentgen/o	Combining form for x-ray
rotat/o	Combining form for revolve
-rhage	Suffix for brushing forth
-rrhage	Suffix for bursting forth
-rrhagia	Suffix for bursting forth
-rrhaphy	Suffix for suture
-rrhea	Suffix for flow or discharge
-rrhexis	Suffix for rupture
rug/o	Combining form for wrinkle or fold
sacc/i	Combining form for sac
sacc/o	Combining form for sac
sacr/o	Combining form for sacrum
saliv/o	Combining form for spit
salping/o	Combining form for tube (uterine or auditory)
-salpinx	Suffix for uterine tube, oviduct, or fallopian tube

san/o	Combining form for sound or healthy
sanguin/o	Combining form for blood
sangu/i	Combining form for blood
sanit/o	Combining form for soundness or health
saphen/o	Combining form for clear or apparent
sapr/o	Combining form for dead or decaying
sarc/o	Combining form for flesh (connective tissue) or malignancy of connective tissue
sarcomat/o	Combining form for flesh (connective tissue) or malignancy of connective tissue
sarco-	Prefix for flesh
-sarcoma	Suffix for malignant tumor
scalp/o	Combining form for carve or scrape
scapul/o	Combining form for shoulder blade
-schisis	Suffix for cleft or divided
schiz/o	Combining form for cleft or divided
scirrh/o	Combining form for hard
scler/o	Combining form for white of the eye (sclera) or hard
-sclerosis	Suffix for abdominal dryness or hardness
scoli/o	Combining form for curved or crooked
-scope	Suffix for instrument for visual examination
-scopic	Suffix for pertaining to visual examination
-scopy	Suffix for process of visually examining
scot/o	Combining form for darkness
scrot/o	Combining form for bag or pouch
seb/o	Combining form for wax or sebum
sect/o	Combining form for cutting
secti/o	Combining form for cut
secundi-	Prefix for second
segment/o	Combining form for in pieces
sell/o	Combining form for saddle
semi-	Prefix for half
semin/i	Combining form for sperm, semen, or seed
sen/i	Combining form for old
sens/i	Combining form for feeling or sensation
sensitiv/o	Combining form for affected by or sensitive to
seps/o	Combining form for infection
sept-	Prefix for seven
septi-	Prefix for seven
sept/o	Combining form for infection or partition

ser/o	Combining form for clear, fluid portion of blood (serum)
seros/o	Combining form for serous
sesqui-	Prefix for one and one-half
sex-	Prefix for six
sial/o	Combining form for saliva or salivary glands
sialaden/o	Combining form for salivary glands
sider/o	Combining form for iron
sigmoid/o	Combining form for sigmoid colon
sin/o	Combining form for hollow, sinus, or tubelike passage
sin/u	Combining form for hollow, sinus, or tubelike passage
sinistr/o	Combining form for left
sinus/o	Combining form for air-filled or fluid-filled cavity within a bone; also means channel
sit/u	Combining form for place
skelet/o	Combining form for bony framework of the body (skeleton)
soci/o	Combining form for companion
solut/o	Combining form for dissolved
solv/o	Combining form for dissolved
soma-	Prefix for body
somat/o	Combining form for body
somn/i	Combining form for sleep
somn/o	Combining form for sleep
-somnia	Suffix for sleep
son/o	Combining form for sound
sopor/o	Combining form for sleep
spad/o	Combining form for draw off
-spasm	Suffix for sudden involuntary contraction or tightening
spasm/o	Combining form for sudden involuntary contractions or tightening
spasmod/o	Combining form for sudden involuntary contraction or tightening
spec/i	Combining form for look at or a sort
specul/o	Combining form for mirror
sperm/o	Combining form for cell or seed
spermat/o	Combining form for sperm cell or seed
sphen/o	Combining form for wedge
spher/o	Combining form for round, sphere, or ball
sphincter/o	Combining form for tight band
sphygm/o	Combining form for pulse
spin/o	Combining form for backbone
spir/o	Combining form for breathing or breath
spirill/o	Combining form for little coil

spirochet/o	Combining form for coiled microorganism
splen/o	Combining form for spleen
spondyl/o	Combining form for vertebra
spontane/o	Combining form for unexplained
spor/o	Combining form for seed
sput/o	Combining form for spit
squam/o	Combining form for scale
-stalsis	Suffix for contraction
staped/o	Combining form for one of the middle ear bones (stapes or stirrup)
stapedi/o	Combining form for one of the middle ear bones (stapes or stirrup)
staphyl/o	Combining form for cluster or bunch of grapes
stas/i	Combining form for controlling or stopping
-stasis	Suffix for controlling or stopping
stat/i	Combining form for controlling or stopping
-static	Suffix for controlling or stopping
steat/o	Combining form for fat
sten/o	Combining form for narrowing or contracted
-stenosis	Suffix for narrowing or stricture of a duct or canal
steno-	Prefix for narrowness
ster/o	Combining form for solid structure
stere/o	Combining form for solid or three dimensional
steril/i	Combining form for barren
stern/o	Combining form for breastbone
stert/o	Combining form for snore
steth/o	Combining form for chest
sthen/o	Combining form for strength
-sthenia	Suffix for strength
stigmat/o	Combining form for point or spot
stimul/o	Combining form for good or incite
stol/o	Combining form for send or place
stomat/o	Combining form for mouth
-stomosis	Suffix for make a new opening
-stomy	Suffix for make a new opening
strab/i	Combining form for squint
strat/i	Combining form for layer
strept/o	Combining form for twisted chain
striat/o	Combining form for stripe or groove
stric-	Prefix for narrowing
strict/o	Combining form for draw tightly together or tie
strid/o	Combining form for harsh sound

stup/e	Combining form for stunned
styl/o	Combining form for pointed instrument or pen
sub-	Prefix for under, less, or below
subluxat/o	Combining form for partial dislocation
sucr/o	Combining form for sugar
sudor/i	Combining form for sweat
suffoc/o	Combining form for choke
suffocat/o	Combining form for choke
sulc/o	Combining form for furrow or groove
super-	Prefix for above, excessive, or higher than
super/o	Combining form for above, excessive, or higher than
superflu/o	Combining form for overflowing or excessive
supin/o	Combining form for lying on the back
supinat/o	Combining form for bent backward or to place on the back
suppress/o	Combining form for press down
suppur/o	Combining form for form pus
suppurat/o	Combining form for form pus
supra-	Prefix for above or excessive
supraren/o	Combining form for above or on the kidney
sutur/o	Combining form for stitch
sym-	Prefix for with or together
symptomat/o	Combining form for falling together or symptom
syn-	Prefix for union or association
synaps/o	Combining form for point of contact
synapt/o	Combining form for point of contact
syncop/o	Combining form for cut short or cut off
-syndesis	Suffix for surgical fixation of the vertebrae
syndesm/o	Combining form for ligament
syndrom/o	Combining form for running together
synovi/o	Combining form for synovial membrane or synovial fluid
syring/o	Combining form for tube
system/o	Combining form for entire body
systemat/o	Combining form for entire body
systol/o	Combining form for contraction
tachy-	Prefix for abnormally fast
tact/i	Combining form for touch
tars/o	Combining form for tarsus or edge of the eyelid
tax/o	Combining form for order or coordination
techn/o	Combining form for skill
techni/o	Combining form for skill

tectori/o	Combining form for covering
tele/o	Combining form for distant or far
tempor/o	Combining form for temple
ten/o	Combining form for tendon, strain, or to extend
tenac/i	Combining form for sticky
tend/o	Combining form for tendon, strain, or to extend
tendin/o	Combining form for tendon
tens/o	Combining form for stretch out, strain, or extend
terat/o	Combining form for malformed fetus
termin/o	Combining form for the end
tert-	Prefix for third
test/i	Combining form for testis or testicle
test/o	Combining form for testis or testicle
testicul/o	Combining form for testis or testicle
tetan/o	Combining form for rigid
tetra-	Prefix for four
thalam/o	Combining form for inner room or thalamus
thanas/o	Combining form for death
thanat/o	Combining form for death
the/o	Combining form for put or place
thec/o	Combining form for sheath
thel/o	Combining form for nipple
therap/o	Combining form for treatment
therapeut/o	Combining form for treatment
-therapy	Suffix for treatment
therm/o	Combining form for heat
thio-	Prefix for sulfur
thorac/o	Combining form for chest
-thorax	Suffix for pleural cavity or chest
thromb/o	Combining form for aggregation of blood in a vessel; "clot"
thym/o	Combining form for thymus gland
thyr/o	Combining form for shield or thyroid gland
thyroid/o	Combining form for shield or thyroid gland
tibi/o	Combining form for tibia or "shinbone"
-tic	Suffix for pertaining to
tine/o	Combining form for gnawing worm or ringworm
tinnit/o	Combining form for ringing
toc/o	Combining form for birth
-tocia	Suffix for labor or birth
tom/o	Combining form for cut or section

-tome	Suffix for instrument to cut
-tomy	Suffix for cutting or incision
ton/o	Combining form for tension or stretching
tone/o	Combining form for stretch
-tonic	Suffix for tone or osmotic pressure
tonsill/o	Combining form for tonsil or throat
top/o	Combining form for place or location
tors/o	Combining form for twist or rotate
tort/i	Combining form for twisted
tox/o	Combining form for poison
toxic/o	Combining form for poison
-toxin	Suffix for poison
trabecul/o	Combining form for beams or little beam marked with crossbars
trache/i	Combining form for windpipe
trache/o	Combining form for windpipe
trachel-	Prefix for neck
tract/o	Combining form for draw, pull, path, or bundle of nerve fibers
tranquil/o	Combining form for quiet
trans-	Prefix for across or through
transfus/o	Combining form for transfer or pour across
transit/o	Combining form for changing
transvers/o	Combining form for across or crosswise
traumat/o	Combining form for injury
trem/o	Combining form for shaking
tremul/o	Combining form for fine tremor or shaking
treponem/o	Combining form for coiled or turning microbe
-tresia	Suffix for opening
tri-	Prefix for three
trich/o	Combining form for hair
trigon/o	Combining form for triangle
-tripsy	Suffix for crushing stone
trit-	Prefix for third
-trite	Suffix for instrument for cutting
trito-	Prefix for third
trochle/o	Combining form for pulley
trop/o	Combining form for turn or change
troph/o	Combining form for nourishment, development, or growth
-trophy	Suffix for formation, development, and increased size of tissue and cell (used to represent size)
-tropia	Suffix for turn

-tropic	Suffix for having an affinity for or turning toward
-tropin	Suffix for having an affinity for or turning toward
tub/i	Combining form for tube or pipe
tub/o	Combining form for tube or pipe
tubercul/o	Combining form for little knot or swelling
tunic/o	Combining form for covering or sheath
turbinat/o	Combining form for coiled or spiral shaped
tuss/i	Combining form for cough
tympan/o	Combining form for tympanic membrane or eardrum
ulcer/o	Combining form for sore or ulcer
uln/o	Combining form for one of the bones in the distal forelimb (ulna)
ultra-	Prefix for beyond, above, or excess
-um	Suffix for structure
umbilic/o	Combining form for navel
un-	Prefix for not
ungu/o	Combining form for nail or hoof
uni-	Prefix for one
ur/o	Combining form for urine, the urinary tract, or urination
-uresis	Suffix for urination
ureter/o	Combining form for ureter
urethr/o	Combining form for urethra
urg/o	Combining form for press or push
-uria	Suffix for urination or urine
urin/o	Combining form for urine, the urinary tract, or urination
urtic/o	Combining form for nettle, rash, or hives
-us	Suffix for thing
uter/i	Combining form for uterus or womb
uter/o	Combining form for uterus or womb
uve/o	Combining form for vascular layer of the eye, iris, choroid, or ciliary body
uvul/o	Combining form for little grape or uvula
vaccin/i	Combining form for vaccine
vaccin/o	Combining form for vaccine
vacu/o	Combining form for empty
vag/o	Combining form for wandering
vagin/o	Combining form for sheath or vagina
valg/o	Combining form for bent or twisted outward
valv/o	Combining form for valve or membranous fold
valvul/o	Combining form for valve or membranous fold
var/o	Combining form for bent or twisted inward

varic/o	Combining form for swollen or dilated vein
vas/o	Combining form for vessel, duct, or vas deferens
vascul/o	Combining form for little vessel
vaso-	Prefix for vessel
vast/o	Combining form for great or extensive
vect/o	Combining form for carry or convey
ven/o	Combining form for vein
vener/o	Combining form for sexual intercourse
venter-	Prefix for the abdomen
ventilat/o	Combining form for expose to air
ventr/o	Combining form for belly side
ventricul/o	Combining form for small chamber or ventricle of the brain or heart
venul/o	Combining form for small vein
verg/o	Combining form for twist or incline
verm/i	Combining form for worm
verruc/o	Combining form for wart
vers/o	Combining form for turn
-version	Suffix for turn
vert/o	Combining form for turn
vertebr/o	Combining form for backbone
vertig/o	Combining form for turning around or revolution
vertigin/o	Combining form for turning around or revolution
vesic/o	Combining form for urinary bladder
vesicul/o	Combining form for seminal vesicle, blister, or little bladder
vestibul/o	Combining form for entrance
vi/o	Combining form for force
vill/i	Combining form for tuft of hair or threadlike projections from a membrane
vir/o	Combining form for poison or virus
viril/o	Combining form for masculine
vis/o	Combining form for seeing or sight
visc/o	Combining form for sticky
viscer/o	Combining form for internal organ
viscos/o	Combining form for sticky
vit/a	Combining form for life
vit/o	Combining form for life
viti/o	Combining form for blemish or defect
vitre/o	Combining form for glassy
voc/i	Combining form for voice
vol/o	Combining form for ventral surface of fore or hind limbs (palm

	or sole)
volv/o	Combining form for turn or roll
vulgar/i	Combining form for common
vulv/o	Combining form for covering or vulva
xanth/o	Combining form for yellow
xen/o	Combining form for strange or foreign
xer/o	Combining form for dry
xiph/i	Combining form for sword
xiph/o	Combining form for sword
zygomat/o	Combining form for yoke or cheekbone
zygot/o	Combining form for joined or yoked together

APPENDIX B

Plural Forms of Medical Terms

Many plural word forms are produced by adding an "s" to the singular term. This is true for medical terms as well. The plural of laceration is lacerations, the plural of bone is bones, etc. However, there are some rules to follow when using plural forms of medical terms. These rules are presented in the table below.

If the singular ending is:	Change or deletion from singular form:	Add plural ending:	Examples (singular)	Plural form
s, ch, or h		es	abscess	abscesses
			stitch	stitches
			cough	coughes
y	delete y	ies	capillary	capillaries
is	delete is	es	diagnosis	diagnoses
um	delete um	a	bacterium	bacteria
us	delete us	i	alveolus	alveoli

(Except: The plural of virus is viruses, and the plural of sinus is sinuses.)

a	delete a	ae	vertebra	vertebrae
ix	delete ix	ices	cervix	cervices
ex	delete ex	ices	cortex	cortices
ax	delete ax	aces	thorax	thoraces
ma		s	carcinoma	carcinomas
ma	delete ma	mata	stoma	stomata
on	delete on	a	spermatozoon	spermatozoa

(Except: The plural of chorion is chorions.)

nx	delete nx	nges	phalanx	phalanges

APPENDIX C

Abbreviations

a	ampere
A.U.	animal unit
A/P	anterior/posterior
AAHA	American Animal Hospital Association
AALAS	American Association of Laboratory Animal Science
ab	antibody
ABO	human blood groups
ABVP	American Board of Veterinary Practitioners
ABVT	American Board of Veterinary Toxicology
ac	before meals (ante cibum)
ACh	acetylcholine
ACLAM	American College of Laboratory Animal Medicine
ACPV	American College of Poultry Veterinarians
ACT	American College of Theriogenologists
ACTH	adrenocorticotrophic hormone
ACVA	American College of Veterinary Anesthesiologists
ACVB	American College of Veterinary Behaviorists
ACVCP	American College of Veterinary Clinical Pharmacology
ACVD	American College of Veterinary Dermatology
ACVECC	American College of Veterinary Emergency and Critical Care
ACVIM	American College of Veterinary Internal Medicine
ACVM	American College of Veterinary Microbiologists
ACVN	American College of Veterinary Nutrition
ACVO	American College of Veterinary Ophthalmologists
ACVP	American College of Veterinary Pathologists
ACVPM	American College of Veterinary Preventative Medicine
ACVR	American College of Veterinary Radiology
ACVS	American College of Veterinary Surgeons
ACZM	American College of Zoological Medicine
AD	right ear
ADH	antidiuretic hormone

ag	antigen
AI	artificial insemination
AKC	American Kennel Club
ALAT	Assistant Laboratory Animal Technician
alb	albumin
ALT	alanine aminotransferase (formerly SGPT)
AMF	anhydrous milkfat
amp	ampere
ANS	autonomic nervous system
AO	Arbeitsgemeinschaft für Osterosyntesesfragen (association for the study of fracture treatment in man founded by a group of Swiss surgeons); used to describe specialized bone plates and instruments used in orthopedic repair
APF	animal protein factor
APHIS	Animal and Plant Health Inspection Service
ARF	acute renal failure
AS	left ear
ASAP	as soon as possible
ASIF	Association of the Study of Internal Fixation
ASPCA	American Society for the Prevention of Cruelty to Animals
AST	aspartate aminotransferase (formerly SGOT)
AU	both ears (also Angstrom unit)
AVDC	American Veterinary Dental College
AVMA	American Veterinary Medical Association
Ba	barium
BAL	broncho alveolar lavage
BAR	bright, alert, responsive
BGH	bovine growth hormone
bid	twice daily (bis in die)
BIF	Beef Improvement Federation
BM	bowel movement
BST	bovine somatotropin
BP	blood pressure
BPM	beats per minute or breaths per minute
BSA	body surface area
BUN	blood urea nitrogen
BVD	bovine viral diarrhea
BVSc	Bachelor of Veterinary Science
Bx	biopsy
C	castrated
c̄	with

cal	calorie
cap	capsule
CAT	computed (axial) tomography
cath	catheter
CBC	complete blood count
cc	cubic centimeter (same as ml)
CDC	Centers for Disease Control
CFT	complement fixation test
ChE	cholinesterase
CHF	congestive heart failure
cm	centimeter
CMT	California Mastitis Test
CNS	central nervous system
CO	carbon monoxide
CO_2	carbon dioxide
conc	concentration
CP	conscious proprioception
CPR	cardiopulmonary resuscitation
crit	hematocrit
CRT	capillary refill time
C-section	cesarean section
CSF	cerebrospinal fluid
CSM	carotid sinus massage
CT	computed (axial) tomography
CVP	central venous pressure
CWE	carcass weight equivalent
D/V	dorsal/ventral
D_5W	5% dextrose in water
DA	displaced abomasum
DE	digestible energy
DEA	Drug Enforcement Agency
DES	diethylstilbestrol
DHIA	Dairy Herd Improvement Association
DHLPP	distemper, hepatitis, leptospirosis, parainfluenza, and parvovirus
DIC	disseminated intravascular coagulation
diff	differential blood count
DLH	domestic long hair (feline)
DNA	deoxyribonucleic acid
DOA	dead on arrival
DOT	Department of Transportation (used for OSHA regulation of hazardous material transfer)

dr	dram; equal to ⅛ oz or 4 ml
DSH	domestic short hair (feline)
DVM	doctor of veterinary medicine
Dx	diagnosis
Dz	disease
EBV	estimated breeding value
ECG	electrocardiogram or electrocardiograph
ED	effective dose
ED_{50}	median effective dose
EDTA	ethylenediaminetetra-acetic acid; type of anticoagulant
EEE	Eastern equine encephalitis
EEG	electroencephalogram or electroencephalograph
EIA	equine infectious anemia
EKG	electrocardiogram or electrocardiograph
EMG	electromyogram
eod	every other day; also abbreviated qod
EPD	expected progeny difference
ESR	erythrocyte sedimentation rate
ET	embryo transfer
F	Fahrenheit
F	female
f	frequency
F/S	female sprayed
FA	fatty acid
FCM	fat-corrected milk
FDA	Food and Drug Administration
Fe	Iron
FeLV	feline leukemia virus; also abbreviated FeLeuk or FeLuk
FFA	Originally, Future Farmers of America; now FFA is used only as an acronym
FIP	feline infectious peritonitis
fl oz	fluid ounce
FLUTD	feline lower urinary tract disease
FSH	follicle stimulating hormone
FUO	fever of unknown origin
FUS	feline urological syndrome
FVRCP	feline viral rhinotracheitis, calicivirus, and panleukopenia
FVRCP-C	feline viral rhinotracheitis, calicivirus, panleukopenia,, and chlamydia
Fx	fracture
g	gram; also abbreviated gm
gal	gallon

GFR	glomerular filtration rate
GH	growth hormone
GI	gastrointestinal
gm	gram; also abbreviated g
gr	grain; unit of weight approximately 65 mg
gt	drop (gutta)
gtt	drops (guttae)
GTT	glucose tolerance test
H & E	hematoxylin and eosin stain
H	hydrogen
H_2O	water
H_2O_2	hydrogen peroxide
hb	hemoglobin
Hb	hemoglobin; also abbreviated Hgb
HBC	hit by car
hCG	human chorionic gonadotropin
HCl	hydrochloric acid
Hct	hematocrit
HDL	high density lipoprotein
Hgb	hemoglobin; also abbreviated Hb
HPF	high power field
HR	heart rate
hr	hour
HW	heartworm
Hx	history
Hz	hertz; a unit of frequency
I	iodine
IA	intra-arterial
IBR	infectious bovine rhinotracheitis
IC	intracardiac
ICSH	interstitial cell-stimulating hormone
ICU	intensive care unit
ID	intradermal
IDA	International Dairy Arrangement
IM	intramuscular or intramedullary
IP	intraperitoneal
IT	intratracheal
IU	international units or intrauterine
IV	intravenous
IVDD	intervertebral disc disease

IVP	intravenous pyelogram
K	potassium
K-9	canine
KCl	potassium chloride
kg	kilogram
km	kilometer
kv	kilovolt
kVp	kilovolts peak
kw	kilowatt
L	left
L	liter
LA	large animal or laboratory animal
LAT	Laboratory Animal Technician
LATG	Laboratory Animal Technologist
lb	pound
LC_{50}	lethal concentration
LD	lethal dose
LD_{50}	lethal dose 50
LDA	left displaced abomasum
LDL	low density lipoprotein
LE	lupus erythematosus
lg	large
LH	luteinizing hormone
LOC	level of consciousness
LPF	low power field
LRS	lactated Ringer's solution
LV	left ventricle
M	male
m	meter
M/C	male castrated
M/N	male neutered
mA	milliamperage
mAs	milliamperage in seconds
mc	millicurie
mcg	microgram; also abbreviated µg
MCH	mean corpuscular hemoglobin
MCHC	mean corpuscular hemoglobin concentration
MCV	mean corpuscular volume
MDB	minimum data base
ME ratio	myeloid-erythroid ratio

MED	minimal effective dose
mE	milliequivalent
mg	milligram
MIC	minimum inhibitory concentration
MID	minimum infective dose
ml	milliliter
MLD	minimum lethal dose
MLV	modified live virus
mm	millimeter; also abbreviation for muscles
MM	mucous membrane
mm Hg	millimeters of mercury
MRI	magnetic resonance imaging
MS	mitral stenosis
MSDS	Material Safety Data Sheet
N	neutered
N	nitrogen; also abbreviation for normal
NA	not applicable
Na	sodium
NAVTA	North American Veterinary Technician Association
NH_3	ammonium
NPN	non-protein nitrogen
NPO	nothing by mouth (non per os)
NRBC	nucleated red blood cell
NS	normal saline
NSAID	nonsteroidal anti inflammatory drug; examples of NSAIDs include aspirin, ibuprofen, phenylbutazone, and flunixin meglumine; pronounced N-said
O_2	oxygen
OB	obstetrics
OD	right eye; also abbreviation for overdose
OFA	Orthopedic Foundation for Animals
OHE	ovariohysterectomy; also abbreviated OVH
OR	operating room
OS	left eye
OSHA	Occupational Safety and Health Administration
OTC	over the counter
OU	both eyes
OVH	ovariohysterectomy; also abbreviated OHE
oz	ounce
P	pulse
P/A	posterior/anterior

pc	after meals (post cibum)
PCV	packed cell volume
PD	polydipsia
PDA	patent ductus arteriosus
PDR	Physician's Desk Reference
PE	physical examination
pg	pregnant
pH	hydrogen ion concentration; indicates acidity or alkalinity
PHF	Potomic horse fever
PI	parainfluenza virus
PLR	pupillary light reflex
PM	postmortem; also abbreviation for evening
PMN	polymorphonuclear neutrophil leukocyte
PNS	peripheral nervous system
po	orally (per os)
POVMR	problem oriented veterinary medical records
ppm	parts per million
PR	per rectum
prn	as needed
PSS	Porcine Stress Syndrome
pt	pint
PT	prothrombin time
PTT	partial thromboplastin time
PU	polyuria
PVC	premature ventricular complex
PZI	protamine zinc insulin
q	every
q12h	every twelve hours
q4h	every four hours
q6h	every six hours
q8h	every eight hours
qd	every day
qh	every hour
qid	four times daily (quater in die)
qn	every night
qns	quantity not sufficient
qod	every other day; also abbreviated eod
qp	as much as desired
qs	sufficient quantity
qt	quart

R	respirations
R	right
R/O	rule out
rad	unit of measurement of absorbed dose of ionizing radiation
RBC	red blood cell
RDA	right displaced abomasum
RNA	ribonucleic acid
RP	retained placenta
rpm	revolutions per minute
RR	respiration time
RT	radiation therapy
RV	rabies vaccine
Rx	prescription
S	spayed
SA	sinoatrial
SA	small animal
SAP	serum alkaline phosphatase
SBM	soybean meal
SCC	somatic cell count
SGOT	serum glutamic oxaloacetic transaminase; now abbreviated AST
SGPT	serum glutamic pyruvic transaminase; now abbreviated ALT
SID	once daily
sig	let it be written as
SLE	systemic lupus erythematosus
SLUDDE	acronym for salivation, lacrimation, urination, defecation, dyspnea, and emesis; used in reference to the classic signs of organophosphate toxicity
SOAP	subjective, objective, assessment, plan (record keeping acronym)
soln	solution
sp.	species (singular)
spp.	species (plural)
sp. gr.	specific gravity
SPF	specific pathogen free
SQ	subcutaneous; also abbreviated SC, subQ, or subc
Staph	staphylococcus bacteria
Stat	immediately (statim)
Strep	streptococcus bacteria
Sx	surgery
T	period of time (used in ultrasound)
T	tablet or tablespoon; tablet is also abbreviated tab
T	temperature

T_3	triiodothyronine (one type of thyroid hormone)
T_4	thyroxine (one type of thyroid hormone)
tab	tablet
TB	tuberculosis
TB	tuberculin
TDN	total digestible nutrients
TE	tetanus
TGE	transmissible gastroenteritis
tid	three times daily (ter in die)
TLC	tender loving care
TNTC	too numerous to count
TPN	total parenteral nutrition
TPO	triple pelvic osteotomy
TPR	temperature, pulse, and respiration
TSH	thyroid stimulating hormone
tsp	teaspoon
TTA	transtracheal aspiration
TTW	transtracheal wash
TVT	transmissible venereal tumor
Tx	treatment
UA	urinalysis
URI	upper respiratory infection
USDA	United States Department of Agriculture
USP	United States Pharmacopeia
UTI	urinary tract infection
v	velocity
v	volt
V/D	ventral/dorsal
VM	vagal maneuver
VMD	veterinary medical doctor
vol	volume
VPB	veterinary pharmaceuticals and biologicals
VSD	ventricular septal defect
W	watt
WBC	white blood cell
WEE	Western equine encephalitis
WL	wavelength
WMT	Wisconsin Mastitis Test
WNL	within normal limits
wt	weight

APPENDIX D

Unit Conversions

LENGTH

Metric base unit = meter
1 meter = 1.0936 yards
1 centimeter = 0.39370 inch
1 inch = 2.54 centimeters
1 kilometer = 0.62137 mile
1 mile = 5280 feet or 1.6093 kilometers
1 foot = 0.3048 meter

MASS

Metric base unit = gram
1 kilogram = 2.2 pounds
1 pound = 453.59 grams
1 pound = 16 ounces
1 grain = 65 milligrams
1 dram = 3.888 grams
1 ounce = 28.35 grams
1 ton = 2000 pounds
1 gram = 0.035274 ounces

VOLUME

Metric base unit = liter
1 liter = 1.0567 quarts
1 gallon = 4 quarts
1 gallon = 8 pints
1 pint = 2 cups = 16 ounces
1 cup = 8 ounces
1 gallon = 3.7854 liters
1 quart = 32 fluid ounces
1 quart = 0.94633 liter
1 minim = 0.06 milliliter
1 fluid dram = 3.7 milliliter
1 ounce = approximately 30 milliliters
1 milliliter = 1 cubic centimeter

TEMPERATURE

$°F = 1.8°C + 32$
$°C = \dfrac{°F - 32}{1.8}$

PREFIXES FOR SI UNITS (INTERNATIONAL SYSTEM OF UNITS)

1,000,000 = mega- = M
1,000 = kilo- = k
100 = hecto- = h
10 = deka- = dk
0.1 = deci- = d
0.01 = centi- = c
0.001 = milli- = m
0.000001 = micro- = μ
0.000000001 = nano- = n
0.000000000001 = pico- = p